Modern Methods of Plant Analysis

New Series Volume 11

Editors
H.F. Linskens, Erlangen/Nijmegen
J.F. Jackson, Adelaide

Physical Methods in Plant Sciences

Edited by
H.F. Linskens and J.F. Jackson

Contributors

C. Buschmann   R. Hampp   G. Heinrich   J. S. Heslop-Harrison
A. J. Hoff   F. Homblé   A. Jenard   M. J. M. Martens   K. Omasa
W. H. Outlaw, Jr.   H. Prehn   D. Rhodes   A. Rieger

With 140 Figures and 15 Tables

Springer-Verlag   Berlin   Heidelberg   New York
London   Paris   Tokyo   Hong Kong

Professor Dr. Hans-Ferdinand Linskens
Goldberglein 7
D-8520 Erlangen

Professor Dr. John F. Jackson
Department of Biochemistry
Waite Agricultural Research Institute
University of Adelaide
Glen Osmond, S.A. 5064
Australia

ISBN 3-540-50332-3 Springer-Verlag Berlin Heidelberg New York
ISBN 0-387-50332-3 Springer-Verlag New York Berlin Heidelberg

Typesetting: International Typesetters Inc., Makati, Philippines

Printing and bookbinding: Brühlsche Universitätsdruckerei, Giessen
2131/3145-543210 – Printed on acid-free paper

Introduction

Modern Methods of Plant Analysis

When the handbook *Modern Methods of Plant Analysis* was first introduced in 1954 the considerations were:
1. the dependence of scientific progress in biology on the improvement of existing and the introduction of new methods;
2. the difficulty in finding many new analytical methods in specialized journals which are normally not accessible to experimental plant biologists;
3. the fact that in the methods sections of papers the description of methods is frequently so compact, or even sometimes so incomplete that it is difficult to reproduce experiments.

These considerations still stand today.

The series was highly successful, seven volumes appearing between 1956 and 1964. Since there is still today a demand for the old series, the publisher has decided to resume publication of *Modern Methods of Plant Analysis*. It is hoped that the New Series will be just as acceptable to those working in plant sciences and related fields as the early volumes undoubtedly were. It is difficult to single out the major reasons for success of any publication, but we believe that the methods published in the first series were up-to-date at the time and presented in a way that made description, as applied to plant material, complete in itself with little need to consult other publications.

Contributing authors have attempted to follow these guidelines in this New Series of volumes.

Editorial

The earlier series *Modern Methods of Plant Analysis* was initiated by Michel V. Tracey, at that time in Rothamsted, later in Sydney, and by the late Karl Paech (1910–1955), at that time at Tübingen. The New Series will be edited by Paech's successor H.F. Linskens (Nijmegen, The Netherlands) and John F. Jackson (Adelaide, South Australia). As were the earlier editors, we are convinced "that there is a real need for a collection of reliable up-to-date methods for plant analysis in large areas of applied biology ranging from agriculture and horticultural experiment stations to pharmaceutical and technical institutes concerned with raw material of plant origin". The recent developments in the fields of plant biotechnology and genetic engineering make it even more important for workers in the plant sciences to become acquainted with the more sophisticated methods, which sometimes come from biochemistry and biophysics, but which also have been developed in commercial firms, space science laboratories, non-university research institutes, and medical establishments.

Concept of the New Series

Many methods described in the biochemical, biophysical, and medical literature cannot be applied directly to plant material because of the special cell structure, surrounded by a tough cell wall, and the general lack of knowledge of the specific behavior of plant raw material during extraction procedures. Therefore all authors of this New Series have been chosen because of their special experience with handling plant material, resulting in the adaptation of methods to problems of plant metabolism. Nevertheless, each particular material from a plant species may require some modification of described methods and usual techniques. The methods are described critically, with hints as to their limitations. In general it will be possible to adapt the methods described to the specific needs of the users of this series, but nevertheless references have been made to the original papers and authors. While the editors have worked to plan in this New Series and made efforts to ensure that the aims and general layout of the contributions are within the general guidelines indicated above, we have tried not to interfere too much with the personal style of each author.

There are several ways of classifying the methods used in modern plant analysis. The first is according to the technological and instrumental progress made over recent years. These aspects were used for the first five volumes in this series describing methods in a systematic way according to the basic principles of the methods.

A second classification is according to the plant material that has to undergo analysis. The specific application of the analytical method is determined by the special anatomical, physiological, and biochemical properties of the raw material and the technology used in processing. This classification will be used in Volumes 6 to 8, and for some later volumes in the series. A third way of arranging a description of methods is according to the classes of substances present in the plant material and the subject of analytic methods. The latter will be used for later volumes of the series, which will describe modern analytical methods for alkaloids, drugs, hormones, etc.

Naturally, these three approaches to developments in analytical techniques for plant materials cannot exclude some small overlap and repetition; but careful selection of the authors of individual chapters, according to their expertise and experience with the specific methodological technique, the group of substances to be analyzed, or the plant material which is the subject of chemical and physical analysis, guarantees that recent developments in analytical methodology are described in an optimal way.

Volume Eleven – Physical Methods in Plant Sciences

Volumes 2, 3, and 5 of the New Series of Modern Methods of Plant Analysis deal with three important physical methods of analysis. Perusal of these volumes shows the enormous impact that Nuclear Magnetic Resonance (NMR; Vol. 2) Gas Chromatography – Mass Spectrometry (GC/MS; Vol. 3) and High Performance Liquid Chromatography (HPLC; Vol. 5) respectively have had on the plant sciences. NMR has been of great use in general metabolism, in estimating intercellular pH or energy status, and in structural analysis, among other things. GC/MS combines the separation power of GC with the highly specific and sensitive detection properties of MS, and is particularly useful for plant hormones, detecting amounts down to as low as 10 pg, for essential oils and various volatiles, etc. Commercial availability of columns for HPLC has brought about a surge in its use for the detection and measurement of plant hormones, aromatic compounds, and alkaloids, etc., as well as for the separation of organic acids, carbohydrates and proteins.

Advances in plant biochemistry and physiology have often followed the development of physical methods, opening the way to new and different sets of analysis of compounds or biological parameters. This can easily be seen by looking back at many of the physical methods described in the first series of Modern Methods of Plant Analysis. Between 1954 and 1964, two of the total of seven volumes in this series dealt with physical methods, and many of the physical methods described were crucial to advances in plant analysis, biochemistry and physiology at the time. Thus, chromatography, developed in the 1940s, was essential to the amino acid analysis of proteins, and descriptions of this technique are to be found in the first volume. Similarly, electrophoresis, essential for protein separations, is also to be found in the first volume, as well as manometry, spectrophotometry, the use of radioisotopes, and differential cell fractionation by centrifugation. All of these techniques played a role in the biochemical advances made in that era. The fifth volume, published in 1962, described the physical techniques which developed later, and which were also instrumental in biochemical advances, including gel filtration, ion exchange, and mass spectrometry.

This eleventh volume in the *New Series of Modern Methods of Plant Analysis* we are again devoting to physical methods, and we hope it will build on the trend set in the first series, as well as that seen in the three volume of the New Series dealing with physical methods (Vols. 2, 3, and 5). Instead of concentrating on one physical method, as each of these later three volumes in the New Series have done, we are reverting to the plan used in the first series and have compiled a series of chapters on different physical methods in the one volume, all of which we feel could play a prominent part in plant sciences in the near future. Thus, this volume includes a chapter on microdissection and the biochemical analysis of small amounts of plant tissue, even down to single cells. The authors make the point that plant tissues are not homogeneous, but consist of cells with different functions. These differences are obliterated by the usual homogenization of tissues that most methods of biochemical analysis utilize, a problem which is addressed in this volume in a chapter which shows that even single cells can be biochemically characterized. When linked to gas chromatography, mass spectrometry is a very powerful technique (Vol. 3, *New Series*); however, the requirement for volatilization limits the applications of this technique. Fast atom bombardment is an alternative ionization technique that can extend the range of the application of mass spectrometry, as detailed in one of the following chapters, to simple nitrogenous compounds and the structural analysis of more complex molecules like oligosaccharides and the glycoproteins. Another method for extending the analytical advantages of mass spectrometry is the subject of a further chapter, which concerns the linking of laser radiation-induced fragment ions to a mass analyzer, enabling identification of complex ions through "fingerprint" procedures. Lasers can be of use in a number of other ways, one of which is described in a chapter by Martens on the vibrational behaviour of a leaf in a sound field. As part of a study on noise pollution abatement through plants, he uses Laser-Doppler vibrometry and Laser-Doppler vibrometer-scanning to study vibrations in leaves, thus gaining more information about sound reflection, diffraction, and absorption.

The electrical properties of plant membranes are also explored in a chapter dealing primarily with one of the methods used for impedance spectroscopy measurements: the Laplace transform analysis. There is often a direct connection between the electrical behaviour of the membrane and that of an idealized model circuit which is representative of the physical process taking place in the membrane. The subject of ion currents in plants systems as studied by vibrating micro-electrode techniques is also dealt with in this volume. An intriguing method of plant analysis presented here as well is that of photoacoustic and photothermal measurement. This involves the measurement of acoustic waves and thermal effects arising from plant material following absorption of continuously amplitude-modulated light or other types of energetic radiation. The method can yield information on complex processes like photosynthesis, metabolic rates and transport phenomena, etc. and could be used for screening in quality control, production, photosynthetic activity and other types of bioactivity.

Nuclear Magnetic Resonance (NMR) can be applied in seemingly ever-expanding ways. Two-dimensional NMR can be used for elucidating the structure of the ologosaccharides in glycoproteins. Optically detected magnetic resonance (ODMR) or absorbance-detected magnetic resonance (ADMR) are used to study the triplet state which arises in the organic molecules of biological material following photoexcitation. A chapter on these techniques shows what promise this approach has for the study of plant pigments and chromophores.

Most books dealing with physical methods are aimed at chemists and biochemists working with bacterial or animal tissues; the editors feel that this volume fills a gap and indicates the way plant scientists can go in achieving further sophistication in their analytical procedures. The book should be useful to both undergraduates and graduate students of the plant sciences, as well as to professional scientists in industry and tertiary institutions. We are greatly indebted to the authors of the various chapters for the way in which they have undertaken the task of presenting each method of analysis for the maximum advantage of each of these groups of readers.

Acknowledgements. The editors express their thanks to contributors for their efforts in keeping to production schedules, and to Dr. Dieter Czeschlik, Ms. S. Böckenhaupt, Ms. T. Krammer and Ms. I. Samide of Springer-Verlag for their cooperation in preparing this and other volumes in the Series *Modern Methods of Plant Analysis.* José Broekmans continuous assistance is also gratefully acknowledged.

Nijmegen and Adelaide, Spring 1990   H. F. LINSKENS
                                                                                                J. F. JACKSON

Contents

Fast Atom Bombardment Mass Spectrometry
D. RHODES (With 5 Figures)

Microdissection and Biochemical Analysis of Plant Tissues
R. HAMPP, A. RIEGER, and W. H. OUTLAW, JR. (With 20 Figures)

Photoacoustic Spectroscopy – Photoacoustic and Photothermal Effects
C. Buschmann and H. Prehn (With 8 Figures)

Membrane Operational Impedance Spectra of Plant Cells

F. HOMBLÉ and A. JENARD (With 18 Figures)

Image Instrumentation Methods of Plant Analysis
K. OMASA (With 30 Figures)

Energy Dispersive X-Ray Analysis
J. S. HESLOP-HARRISON (With 9 Figures)

List of Contributors

BUSCHMANN, CLAUS, Botanisches Institut II, Universität Karlsruhe, Kaiserstrasse 12, 7500 Karlsruhe, FRG

HAMPP, RÜDIGER, Institut für Botanik, Biochemie der Pflanzen, Universität Tübingen, Auf der Morgenstelle 1, 7400 Tübingen, FRG

HEINRICH, GEORG, Institut für Pflanzenphysiologie, Universität Graz, Schubertstrasse 51, 8010 Graz, Austria

HESLOP-HARRISON, JOHN S., Cambridge Laboratory, Institute of Plant Science Research, Colney Lane, Norvich NR4 7UH, United Kingdom

HOFF, ARNOLD J., Department of Biophysics, Huygens Laboratory, University of Leiden, P.O. Box 9504, 2300 RA Leiden, The Netherlands

HOMBLÉ, FABRICE, Laboratoire de Physiologie Végétale, Université Libre de Bruxelles, Faculté des Sciences (CP 160), Avenue F. Roosevelt 50, 1050 Bruxelles, Belgique

JENARD, A., Laboratoire de Physiologie Végétale, Université Libre de Bruxelles, Faculté des Sciences (CP 160), Avenue F. Roosevelt 50, 1050 Bruxelles, Belgique

MARTENS, M. J. M., Department of Biology, Toernooiveld, 6525 ED Nijmegen, The Netherlands

OMASA, KENJI, The National Institute for Environmental Studies, Onogawa 16-2, Tsukuba, Ibaraki 305, Japan

OUTLAW, WILLIAM, H., JR., The Florida State University, Biology Unit 1, B-157, Tallahassee, FL 32306-3050, USA

PREHN, HORST, Institut für Biomedizinische Technik, Hochschule für Technik und Wirtschaft, FH Giessen, Wiesenstrasse 14, 6300 Giessen, FRG

RHODES, DAVID, Department of Horticulture, Purdue University, West Lafayette, IN 47907, USA

RIEGER, ANDREAS, Institut für Botanik, Biochemie der Pflanzen, Universität Tübingen, Auf der Morgenstelle 1, 7400 Tübingen, FRG

Laser-Doppler Vibrometer Measurements of Leaves

M.J.M. MARTENS

1 Introduction

In recent years, interest in the effects of vegetation on sound absorption has increased because of the possible use of vegetation in noise pollution abatement (e.g. Martens 1986). Furthermore, there is an increasing interest in the effects of vegetation on sound produced by animals during vocalisation (Foppen and Martens 1986, Martens and Huisman 1987). Sound waves induce resonances (Embleton 1966) and vibrations in plant material (Attenborough and Hess 1988, Martens et al. 1982), and part of the incident energy is absorbed by the plant material by converting the kinetic energy of the vibrating air molecules in a sound field into a pattern of vibrations of, for example, a plant leaf. Part of the vibration energy of the air molecules is lost as heat, since friction occurs in a vibrating leaf (Martens and Michelsen 1981).

To understand the physics concerning sound reflection, diffraction and absorption in leaves, a better understanding is necessary concerning the interaction between the sound field and leaves. Therefore it can be helpful to study the vibrational behaviour of a leaf in a sound field.

Although classical methods to determine vibrations, for example, using accelerometers, have been used to study sound-induced vibrations in leaves (Tang et al. 1986), we preferred to study vibrations in leaves without loading the leaves with a mass. Laser-Doppler-vibrometry and interferometry are recently developed techniques for measuring vibrations. These techniques do not load the vibrating object by a mass or a vibration pick-up that can influence the vibration characteristics of relatively weak biological materials like plant organs. The method has also been proven to be useful to study vibrations in other biological materials such as tympanic membranes of insects (Michelsen and Larsen 1978) and humans (Vlaming and Feenstra 1988). The technique can be made more powerful by connecting a scanning system to the equipment (Bank and Hathaway 1980, Martens et al. 1982). Here we will show the use of Laser-Doppler-vibrometry (LDV) and Laser-Doppler vibrometer scanning (LDVS) in studying vibrations in leaves.

2 Materials and Methods

2.1 Laser-Doppler Vibrometer Scanning System

The LDV is an instrument that has been developed by the Institute of Biology, University of Odense, Denmark and DISA Elektronik, Denmark (Buchhave 1975). The instrument has a wide range of applications to measure frequencies and amplitudes of vibrations. The frequency range extends from DC to about 100 kHz. The displacements to be measured vary from 10 nm to 10 cm. The instrument is based on the principle of optical heterodyne detection of the Doppler shift of light scattered from a small area of the vibrating surface. The scattered light is collected and collimated by a lens system and mixed with a reference beam that is frequency shifted by a Bragg cell. The system therefore delivers an analogue voltage proportional to the instantaneous velocity of the vibrating object. An extended description of the LDV and the optical basics of the instrument is given by Buchhave (1975).

Figure 1 shows a scheme of the equipment we used. The LDV system consists of a 5 mW He-Ne laser NEC GLC 2033, a laser adapter DISA 55X19, the mounting

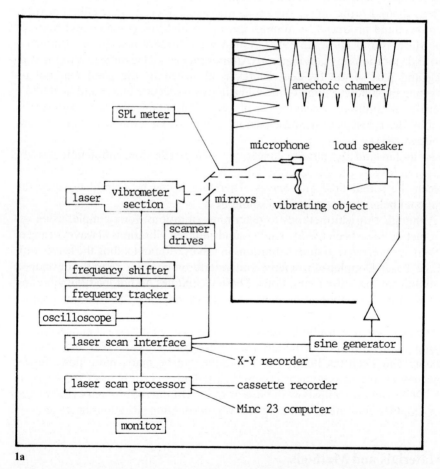

1a

Fig. 1a-c. The Laser-Doppler-vibrometer-scanning system and the equipment necessary to carry out the sound-induced vibration measurements of leaves. A detailed description of the equipment is given in the text. **b** A photograph of the laser-Doppler-vibrometer equipment outside the anechoic chamber: *a* the laser-Doppler-vibrometer section; *b* scanning system with drives and mirrors enclosed in a steel protecting box; *c* the opening in the wall of the anechoic chamber; *d* the frequency tracker and shifter; *e* laser scan inter-face and processor; *f* the monitor; *g* the oscilloscope; *h* the sine generator; *i* the Quad

1b

1c

amplifier; *k* the X-Y recorder; *l* the concrete console with the granite balanced table; *m* the protecting grid for the NEC lasertube. To avoid damage to the equipment a protecting fence is installed in front.
c A leaf part of *Sparmannia africana* clamped in between two aluminium rings with inner diameter of 25 mm. A Brüel & Kjear microphone is visible next to the leaf sample. The white spot at the *left* is the entering laser beam, projected to the leaf

section DISA 55L63, the beam splitter and modulator section DISA 55L83, the vibrometer section DISA 55L66, the frequency shifter DISA 55N10 and the Doppler frequency tracker DISA 55N20.

The scanning system is made of two scanner drives (General Scanning Inc.), AX-200 equipped with two Melles Griot mirrors 02 MFG 005, a laser scan interface, and a processor developed and built by the department of Electronics Research of the Faculty of Sciences of the University. This scanning system scans a matrix of 64×64 points. Furthermore, we use a Philips monitor and a Brüel & Kjaer sine generator 1023.

The other equipment consists of a Philips 10 MHz dual-beam storage oscilloscope PM 3234, a Houston X-Y recorder 2000, a Philips casette recorder D 6350 and a Digital Computer Minc-23. The sound is generated by the Brüel & Kjaer sine generator, amplified by a Quad 303 amplifier and produced by a Philips AD 400 M or a Dynacord D 300 loudspeaker. To measure the sound pressure level a Gen Rad 1933 sound pressure level meter with microphone is used. Most of the experiments have been carried out in an anechoic chamber in the laboratory. A photographic impression of the equipment is given in Fig. 1b,c. When necessary, the objects were sprayed with Spotcheck-Zyglo developer (Magnaflux Ltd, Swindon), to obtain a better reflection of the laser light.

2.2 Specific Materials

In the different experiments which we describe in the next chapters, the leaves of several plant species have been investigated. The specific plant material used in those experiments is indicated in the specific section.

We also investigated the possibility of using LDV-scanning to detect the vibration characteristics of fruit and an organic foam, i.e., whipped cream. As a start, before each of the experiments we investigated the vibration characteristics of some test materials, like paper and plastic sheets. The results of these testings are not described here.

For the experiments described in Chapter 4 the leaf cuttings were clamped in between two aluminium rings. The inner radii of these rings were 25 or 68 mm.

3 Two Simple Models for Vibrating Leaf Tissue

A fully developed theory on the vibration characteristics of leaves does not exist in literature. Therefore, we tried to find an analogy in acoustics to compare leaves with. A structural description of a leaf shows that the main parts are the epidermic layers at both sides of the leaf, the pallisade and parenchyma cell layers in between the epidermic layers, the nerves and veins, some cavities in the neighbourhood of the stomata, water and, in some plant species, viscous fluid, for example, latex, in the cells.

Some well-defined structures are described in the acoustic literature that can be considered as analogous with (the parts of) a leaf. For example a nerve can be compared to a bar and the surface of a leaf can be compared to a membrane or a plate. For vibrating bars, isotropic membranes and isotropic plates models and theory are available (e.g. Morse 1983). For clamped circular isotropic membranes and plates the following theoretical considerations are useful for understanding more or less the behaviour of (clamped) vibrating plant leaves.

3.1 Clamped, Isotropic Membranes

Membranes are surfaces with a stiffness that is negligible compared to the restoring force due to tension. In other words a membrane is the two-dimensional analogon of a flexible string. For a membrane with tension T and density σ, fastened along a boundary circle of radius r, the resonance frequencies ν_{mn} are defined by:

$$\nu_{mn} = \frac{1}{2r} \sqrt{\frac{T}{\sigma}} \, b_{mn}, \tag{1}$$

where m and n are integer numbers, $m \geq 0$ and $n \geq 1$ and b_{mn} are constants (Morse 1983). For the fundamental frequency ν_{01}, b_{01} will have the value of 0.7655. The fundamental frequency (f_0) of a round clamped membrane of radius 25 mm is 2.7 times larger than the fundamental frequency of a comparable membrane of radius 68 mm.

3.2 Clamped, Isotropic Plates

The most important property of a plate is stiffness. Therefore a plate is a two-dimensional analogon of a bar. The fundamental frequency and the overtones of a circular plate of radius r and density σ, clamped at its edges are defined by:

$$\nu_{mn} = \frac{\pi h}{r^2} \sqrt{\frac{Q}{3\sigma (1-s^2)}} \, b_{mn}, \tag{2}$$

where h is the half-thickness of the plate, Q its modulus of elasticity and s the Poisson ratio; m, n and b_{mn} are constants identical to those in equation (1). At the fundamental frequency ν_{01} the value of b_{01} equals 1.015. The ratio of the fundamental frequencies ($f_{0,25}/f_{0,68}$) of two clamped, circular plates with radii of 25 mm and 68 mm respectively equals 7.3.

4 Experiments and Results with Clamped Leaves

The vibration frequencies of all the leaf cuttings clamped in between the metal rings were equal to the frequency of the impinging pure tone sound waves. Vibration frequencies other than these sound frequencies were not detected. This confirmed the results of the earlier experiments that have been carried out by Martens and Michelsen (1981 and described in the next chapter). The vibration amplitude was linearly related to the effective sound pressure in the whole measurable vibration velocity range (Geveling 1986, Roebroek 1984).

4.1 Point Measurements of the Vibration Velocity with LDV

Both Eqs. (1) and (2) from Morse (1983) will be used further on to be compared with the results from measurements carried out at clamped circular parts of leaves of *Sparmannia africana* L.f. and *Ficus elastica* L.. The first plant has broad leaves with few nerves, whereas the leaves of the second plant are stiff because of the viscosity of the latex content. During the experiments cuttings of freshly picked leaves of the plants were clamped in between aluminium circular rings with inner radii of 25 and 68 mm. The mass of the leaf cuttings was weighted with a balance and the leaf area was measured with a surface area meter.

Pure tone sound was impinging perpendicularly on the clamped leaf cuttings and the vibration velocity was measured with the laser-Doppler-vibrometer in the centre of the leaf part. The vibration velocity was monitored on the oscilloscope. The first velocity vibration maximum above 50 Hz was determined, and considered to be the fundamental frequency (f_0). The sound forced the clamped leaf cutting into harmonic vibration with the frequency of the sound wave. The displacement amplitude of the centre of a *Sparmannia* leaf at the fundamental frequency was about 20 μm at 80 dB SPL (re 20 μPa). The experimental results of the measurements published elsewhere (Roebroek 1984, Geveling 1986) have been summarized in Table 1 (e.g., Geveling et al. 1986, Martens 1988).

In this table the ratios have been presented of the measured fundamental frequencies of the leaf cuttings at the two radii used. As has been theoretically stated in Chapter 3, these ratios are for an isotropic membrane 2.7 and for an isotropic plate

Table 1. Ratios of two fundamental frequencies at different radii of clamped leaf cuttings of two plant species

Plant species	Mass (mg/cm^2)	Fund. freq. (Hz) r = 68 mm	Fund. freq. (Hz) r = 25 mm	$\dfrac{f_o\,(r=25)}{f_o\,(r=68)}$
Sparmannia africana	15.3	60	157	2.6
	15.7	70	193	2.8
	15.0	61	157	2.6
	14.7	56	151	2.7
	16.3	66	175	2.7
Ficus elastica	64	75	399	5.3
	64	79	450	5.7
	63	75	395	5.3
	63	72	385	5.3
	63	79	387	4.9

7.3. The calculated ratio for the *Sparmannia* leaf cuttings equals the theoretical ratio for an isotropic membrane. It can therefore be concluded that *Sparmannia* leaf cuttings behave in this respect like membranes. The calculated ratio for the *Ficus* leaf cuttings is in between the theoretical ratios for a membrane and a plate. The vibrational behaviour of the *Ficus* leaf cuttings is therefore not membrane-like nor plate-like, but somewhere in between.

4.2 Optical Scanning over the Surface with LDV-Scanning

By using the LDV scanning system it is possible to visualize the vibration velocity pattern of a whole clamped circular leaf cutting, since the scanning system makes 4096 point-measurements·in a 64×64 matrix. The vibration velocities of these 4096 points are stored in the laser scan interface and can be displayed in a semi-3-dimensional way on a monitor. The pictures of 12 regularly divided moments in a vibration cycle, $0°-30°-60°$ and so on, can be shown one after the other, with a speed of up to fifty pictures per second, resulting in an animation of the motion of the scanned surface. The extremes from scans that have been made of some clamped leaf cuttings are shown in Figs. 2 and 3.

Figure 2 shows two scans of two different leaf cuttings of *Sparmannia* at what will be shown to be the fundamental frequency. The (a) series gives the measuring results of a leaf cutting without a nerve, while the (b) series shows those of a leaf cutting with a nerve. In Fig. 2a and a′ and 2b and b′ the two extremes of a vibration cycle have been drawn, whereas in 2a″ and a‴, and in 2b″ and b‴ the vibration velocity amplitudes and the phase of the different leaf parts compared to each other have been given from the centre-lines in the X and the Y directions.

From the phase information (dotted lines) it is evident that all leaf parts vibrate in the same direction. Therefore, it is concluded that the leaf cuttings a and b both vibrate at their fundamental frequencies. The influence of the bar-like nerve is shown in series (b) of Fig. 2. Since the nerve is stiffer than the leaf tissue, the inhomogeneity of a leaf cutting with a nerve is clearly visible: firstly, the fundamental frequency is shifted to the higher frequencies, and secondly the vibration velocity is less at the nerve itself than next to the nerve in the leaf surface. More detailed results have been published by Geveling (1986).

Next to scans at the fundamental frequency as shown in Fig. 2, it is possible to obtain scans at overtones. From the same *Sparmannia* leaf cutting without a nerve, of which the fundamental frequency is visualized in Fig. 2a, that fundamental one and three higher modes are shown in Fig. 3. These higher modes are comparable to those found theoretically for clamped membranes (Morse 1983).

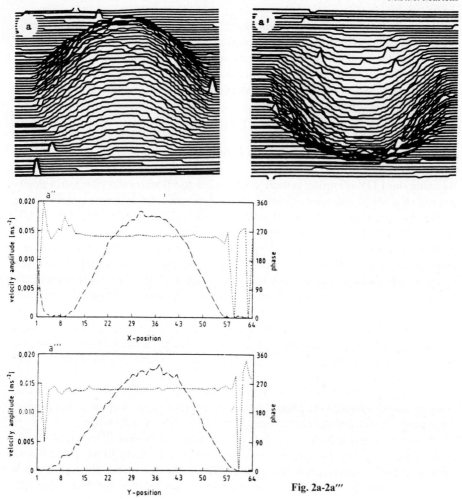

Fig. 2a-2a'''

Fig. 2a,b. Vibration velocity patterns of *Sparmannia africana* leaf-cuttings clamped in between two metal rings with an inner radius of 25 mm. Series **a-a'''** shows the vibration pattern of a homogeneous leaf-cutting without a nerve at its fundamental frequency of 110 Hz, 80 dB SPL; series **b-b'''** shows the vibration pattern of a leaf-cutting with a nerve at its fundamental frequency of 167 Hz, 80 dB SPL. Scans **a** and **a'** and **b** and **b'** respectively show the extreme positions of the vibration velocity pattern out of a full cycle. The *spikes* in the pictures are artifacts caused by an insufficiently reflected signal. The *curves* in **a''** and **a'''** and **b''** and **b'''** respectively show the vibration velocity amplitude and phase of a *horizontal* (x-position) and *vertical* (y-position) line through the middle of the scans

Fig. 3. Vibration velocity patterns of a circularly clamped *Sparmannia africana* leaf-cuttings, radius 25 mm at the fundamental frequency and expected higher modes: **a** 110 Hz, 80 dB; **b** 189 Hz, 80 dB; **c** 275 Hz, 85 dB; and **d** 432 Hz 100 dB SPL. All shown patterns are extreme positions of a vibration cycle. *Sharp spikes* are artifacts

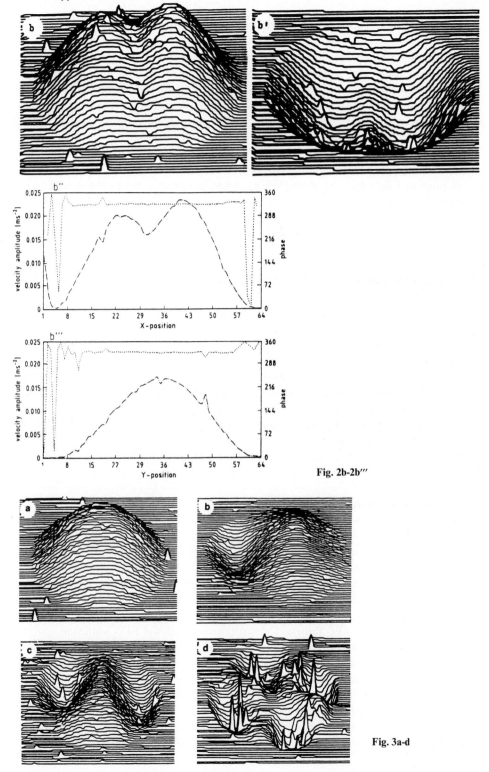

Fig. 2b-2b'''

Fig. 3a-d

5 Experiments and Results with Free Hanging Leaves

As is shown in the preceding chapter, the work with clamped leaf cuttings offers a
tool to compare the vibration characteristics of plant leaves with well-defined
models in acoustic literature. For realism, however, the vibrational behaviour of
free hanging plant leaves is of much more interest. A model for free hanging plant
leaves does not exist. Laser-Doppler-vibrometry and interferometry are well suited
to do in situ experiments at free hanging plant leaves exposed to sound waves.

Free hanging leaves have a much smaller vibration velocity amplitude compared to clamped leaves. All leaves studied vibrate at the same frequency as the
frequency of the inpinging sound waves. A generation of lower or higher vibration
frequencies has not been observed up till now, although this effect has been
suggested as a hypothesis to explain a sound amplification in experiments with
model forests in an anechoic chamber (Martens 1980).

5.1 Point Measurements of the Vibration Velocity with LDV

A pilot study has been carried out to investigate with LDV the sound-induced
vibrations of free hanging plant leaves (Martens and Michelsen 1981). The experiments have been carried out with four plant species: *Ligustrum regelianum, Betula
verrucosa* Ehrh., *Corylus avellana* L. and *Quercus robur* L., and the leaves remained
attached to the plants during the measurements. In a sound field up to 100 dB SPL
(re 20 μPa) vibration velocities have been measured up to $3*10^{-4}$ m/s. An example
has been shown in Fig. 4, using a slowly sweeping sine-wave synchronized with the
frequency analysis in the spectrum analyser (Fig. 4a,b) or white noise (4c). The
complexity of the vibration velocity distribution over a leaf can be observed when
studying the vibration velocities of different spots of the leaf as shown in Fig. 5.

5.2 Optical Scanning over the Surface
with LD-Interferometry and LDV-Scanning

In the preceding paragraph the complexity of the vibration velocity distribution
over the leaf is mentioned. Laser-Doppler interferometry (LDI) and Laser-Doppler
vibrometer scanning (LDV scanning) offer the opportunity to study the vibrations
of free hanging leaves. Some results from measurements with LDI are shown in Figs.
6 and 7.

Some results are also shown of measurements with the LDV scanning system
described in Chapter 2. Figure 8 shows an example of *Euphorbia pulcherrima*. Six
moments out of a vibration cycle at 400 Hz and 100 dB (re 20 μPa) are visualized.
The vibration pattern is shown to be very complicated.

To obtain information concerning the vibration velocity amplitudes of the leaf
surface area, a computer program has been developed to calculate these amplitudes
from the measured data and to plot the amplitudes like sonograms. Light regions
have a small vibration velocity amplitude ($0-30*10^{-5}$ m/s) and the dark regions have

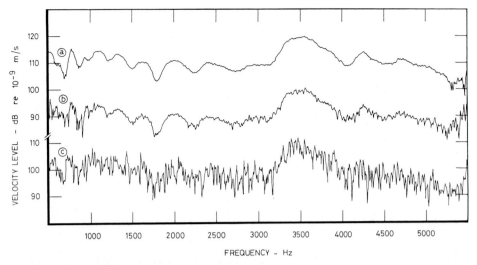

Fig. 4. Vibration velocity spectrum versus sound frequency of one spot of a free hanging *Corylus avellana* leaf in situ. In **a** the leaf was activated by pure tones of about 100 dB and in **b** of about 80 dB SPL. The velocity amplitudes are linearly related to the magnitude of the sound pressure. In **c** the leaf was activated by white noise of about 100 dB SPL. The leaf was 100 mm in length and 85 mm broad, and the angle between the plane of the leaf and the direction to the loudspeaker was 60°. The measuring spot was situated on the mesophyll between the main nerve and the margin

a relatively large one (3–8*10⁻³ m/s). Figure 9 shows such an amplitude plot of a vibrating *Euphorbia pulcherrima* leaf at 400 Hz and 100 dB (re 20 μPa). The maximum displacement amplitude of the leaf is 2 μm. From this vibration velocity amplitude plot the region of the nerves and veins is light, which means that the nerves have a much smaller amplitude than the leaf surface.

The effect of a relatively less vibrating nerve is not always found, as is demonstrated in Fig. 10, where we show comparable vibration velocity amplitude plots of a *Carpinus betulus* leaf. This free hanging leaf has been scanned at two pure tone frequencies of 300 Hz (Fig. 10a) and 400 Hz (Fig. 10b), both at 100 dB SPL. We therefore investigated more carefully the vibration velocity pattern of some single points at the surface of the *Carpinus* leaf by studying the vibrations versus discrete frequencies. The results of these measurements are shown in Fig. 11. These spectra indicate that the vibration pattern of the nerve strongly varies at different frequencies, not only in the high frequencies, as has been shown earlier (e.g., Fig. 5), but also in the low frequency region. It is also clear from these spectra that at some frequencies the nerve has a larger vibration amplitude than the internerval regions of the leaf, which was already mentioned earlier in discussing the vibration velocity amplitude plots.

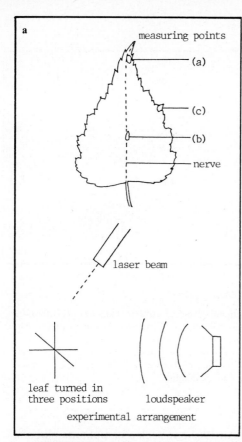

a

measuring points

(a)

(c)

(b)

nerve

laser beam

leaf turned in
three positions

loudspeaker

experimental arrangement

Fig. 5a,b. Vibration velocity spectrum versus frequency of a *Betula pendula* leaf in situ measured at three spots on the leaf indicated as (*a*), (*b*) and (*c*) and with different orientation to the sound source. **a** shows the experimental arrangement and **b** the spectra. The three curves per measuring point, ⓐ, ⓑ and ⓒ, corresponding to the three spots (*a*), (*b*) and (*c*), show the results related to the three different angles of incidence of the sound. *Dotted lines* 90°; *broken lines* 45°; *solid lines* 0°. The angles of incidence of the laser beam are 30° for the *dotted lines*, 60° for the *solid lines* and 75° for the *broken lines*. Therefore, the real vibration velocities perpendicular to the plane of the leaf were 6 dB higher for the *dotted curves* and about 1 dB higher for the *solid curves*

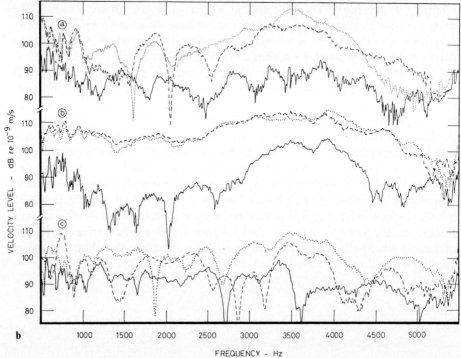

VELOCITY LEVEL – dB re 10⁻⁹ m/s

ⓐ

ⓑ

ⓒ

110
100
90
80

110
100
90
80

110
100
90
80

b

1000   1500   2000   2500   3000   3500   4000   4500   5000

FREQUENCY – Hz

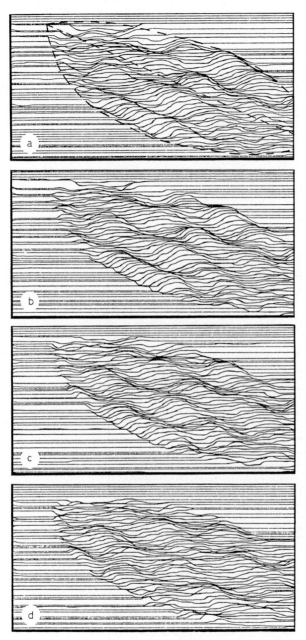

Fig. 6a-d. Vibration velocity pattern of a *Vinca minor* leaf in a sound field of 3290 Hz, 120 dB SPL, measured with laser-Doppler interferometry. **a-d** Four successive scans of a vibration cycle. An oscillating pattern is evident. In **a** the shape of the leaf is schematically drawn

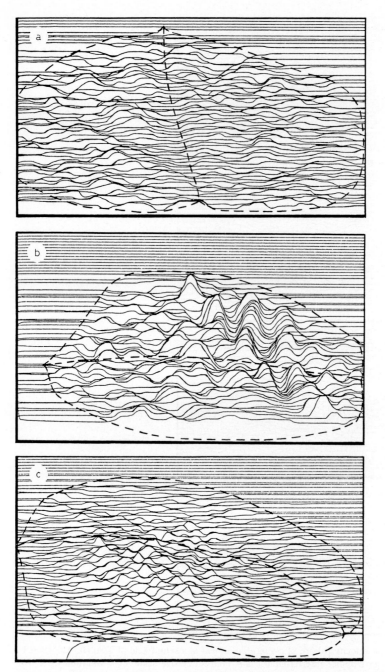

Fig. 7a-c. Vibration velocity pattern of *Tilia platyphyllos* leaves in pure tone sound fields, measured with LDI; **a** 444 Hz, 108 dB; **b** 1200 Hz, 122 dB; and **c** 6700 Hz, 104 dB SPL. The shape of the leaves is indicated

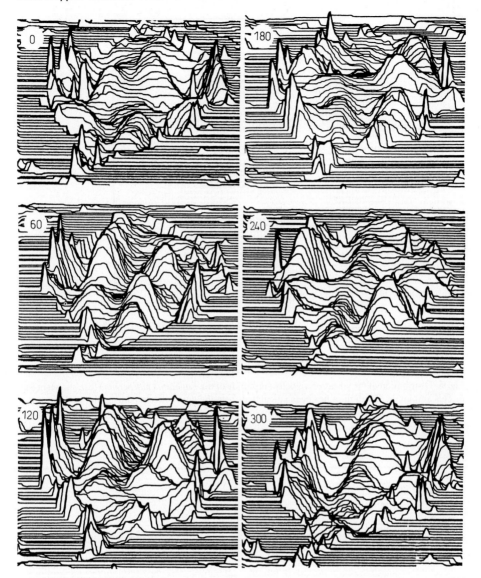

Fig. 8. Vibration velocity pattern of a free hanging *Euphorbia pulcherrima* leaf in a pure tone sound field of 400 Hz at 100 dB SPL. From a full vibration cycle six regularly divided scans are pictured. The phase is indicated. The main nerve can be clearly recognized, and the surface and margins of the leaf show the largest velocity amplitudes

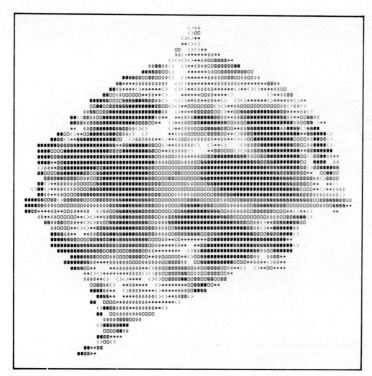

Fig. 9. Distribution of the vibration velocity amplitude of the *Euphorbia pulcherrima* leaf, shown in Fig. 8. This amplitude distribution has been calculated from the data collected during the LDV-scanning experiment. The velocity amplitudes vary from $0-30 \times 10^{-5}$ m/s to $3-8 \times 10^{-3}$ m/s. The maximum displacement is 2 μm

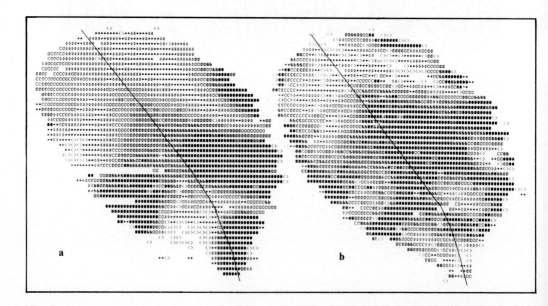

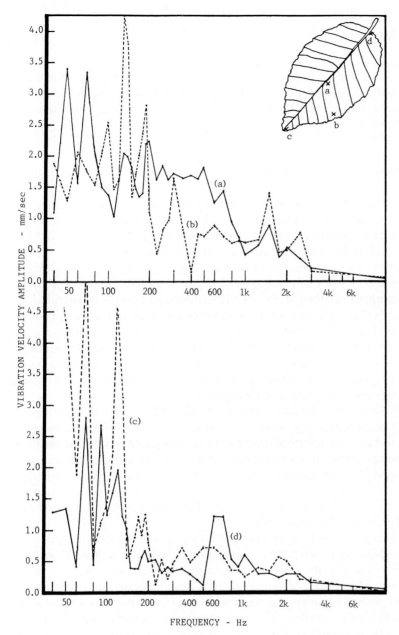

Fig. 11. Vibration velocity amplitude spectrum versus frequency of four different spots of a *Carpinus betulus* leaf at pure tone sounds of 100 dB SPL. The four measured spots are indicated in the *upper right* drawing of the leaf (*a-d*)

Fig. 10a,b. Distribution of the vibration velocity amplitude of a *Carpinus betulus* leaf. **a** 300 Hz, 100 dB and **b** 400 Hz, 100dB SPL. The main nerve is indicated. The values of the velocity amplitudes are equal to those indicated in Fig. 9

6 Discussion

In this chapter we will try to give an overall discussion of the findings with these measuring methods concerning sound-induced vibrations occurring in leaves. We will also discuss the use of LDV and LDV-scanning in measuring vibrations in biological materials.

6.1 Sound-Induced Vibrations of Leaves

LDV, LDI and LDVS have been shown to be useful methods to detect and measure vibrations in weak biological materials. Concerning plant leaves, we have shown that sound indeed can induce vibrations and that these vibrations can be visualised rather easily and studied very well using laser-Doppler techniques.

To understand the vibration patterns, we have tried to compare these patterns in leaves with some vibration models, well-described in literature (e.g. Morse 1983). From these comparisons it can be concluded that the studied models are useful when comparing to leaf cuttings that are homogeneous and without nerves and veins. When studying leaf cuttings comprising a nerve, the model of a membrane or that of a plate are not adequate. This is evident in Fig. 2b–b'''. As well as this visible effect, Geveling (1986) also calculated irregular modes instead of the circular symmetrical ones that can be expected from homogeneous membranes and plates. The nerves seem to correspond to mechanical inhomogeneities.

The LDV-scanning technique is highly suitable for investigating vibrations in free hanging plant leaves, e.g. in situ. This still unexplored field of research is developing, and we have described here the first descriptive studies carried out. One of the surprising observations we have made is that at some sound frequencies the nerve vibrates stronger than the surface of the leaf. The state of the research at this moment has not yet advanced so far that it is possible to calculate exactly the sound energy that is absorbed by a single plant leaf. This can only be estimated at the moment (Martens and Michelsen 1981).

On the other hand, the laser-Doppler technique seems to give the possibility of studying sound-induced vibrations in plant leaves in simple as well as in complex situations. For example, the described measurements with single leaves can be extended to a more complex and natural situation by introducing more plants in the sound field between the loudspeaker and the studied leaf. In that way it will also be possible to investigate variations in a sound field introduced by other structures. Knowledge of these absorbing, diffracting, reflecting and scattering characteristics of vegetation may help us to understand sound attenuation in natural and urban circumstances, and the adaptation of animals to the acoustic environment in complex environmental circumstances, where still is a need for adaptation (Foppen 1988).

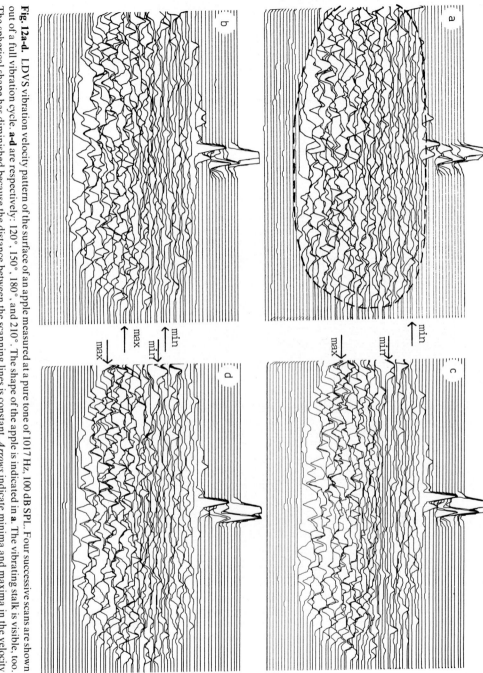

Fig. 12a-d. LDVS vibration velocity pattern of the surface of an apple measured at a pure tone of 1017 Hz, 100 dB SPL. Four successive scans are shown out of a full vibration cycle. **a-d** are respectively: 120°, 150°, 180°, and 210°. The shape of the apple is indicated in **a**. The vibrating stalk is visible, too. The spherical shape has diminished because the distance between the scanning lines is constant. *Arrows* indicate minima and maxima in the velocity pattern, indicating a surface wave running over the surface

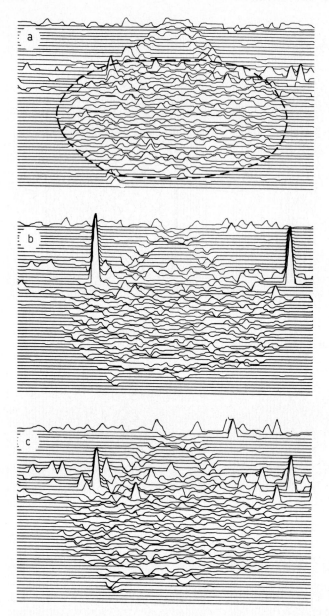

Fig. 13a-c. LDVS vibration velocity pattern of the surface of a tomato measured at a pure tone of 215 Hz at 100 dB SPL. Three scans are shown of a full vibration cycle: **a** 90°, **b** 240°, and **c** 270°. **a, c** The extreme positions of the vibration cycle. The shape of the tomato is indicated in **a**; the rubber bands in which the tomato was hung are visible at the *top* of the scans

6.2 The Usefulness of LDV and LDV-Scanning

The vibrations of weak biological materials can be measured with laser-Doppler techniques. We have referred to the work of the Michelsen group on tympanum vibrations in *Tettigonidae* (e.g. Michelsen and Larsen 1978). The vibrations of the human tympanum have also been studied using laser-Doppler techniques (Vlaming and Feenstra 1988). In addition to plant leaves, the study of other plant organs is possible.

A pilot study has been initiated quite recently, to investigate whether LDV scanning can be used to measure sound-induced vibrations in fruit. LDV-scanning appeared to be suitable for detecting very easily sound-induced vibrations in spherical fruit like apples and tomatoes. We had expected many problems that did not eventuate. For example, we had expected a poor laser-beam reflection: firstly, because of the non-perpendicular angle of incidence. However, the whole spherical surface, even at shearing incidence, still reflected so much laser light that the LDV-scanning system acted perfectly. We only detected more artificial spikes than usual in the scanning plots; secondly, we had expected that the surface of apples and tomatoes would not be bright enough to be scanned, but it appeared not to be necessary to spray the surfaces with Spotcheck-Zyglo developer. Figures 12 and 13 show some of the results obtained by scanning sound-induced vibrations in apples and tomatoes.

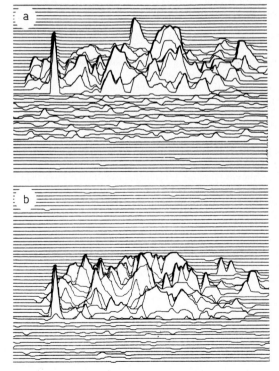

Fig. 14a,b. LDVS vibration velocity patterns of whipped cream on a small plate. **a** 440 Hz, 100 dB; and **b** 215 Hz, 100 dB SPL. The two *sharp spikes* in the *left* part of the scans are artifacts. The vibration velocity pattern of the small plate is much less

Finally, the method also seems to be useful to study some physical characteristics of food like whipped cream, as is shown in Fig. 14. Therefore, it can be
concluded that LDV-scanning can be a powerful tool to investigate vibrations in all
kinds of biological materials and other weak substances.

References

Attenborough K, Hess HM (1988) The uses of sound in the study of soils and plants. Neth Akoest
    Genootschap 93:31–49
Bank G, Hathaway GT (1980) A revolutionary 3-D interferometric vibrational mode display. AES
    preprint presented at the 66th Convention 1980 May 6 through 9, Los Angeles, 32 pp
Buchhave P (1975) Laser-Doppler vibration measurements using variable frequency shift. DISA Inf
    18:15–20
Embleton TFW (1966) Scattering by an array of cylinders as a function of surface impedance. J Acoust
    Soc Am 40:667–670
Foppen RPB (1988) Animals, sounds and noise. Neth Akoest Genootschap 93:51–57 (In Dutch with
    English summary)
Foppen RBP, Martens MJM (1986) Investigating bird communication in relation to the acoustic
    environment. In: Martens MJM (ed) Proceedings of the workshop on Sound Propagation in Forested
    Areas and Shelterbelts. 3–6 March 1986, Fac Sci, Nijmegen, pp 207–214
Geveling SM (1986) Sound-induced vibrations of plant leaves studied using laser-Doppler vibrometry.
    Int Rep Dep Bot, Catholic Univ, Nijmegen, p 42
Geveling SM, Martens MJM, Roebroek JGH (1986) Sound-induced vibration-patterns of plant leaves.
    In: Martens MJM (ed) Proceedings of the workshop on Sound Propagation in Forested Areas and
    Shelterbelts. 3–6 March 1986, Fac Sci, Nijmegen, pp 153–160
Martens MJM (1980) Foliage as a low-pass filter: Experiments with model forests in an anechoic
    chamber. J Acoust Soc Am 67:66–72
Martens MJM (ed) (1986) Proceedings of the workshop on Sound Propagation in Forested Areas and
    Shelterbelts. 3–6 March 1986, Fac Sci, Nijmegen, p 214
Martens MJM (1988) Laser-Doppler-vibrometry as a measuring tool for vibrating plant leaves. Neth
    Akoest Genootschap 93:3–13 (In Dutch with English summary)
Martens MJM, Huisman WHT (1987) Sound propagation in the natural environment. In: Van Gelder
    JJ, Strijbosch H, Bergers PJM (eds) Proceedings of the Fourth Ordinary General Meeting of the
    Societas Europaea Herpetologica. 17–21 August 1987, Fac Sci, Nijmegen, pp 217–274
Martens MJM, Michelsen A (1981) Absorption of acoustic energy by plant leaves. J Acoust Soc Am
    69:303–306
Martens MJM, Van Huet JAM, Linskens HF (1982) Laser interferometer scanning of plant leaves in
    sound fields. Proc Kon Ned Akad Wet, Amst, C85:287–292
Michelsen A, Larsen ON (1978) Biophysics of the ensiferan ear. I. Tympanal vibrations in bushcrickets
    (Tettigoniidae) studied with laser vibrometry. J Comp Physiol A123:193–203
Morse PM (1983) Vibration and sound. Am Inst Phys Acoust Soc Am, MacGraw Hill, NY
Roebroek JGH (1984) Onderzoek naar de trillingen van bladeren in een akoestisch veld. Int Rep,
    Catholic Univ, Nijmegen, 496; 107 pp
Tang SH, Ong PP, Woon HS (1986) Monte Carlo simulation of sound propagation through leafy foliage
    using experimentally obtained leaf resonance parameters. J Acoust Soc Am 80:1740–1744
Vlaming MSMG, Feenstra L (1988) The measurement of vibrations in the normal and reconstructed
    middle ear. Neth Akoest Genootschap 89:9–20

Triplet States in Photosynthesis: Linear Dichroic Optical Difference Spectra via Magnetic Resonance[1]

A.J. HOFF

1 Why Triplet States Are of Interest

Triplet states are, with few exceptions (notably dioxygen), (photo) excited meta-stable states, usually of aromatic, organic molecules. These types of molecules abound in all biological material. Whenever a photon impinges on a substance containing such molecules, there is a finite, and often quite appreciable, probability that such a molecule is converted into the triplet state.

In some important instances, triplet states are formed by radical recombination reactions, and the radicals may be generated from a singlet excited state that is created by illumination, or by ionizing radiation. In photosynthesis in which normal, forward electron transport is blocked, such a recombinational triplet state is formed with high efficiency.

The triplet state is an important probe of molecular structure. It has certain physical properties that make it eminently useful for identifying reactants in photoreactions, for probing structural environment and binding, etc. and for serving as a gentle perturber of the system. Studying the system's response then gives important clues to its structure and function.

In this chapter I will first introduce, in a general way, the physical characteristics of the triplet state. Avoiding mathematics where not absolutely needed, this introduction aims at developing an intuitive feeling for those properties of the triplet state that make it such a useful probe. I will then briefly discuss a variant of the technique of electron paramagnetic resonance (EPR): optically detected magnetic resonance (ODMR) in zero magnetic field, which is one of the most important spectroscopic tools for studying the triplet state.

---

[1] *Abbreviations*:

ADMR absorbance-detected magnetic resonance; *BC, BChl* bacteriochlorophyll; *B_A* accessory BC of the active or L chain; *B_B* accessory BC of the inactive or M chain; *Chl* chlorophyll; *CT* charge transfer; *D* generic label of the primary donor; *D_A* BC of the primary donor D of L chain; *D_B*, BC of D of M chain; *EPR* electron paramagnetic resonance; *FDMR* fluorescence-detected magnetic resonance; Φ, bacteriopheophytin; *HOMO* highest occupied molecular orbital; *ISC* intersystem crossing; *LED* light emitting diode; *LD* linear dichroic; *LUMO* lowest unoccupied molecular orbital; *MIA* microwave-induced absorbance; *MIF* microwave-induced fluorescence; *MIP* microwave-induced phosphorescence; *ODMR* optically detected magnetic resonance; *PDMR* phosphorescence-detected magnetic resonance; *PEM* photoelastic modulator; *PS I* photosystem I of plants; *PS II* photosystem II of plants; *P680* primary donor of PS II; *P700* primary donor of PS I; *P860* primary donor of bacteria containing BChl *a*; *RC* reaction center; *T–S* triplet-minus-singlet; *tm* transition moment; *ZFS* zero field splitting; *C. Chromatium; Cfl. Choroflexus; P. Prosthecochloris; R. Rhodospirillum; Rb. Rhodobacter; Rps. Rhodopseudomonas.*

The rest of this chapter is devoted to the applications of ODMR in photosynthesis. Most attention will be devoted to absorbance-detected magnetic resonance, ADMR, a technique which has taken a real flight in the past few years. With ADMR it is possible to monitor low temperature triplet absorbance difference spectra with unparalleled accuracy and sensitivity. For coupled pigment systems, such as those encountered in photosynthetic membranes, this has opened up a whole new field for the study of pigment interaction and the interpretation of optical spectra, etc.

1.1 Some Physics

Triplet states are characterized by having two unpaired electrons whose electron spins, s_1 and s_2, are parallel. Hence, their total spin magnetic angular momentum S (their 'magnetism') is $S = \bar{s}_1 + \bar{s}_2 = \frac{1}{2} + \frac{1}{2} = 1$. Quantum mechanics teaches us that systems with spin vector S possess $2S + 1$ magnetic energy levels, called the system's *multiplicity*. Here, the multiplicity is $2 \times 1 + 1 = 3$, hence the name triplet state.

Although some molecules have a triplet ground state (e.g. dioxygen), most molecular triplet states are produced by photoexcitation from a singlet ($S = 0$, all electrons paired, the system is diamagnetic) ground state to the first electronically excited singlet state, which converts into an excited triplet state by spin reversal of one of the two highest energy electrons. This process is illustrated in Fig. 1.

The triplet state can revert to the singlet ground state by the emission of light, *phosphorescence,* or by radiationless processes. This decay is 'spin forbidden': the law of conservation of angular momentum does not allow destruction of the triplet state's spin moment by its conversion to an angular momentumless singlet state. The spin angular momentum has to be taken up by the angular momentum of the orbital motion of the electrons. Because the two angular momentum reservoirs do not communicate easily, especially in molecules consisting of light atoms, the rate of conversion is much slower than the decay of the singlet excited state to the singlet ground state ('spin allowed' because S remains zero). Thus, the triplet state has a relatively long lifetime, of the order of microseconds to seconds; it is *metastable.*

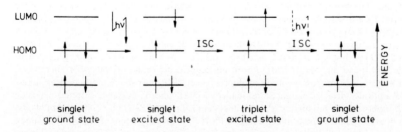

Fig. 1. Distribution of electrons over the highest occupied (*HOMO*) and lowest unoccupied (*LUMO*) molecular orbital of singlet and triplet states. *ISC* Intersystem crossing [76]

When a molecule is excited to the singlet or triplet excited state, it has two single electrons moving in different orbitals. In the singlet excited state, S_1, they may approach each other very closely without violating the Pauli principle, which states that no two electrons may be in the same spin and orbital state. In the triplet state, however, they have to keep apart, since the two have the same spin state (both 'up' or both 'down'). Their coulombic repulsion is therefore considerably less than in the singlet excited state, so that the triplet states lies lower in energy than the S_1 state by an amount that is called the *exchange energy*.

The two unpaired electrons of the triplet state can be viewed as tiny magnets. Each magnet generates a magnetic field that exerts a force on the other magnet. This mutual action is called the magnetic *dipole-dipole interaction*. The force depends on the distance and on the angle the dipoles make with each other and with their distance vector. Since forceful action and energy are interrelated, the two unpaired electrons will have a certain dipolar magnetic energy. For two electrons moving in a molecule, the time-averaged magnetic energy will be the average of the dipolar energy over time and space (the molecular coordinate frame). Thus, the dipolar energy of the triplet state depends on the structure of the molecule, its symmetry, etc.

A basic tenet of quantum mechanics is that in an external magnetic field $\bar{B}$ (e.g. of an electromagnet) a magnetic moment (thus also the spin magnetic moment) cannot have an arbitrary position with respect to the magnetic field vector $\bar{B}$. It is *quantized*, i.e. it can only have certain positions, numbering precisely $2S + 1$. For example, a single electron with spin ½ can only be parallel (up, $m_S = +½$) or antiparallel (down, $m_S = -½$). Here m_S is the projection of the spin angular momentum vector on $\bar{B}$; it is called the spin magnetic quantum number. For a triplet state with $S = 1$ we have three possible positions with $m_S = +1, 0$ or -1.

If there is no external magnetic field, it can be shown that the magnetic moment of the triplet state is still quantized, not along a vector but in a plane. There are three such planes that are mutually perpendicular and whose position is determined by molecular symmetry. The dipolar energy for each plane of quantization is generally different: For the plane defined by $x = 0$, it is X, for the plane $y = 0$ it is Y and for the plane $z = 0$ it is Z. Here the vectors $\bar{x}, \bar{y}$ and $\bar{z}$ span a cartesian molecular coordinate frame. Conventionally $|Z| > |X|, |Y|$, whereas $X + Y + Z = 0$. The latter relation follows from the law of conservation of energy: As long as there are no external forces acting on a system its energy is constant, and the sum of the quantum energies of an internal interaction must be zero. For a spherically symmetric system there is no distinction between the $\bar{x}, \bar{y}$ or $\bar{z}$ direction so $X = Y = Z$ and from the sum rule it follows that then all are zero: The three energy levels are said to be *degenerate*. For axially symmetric systems conventionally $X = Y$; for rhombic systems $X \neq Y \neq Z$, as illustrated in Fig. 2.

Since X, Y and Z are mutually dependent (their sum is zero) it is customary to express them into two independent, so-called fine structure or zero field parameters: $D = -\frac{3}{2} Z$ and $E = -\frac{1}{2} (X-Y)$. The energies of the three levels corresponding to Z, Y and X are then given by $-\frac{2}{3}D, \frac{1}{3}D + E$ and $\frac{1}{3}D - E$ respectively. The levels as depicted in Fig. 2 thus correspond to a rhombic triplet state with $D > 0$ and $E < 0$. When $X = Y$, $E = 0$, so this corresponds to axial symmetry. By convention $|D| \geq 3|E|$. The physical meaning of the zero field splitting parameters is that they

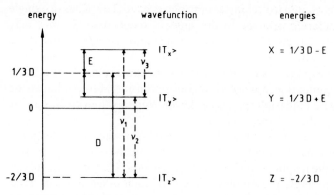

Fig. 2. Triplet sublevel energies X, Y, Z and wave functions $|T_x>$, $|T_y>$ and $|T_z>$. The levels are drawn for the zero field splitting parameters $D>0$, $E<0$; their values are indicated by *arrows. Dashed arrows* ν_1, $(|D| + |E|)/h$; ν_2, $(|D| - |E|)/h$; ν_3, $2|E|/h$ transition

represent averages over the spatial coordinates x′, y′, z′ of the distance vector $\bar{r}$ of the two unpaired electrons:

$$D = \frac{3}{4} \cdot \frac{g^2\beta^2\mu_o}{4\pi} < \frac{r^2 - 3z'^2}{r^5} > \text{ and } E = -\frac{3}{4} \cdot \frac{g^2\beta^2\mu_o}{4\pi} < \frac{x'^2 - y'^2}{r^5} >,$$

where g is the ratio between the electronic orbital and spin magnetic moment (for a 'free' electron, $g = 2.0023$), β the electronic Bohr magneton and μ_o the permeability of vacuum. For a flat molecule such as chlorophyll, one would expect D to be positive. (The $\bar{z}$-axis is the axial symmetry axis and is perpendicular to the plane of the molecule, so that the z′ component of $\bar{r}$ is on the average much smaller than $|\bar{r}|$.) For a rod-like molecule, such as a biradical, D will be negative.

1.2 Magnetic Resonance

At a given temperature the ensemble of triplet states will stay in heat contact with the environment and will consequently be in Boltzmann equilibrium: Most triplet states will have the lowest energy, a minority the intermediate energy and a few the highest energy. At very low temperatures (say 10 K or lower), the environment is effectively isolated from the ensemble of triplet states, the system is not in Boltzmann equilibrium and, under continuous illumination, the relative population n_i of each of the three sublevels will be determined by the probability p_i to transit from the singlet excited state to the i-th sublevel and the decay rate k_i that governs de-excitation from the ith sublevel back to the singlet ground state: $n_i = p_i K/k_i$ ($i = x,y,z$), where K is the rate of population of the triplet state. Both p_i and k_i are determined by molecular symmetry.

As in optical spectroscopy, an oscillating electromagnetic field of the proper frequency may elicit transitions from one sublevel to another. Since the levels represent the quantum energies of a magnetic moment, it is the magnetic vector of the field, not the electric, that does this trick. If, in the transition, population is transferred from a level of lower to higher energy, electromagnetic radiation is absorbed. Conversely, stimulated transfer of population from a higher to a lower energy level leads to emission of radiation. Both absorption and emission may be detected in a conventional way with some radiation detector. Since the frequencies corresponding to the energy differences of the triplet sublevels generally lie in the microwave region, one could use a microwave diode. It is often advantageous, however, to employ a different detection method, making use of the optical properties of the triplet state.

1.3 Optical Detection of Magnetic Resonance, ODMR

Continuous illumination will generate an equilibrium population of the triplet sublevels given by n_i, $\Sigma_i n_i = N$, the total triplet population. The n_i are represented by dots in Fig. 3. We have assumed that the p_i are about equal and that $k_x, k_y \gg k_z$. Let us now switch on a microwave field of a frequency corresponding to a transition between the y and the z level. The field will transfer population from the heavily populated, slowly decaying z level to the much less populated, fast decaying y level. Obviously, this transferred population will not remain there but will quickly decay to the singlet ground state. This will lead to enhanced phosphorescence if the triplet decays radiatively. Furthermore, a new equilibrium will be established that for a strong enough microwave field is given by $n'_y = p_y K / [\frac{1}{2}(k_y + k_z)]$ and $n'_z = p_z K / [\frac{1}{2}(k_y + k_z)]$. If we take $p_y \approx p_z$ it is then immediately seen that $(n'_y + n'_z) < (n_y + n_z)$, because for $k_y > k_z$, $2/(k_y + k_z) < 1/k_y + 1/k_z$. Although the x level will sense the new equilibrium via the photogeneration cycle of the triplet state (K contains the light flux *and* the concentration of singlet ground states), this is a second order effect, and we have $\Sigma n'_i = N' < N$. Because the concentration of the singlet excited state will be negligibly low when conventional light sources are used, this means that the concentration of the singlet ground state is enhanced. This will lead

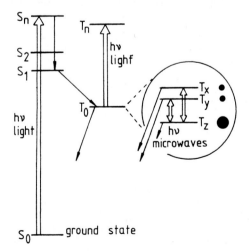

Fig. 3. Energy level diagram of triplet formation. Depopulation rates of the three triplet sublevels are indicated by *arrows*; microwave transitions by *double arrows*; and equilibrium population by *black dots* [77]

to enhanced fluorescence, and enhanced singlet ground state absorbance as both these phenomena are proportional to the ground state population.

In the above we have the essence of ODMR. A sample is continuously illuminated at liquid helium temperatures, preferably below 2.1 K, and simultaneously irradiated by microwaves of a frequency ν not far from that corresponding to one of the triplet sublevel spacings: $\nu_{1,2} = (|D| \pm |E|)/h$ and $\nu_3 = 2E/h$ (Fig. 3). The frequency ν is scanned across one of the frequencies $\nu_{1,2,3}$ while either the phosphorescence, fluorescence or absorbance of the sample is monitored. When ν is close to or precisely equal to $\nu_0 = \nu_{1,2,3}$ the ensemble of triplet states is in resonance with the microwave field and the fluorescence (FDMR), phosphorescence (PDMR) or absorbance (ADMR) will be enhanced or diminished, depending on the relative values of p_i and k_i ($i = x,y,z$). An example of such a resonance is shown in Fig. 4.

An important advantage of ODMR compared to conventional EPR in zero field, where the absorption of microwaves is monitored, is that the optical probing occurs with quanta of much higher energy than the microwave quantum. (E(orange light, 600 nm $= 6 \times 10^{-5}$ cm)/E(microwaves, 3 cm) $= 5 \times 10^4$.) This enhances detector sensitivity enormously. Another advantage is the possibility to probe the resonance at various wavelengths (Sect. 2.2), which, especially for ADMR, gives rise to much new information.

The above gives the basics of ODMR spectroscopy. For further details one should consult 'Triplet State ODMR Spectroscopy' (R.H. Clarke, ed.) [1].

1.4 Instrumentation

It is clear that the minimal requirements for an ODMR experiment are a source of light, one of microwaves, a cryostat and a detector. The best excitation light source in the visible or infrared region is an incandescent lamp (e.g. a tungsten-iodine projection lamp) powered by a current-stabilized DC power supply. Mercury or xenon lamps are more intense, certainly in the UV region, but suffer from instabilities in the intensity I, often to a level of $\Delta I/I = 10^{-3}$ or worse. The same holds

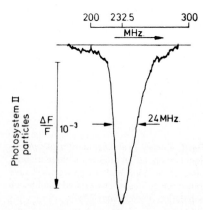

Fig. 4. Fluorescence-detected magnetic resonance line of the $2|E|/h$ transition at 232 MHz of subchloroplast particles enriched in photosystem I. For experimental details see [78]

for laser excitation. When the excitation beam is broad-banded, it may also serve as probe beam at various wavelengths.

Microwaves are supplied preferably by a sweep unit with provisions to scan the frequency over the desired range. Usually, the output ($\sim$50 mW) is enough for regular ODMR at liquid helium temperatures without amplification. For measuring the ODMR lines the microwaves are fed into a broad-band resonator, e.g. a helix, that can admit the whole frequency range. If one is interested in the (probe) wavelength dependence, the microwave frequency is set exactly at resonance and a narrow-band cavity may be used.

The cryostat is a double-walled helium bath cryostat equipped with at least two windows. Helium boils at 4.2 K, and scattering by the bubbles then prohibits optical detection. Fortunately, by lowering the pressure above the helium bath, the temperature of the helium can be lowered to below 2.1 K, the so-called lambda point, below which helium is superfluid and bubbles disappear completely. Thus, adequate pumping facilities should be provided.

The detector system depends on the optical mode. For phosphorescence and fluorescence a photomultiplier is used. For ADMR, one may employ a strong probe beam (in fact the excitation beam may serve as such), providing a sufficiently high number of transmitted quanta that a photodiode may be used (up to 1100 nm, silicium; 1100 nm–2 μm, germanium). To observe the resonance by fluorescence or phosphorescence, the wavelength of detection should be separated from the wavelength of excitation by adequate filtering. For ADMR, either a filter or, when the probe wavelength dependence is monitored, a monochromator is used.

The sensitivity of the ODMR spectrometer can be considerably enhanced by modulating the amplitude of the microwaves and applying frequency-selective amplification of the photodetector signal combined with lock-in detection. Noise is then reduced to that corresponding with the passband of the amplifier-lock-in detector combination, leading to an increase in the signal-to-noise ratio of several orders of magnitude. For example, in our ADMR spectrometer a $\Delta A/A$ ratio of better than 5×10^{-7} is routinely achieved.

Finally, as in all modern spectrometers, the instrument is interfaced to a small, dedicated computer that handles monochromator setting, data collection and storage, and carries out simple operations such as taking the $\Delta I/I$ ratio (important when recording the probe wavelength dependence).

A schematic diagram of our present ADMR setup is shown in Fig. 5. With small modifications it can be used for PMDR and FMDR as well.

2 Information Obtainable from ODMR Spectroscopy

The triplet state is a probe of molecular structure and of the interaction between probe molecule and its environment. The information content of an ODMR spectrum is twofold: (1) information derived from the magnetic resonance line, and (2) information derived from the dependence on probe and excitation wavelength. Both types of information will be discussed in some detail in the next two sections.

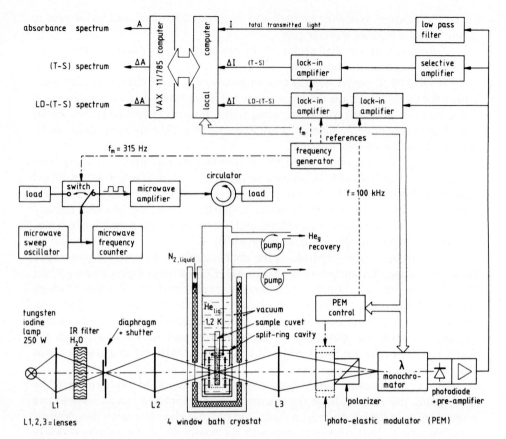

Fig. 5. Setup for absorbance-detected magnetic resonance *ADMR*, and linear dichroic *ADMR* [77]

2.1 Determination of Zero Field Splitting Parameters, Sublevel Population Probabilities and Decay Rates

Zero Field Splitting Parameters. As mentioned in Section 1.3, the possible resonance frequencies are given by $\nu_{1,2} = (|D| \pm |E|)/h$, $\nu_3 = 2|E|/h$ (Fig. 3). Thus, from two observed resonances the values of $|D|$ and $|E|$ can be calculated. (Note that only the absolute values can be determined.) This can be done with a precision of less than 0.5%. $|D|$ and $|E|$ are a fingerprint of the molecule, since they depend on spatial averages over the electronic coordinates (Sect. 1.1), which depend in turn on molecular structure, size and symmetry. In fact, they reflect the detailed structure of the wavefunction of the triplet excited state. Thus, the determination of $|D|$ and $|E|$ may serve to identify the molecule on which the triplet state is residing. Secondly, $|D|$ and $|E|$ are sensitive to the environment. For example, their precise value will reflect binding properties, structural changes of the matrix, etc. This is especially important in photosynthesis, where the triplet state of the primary may serve as a probe of its precise configuration and of its interaction with the reaction center protein matrix.

Often, in biological systems more than one triplet state is photoinduced (e.g. different pigments in a pigment-protein complex). In that case, one generally sees more than three resonance lines. To sort out which ones belong to the same triplet, it is useful to carry out a double resonance experiment. For instance, saturating a $(|D|+|E|)/h$ transition will affect the intensity of the $(|D|-|E|)/h$ and $2E/h$ transitions of the same triplet state, but not the transitions of all the other triplets.

Populating Probabilities. The equilibrium population of sublevel i is given by $n_i = Kp_i/k_i$ (Sect. 1.2), with K a constant depending on light flux, absorption cross section, etc., p_i the populating probability and k_i the decay rate. p_i depends critically on molecular symmetry if the triplet state is populated via spin-orbit induced intersystem crossing. For example, in free base porphin p_z is very much lower than p_x, p_y [2,3]. p_i may be determined via a double resonance experiment, in which the three triplet sublevels are two by two connected by microwaves of different frequency. Measuring the photodetector signal for the various possible ways of connecting the three triplet sublevels: x,y,z, x~y,z, x,y~z and x~y~z, where the ~ sign means 'transition saturated with microwaves', yields the three p_i's [3]. For fast decaying, randomly oriented triplets, however, the methodology of p_i determination is not without its pitfalls, and should be used with care [4]. Reliable values of p_i should provide information on symmetry, changes in symmetry induced by changes in the environment, etc. Note that in some important cases, notably in photosynthesis, triplet states are generated via recombination from photoinduced radical states (see Sect. 3). In that case, the p_i's are not directly related to molecular symmetry but depend on details of the recombination reaction, which are often not well understood.

Decay Rates. The values of the k_i's depend on the details of spin-orbit induced intersystem crossing from the excited triplet to the singlet ground state. As such they are a useful additional 'fingerprint' of molecular structure and environmental influences. The k_i's are much easier to determine reliably than the p_i's. Two methods have been advocated: (1) The saturation-desaturation method in which resonant microwaves of saturating intensity are switched on, and the system left for some time to equilibrate, after which the microwaves are switched off, and the return to equilibrium is followed in time [5]. For fast decaying and/or randomly oriented triplet states, however, it has been demonstrated [4] that this method can yield quite erroneous results. (2) The pulse method in which a brief pulse of microwaves is applied, much shorter than the relevant decay times, and the response of the system is optically monitored [3]. This method is entirely free from the pitfalls of the saturation method, and should be used with preference [4]. Since the response signal is proportional to the number of spins transferred, hence to the intensity of the microwave pulse, it is of advantage to use a cavity with a high quality factor and/or microwave amplification after the sweep unit.

2.2 Spectral Information: Microwave-Induced Phosphorescence, Fluorescence and Absorbance Spectra

Once the ODMR lines of a triplet have been determined, the resonance frequencies are known precisely, and one can investigate the dependence of the intensity of a particular resonance line on the probing wavelength. Thus, one irradiates the sample with resonant microwaves of sufficient, preferably saturating intensity, and monitors via a monochromator the photodetector output as a function of the probe beam wavelength. The resulting spectra are called microwave-induced phospho-rescence, fluorescence or absorbance spectra, abbreviated by MIP, MIF and MIA spectra respectively. For one particular triplet state, the shape of these spectra does not depend on which resonance line, $\nu_{1,2}$ or ν_3, is selected. Obviously, if more than one triplet state is present, the MI spectra provide another means to sort out which resonances belong to the same triplet state.

In biological systems, the phosphorescence is often weak, and only MIF and MIA spectra can be recorded. Besides their use for discriminating between various triplet states, the MIF spectra are useful for identifying the triplet carrying molecule, as exemplified by the elegant studies of Beck et al. [6,7] on photoinduced triplet states in bacterial photosynthetic membranes.

MIA spectra are a case apart, since they provide much more information than the MIP or MIF spectra. As will be discussed in the next section, they represent the difference of the singlet ground state, 'normal' absorbance spectrum and the spectrum for the system when a triplet state is present. They are therefore known as triplet-minus-singlet absorbance difference (T–S) spectra, rather than being labelled by the MIA acronym.

2.3 Triplet-Minus-Singlet (T–S) Absorbance Difference Spectra

When a triplet state is present, the absorbance spectrum contains the following contributions:

1. The *unperturbed* singlet ground state absorbance spectrum ($S_n \leftarrow S_o$ transitions) of all molecules that are *not* in the triplet state and that do not interact with the molecule that *is* in the triplet state;
2. The *perturbed* singlet ground state spectrum of the molecules that are not in the triplet state but *do* interact with the triplet carrying molecule. When this particular molecule is in the triplet state the interaction will generally be different from that when the molecule is in the singlet ground state;
3. The absorbance spectrum of the triplet state itself, consisting of $T_n \leftarrow T_o$ transitions.

With square wave, on-off amplitude-modulated microwaves, the MIA spectrum represents the difference in absorbance of the sample for microwaves *on* and microwaves *off* (Fig. 6). It can be shown [8] that this difference is proportional to the difference in absorbance *with* and *without* the triplet state present. In other words, the MIA spectrum represents the difference of the absorbance of the sample with all

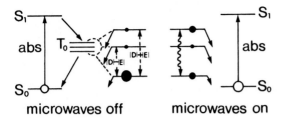

Fig. 6. Scheme of the influence of microwaves resonant with a triplet *ODMR* transition on the singlet ground state population. In actual practice the microwaves are on-off square wave-modulated at a frequency of 300–400 Hz [10]

molecules in the singlet ground state and that when all molecules of one particular type are excited into the triplet state whose ODMR resonance is being monitored.

It is important to note that other triplet states with different values of $|D|$ and $|E|$, and consequently different ODMR resonance frequencies, may be present without showing up in the MIA$\equiv$T–S spectrum. Their absorbance is not changed by the microwaves and, therefore, their contribution to the absorbance cancels in the ADMR-monitored T–S difference spectrum. On the other hand, recording T–S spectra for different ADMR resonance frequencies provides a means to discriminate the resonances belonging to one and the same triplet states, since, in general, contributions 2 and 3 will be different for different triplet states.

The ADMR-monitored T–S spectrum is of multifold interest. For *non-interacting* triplet states it provides a very accurate triplet absorbance spectrum, since that is given by adding the 'normal' singlet ground state spectrum to the T–S spectrum. For *interacting* triplet states present, for example, in a photosynthetic pigment-protein complex, it records these interactions very sensitively and thus provides a unique means to study pigment configuration. This will be illustrated in Sect. 3, where we will also discuss applications of *linear dichroic* (LD) T–S spectroscopy, the principle of which is explained in the next section.

2.4 Linear Dichroic T–S Spectroscopy

The microwave transitions between the triplet sublevels that were mentioned in Section 1.2 are *polarized*: the x~y transition is $\bar{z}$-polarized, the x~y transition $\bar{y}$- and the y~z transition $\bar{x}$-polarized. This is quite analogous to an optical transition, whose transition dipole moment usually has a well-defined direction in the molecular frame. For example, the long wavelength Q_y absorbance bands of chlorophylls is y-polarized, which means that their transition dipole moment is a vector lying approximately in the N_1-N_3 direction (Fig. 7). Often, the direction of the triplet magnetic resonance transition moments are not well-known. In chlorophylls, for example, one may be reasonably certain that the $\bar{z}$-transition moment is perpendicular to the molecule and that the $\bar{x}$- and $\bar{y}$-transition moments lie in the plane of the macrocycle. The precise direction in the plane of the latter, however, was until recently (Sect. 3.3) not known. As will be shown below, LD-(T–S) spectroscopy provides a means to ascertain the directions of the magnetic transition moments. With this knowledge one may then derive precise structural information from the LD-(T–S) spectra.

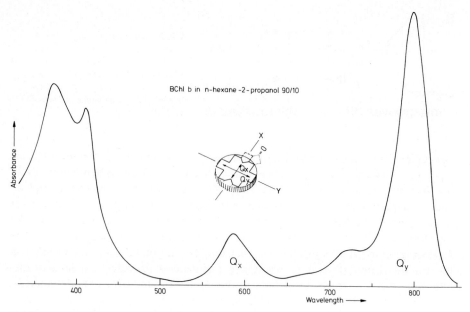

Fig. 7. The absorption spectrum of bacteriochlorophyll *b* (for structure see Fig. 16). *Inset* Direction of the transition moments of the bands labelled Q_x, Q_y

In optical spectroscopy, the transition probability for a transition with transition dipole $\bar{p}$ is proportional to $|\bar{E}|^2|\bar{p}|^2\cos^2\beta$, where β is the angle between $\bar{p}$ and the $\bar{E}$ electric vector of the (polarized) incident light. A similar relation holds for the magnetic microwave transitions between the triplet sublevels. Thus, for a microwave transition moment $\bar{\mu}_{mw}$ and an angle β between $\bar{\mu}_{mw}$ and the $\bar{B}_1$ magnetic vector of the (polarized) microwave field, we have a transition probability $|\bar{B}|^2|\bar{\mu}_{mw}|^2\cos^2\beta$. It follows that molecules oriented with $\bar{\mu}_{mw}$ more or less parallel to $\bar{B}_1$ have a much higher transition probability than those oriented about perpendicular to $\bar{B}_1$. (Of course, for $\beta = 90°$, the transition probability is exactly zero.) Hence, for random excitation to the triplet state, molecules oriented in an angular interval $d\beta$ close to $\beta = 0$ will experience a much higher change in their relative triplet concentration upon the application of (polarized) resonant microwaves than molecules in an interval $d\beta$ close to $\beta = 90°$. Consequently, the *distribution of triplet states*, which was isotropic before the application of the microwaves, becomes *axially anisotropic* with the axis parallel to $\bar{B}_1$ when resonant microwaves are switched on.

The microwave-induced anisotropy in the triplet state distribution can be interrogated with a beam of polarized light. For example, let us assume that the optical transition moment $\bar{p}$ is parallel to $\bar{\mu}_{mw}$, that the microwaves as in the example of Section 1.3 decrease the triplet concentration, and that we interrogate at a wavelength where the singlet ground state has an absorption band and the triplet state does not absorb. For light polarized parallel to $\bar{B}_1$ we will then measure a lower transmittance than for light polarized perpendicular to $\bar{B}_1$. (Along $\bar{B}_1$ there are fewer triplets, hence more singlet ground states, than perpendicular to $\bar{B}_1$.) Obviously, the difference in transmittance (which for small changes can be taken equal to the difference in absorbance ΔA [8]) will depend on the angle α between $\bar{p}$ and $\bar{\mu}_{mw}$. In the above example, the sign of $\Delta A = A_{//} - A_\perp$ would be reversed if not $\alpha = 0$ as assumed, but $\alpha = 90°$. Going from $\alpha = 0$ to $\alpha = 90°$, at a given angle ΔA must become zero. This is the magic angle $\alpha = 54.7°$ [for which $3 \cos^2\alpha - 1 = 0$, see Eq. (1)]. Thus, from the magnitude of ΔA relative to the magnitude of A we should be able to directly derive α.

It will be recognized that the above description of the microwave induced selection in the triplet state distribution is very similar to that of *photoselection*. We can therefore partake of the formalism derived for that technique (see e.g. [9]) to calculate the functional relationship between ΔA and α. In doing so, we must of course average over all positions of the molecules with respect to $\bar{B}_1$, assuming a random initial distribution. To simplify this averaging we further assume that the triplets are isotropically excited. This is not strictly true, as in our setup the light beam does not excite molecules oriented such that their optical transition moment for triplet excitation is parallel to the direction of propagation of the light. Nevertheless, because of energy transfer among differently oriented pigments before the excitation is trapped onto the triplet state, and because of scattering at sample cell walls and at impurities and cracks inside the sample, isotropic excitation proved to be a good approximation. Finally, we have employed unpolarized probe light and interrogate the difference $A_{//} - A_\perp$ *after* the light has passed the sample, using a photoelastic modulator and a polarizer. Taking all this into account it can be shown that the ratio R in intensity of the LD-(T–S) spectrum (which represents the difference $\Delta A_{//} - \Delta A_\perp = (T-S)_{//} - (T-S)_\perp$) and the T–S spectrum is given by [10,11]:

$$R = \frac{LD\text{-}(T\text{–}S)}{T\text{–}S} = \frac{3\cos^2\alpha - 1}{\cos^2\alpha + 3}. \tag{1}$$

Equation (1) is plotted in Fig. 8, together with plots of the LD-(T–S) and T–S intensities versus α. It is seen that R is quite sensitive to α, so that with proper calibration of the T–S and LD-(T–S) spectra, α can be determined quite accurately. Of course, all this applies rigourously only for single absorbance bands. If bands with different directions of $\bar{p}$ vis-à-vis $\bar{\mu}_{mw}$ overlap, R will have some intermediate value and one will have to simulate the complete LD-(T–S) and T–S spectra to obtain values for the various α's.

In photoselection, one customarily extrapolates to zero intensity of the exciting, selecting light beam. This is because the transition, and therefore photoconversion, probability is proportional to $|\bar{E}|^2\cos^2\beta$. Even for β close to $90°$, photoconversion will be appreciable if the field strength $|\bar{E}|$ is high enough. In that case, almost all molecules will be photoconverted, regardless of their orientation, and selection is lost. In microwave selection, the relevant field strength is that of the $\bar{B}_1$ field. Thus, to ensure proper selection one has to measure R as a function of $|\bar{B}_1|$ and extrapolate for $|\bar{B}_1| \to 0$. This is best done by taking the slope of a graph of the LD-(T–S) versus the T–S intensity as a function of $|\bar{B}_1|$ (see Sect. 2.5 and Fig. 10).

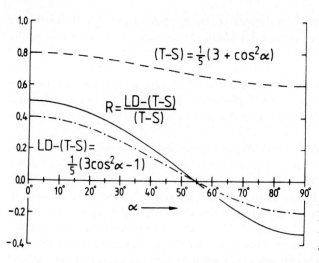

Fig. 8. Plot of Eq. (1). The value $\alpha = 54.7°$ corresponds to the magic angle at which the anisotropy disappears [77]

The angle α in Eq. (1) refers to one particular $\bar{\mu}_{mw}$, say that corresponding to the $\bar{x}$ polarized y~z transition at frequency $(|D|-|E|)/h$. Tuning the microwaves to the $(|D|+|E|)/h$ frequency, that is the x~z or $\bar{y}$ polarized transition, allows the recording of LD-(T–S) spectra and the determination of R for a different α, viz. the angle α_y between $\bar{p}$ and $\bar{y}$. When $\bar{p}$, $\bar{x}$ and $\bar{y}$ lie in one plane, $\alpha_y = 90° - \alpha_x$, since the triplet spin axes, $\bar{x}$, $\bar{y}$ and $\bar{z}$ span a cartesian coordinate frame. If $\bar{p}$, $\bar{x}$ and $\bar{y}$ are not coplanar, the two measurements uniquely define the orientation of $\bar{p}$ in the $\bar{x}, \bar{y}, \bar{z}$ coordinate frame. This is a great advantage over the ordinary photoselection experiment, where one determines just one angle between two transition moments, which leaves one with a conical ambiguity. Note that it suffices to record LD-(T–S) spectra for just two of the three possible ADMR transition frequencies. Since the orientations of all $\bar{p}$'s are determined in one and the same coordinate frame, their mutual angular dependence immediately follows.

2.5 Instrumentation for LD-(T–S) Spectroscopy

The instrumentation for LD-(T–S) spectroscopy is quite similar to that of isotropic T–S spectroscopy, and we can refer to Fig. 5 for a description. First of all we need polarized microwaves, and therefore a simple helix is not suitable. We use a split-ring cavity as described by Hardy and Whitehead [12]. This design has the advantage that for the rather long wavelength of the microwaves (between 50 and 100 cm) one still has a cavity of manageable dimensions, which fits easily into a four-window liquid helium cryostat. The $\bar{B}_1$ field is polarized along the vertical, and therefore is perpendicular to the horizontal light beam. The probe light is unpolarized. The $\bar{B}_1$ field induces an ellipticity ($\equiv T_{//} - T_{\perp}$) in the transmitted light, which is detected via a photoelastic modulator after the sample followed by an analyzer. The PEM rotates the ellips spanned by the unequal transmitted light vectors $\bar{E}_{//}$ and $\bar{E}_{\perp}$ (with respect to $\bar{B}_1$) by 180° at a frequency of 50 kHz. The analyzer converts this polarization modulation into an amplitude modulation at 100 kHz that is proportional to $T_{//}-T_{\perp} \approx A_{//}-A_{\perp}$ (T, transmission; A, absorbance). The microwaves are as in T–S spectroscopy modulated at low frequency (say 315 Hz), so that the light intensity falling on the photodiode is doubly modulated at 100 kHz and 315 Hz. Demodulation at 315 Hz combined with suitable electronic filtering gives the normal T–S signal. Double demodulation at 100 kHz and 315 Hz gives the $(T–S)_{//}-(T–S)_{\perp} \equiv LD-(T–S)$ difference signal. This procedure is illustrated in Fig. 9. The signal-to-noise ratio is enhanced by inserting selective amplifiers in both modulation channels. Scanning the monochromator yields simultaneously the T–S and the LD-(T–S) spectra. As before, the signals are divided by the intensity I to correct for changes in lamp output, monochromator sensitivity, etc. as a function of wavelength.

To correctly evaluate the ratio R [Eq. (1)] we must mutually calibrate the T–S and LD-(T–S) signals. This calibration includes amplification factors, lock-in sensitivities, etc. It is best done by simulating the transmitted modulated light by a modulated light emitting diode (LED). The amplitude of the 100 kHz modulation at specific instrument settings can then be accurately compared with that of the 315 Hz modulation. As much as possible care has to be taken to avoid ellipticities induced by extraneous sources, such as the lamp, cryostat windows, sample cell, etc. In our setup these extraneous ellipticities amounted to less than 1% of the LD-(T–S) signal.

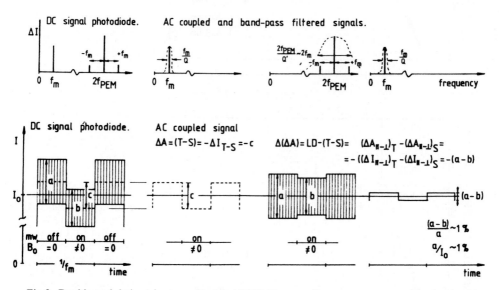

Fig. 9. Double modulation scheme used in *LD-ADMR*. *Upper part* Frequency spectrum of the signal and its side bands before and after AC coupling and filtering. *Lower part* The photodiode signal is modulated at $f_{PEM} = 100$ kHz and $f_m \sim 350$ Hz; filtering at f_m and low frequency demodulation yields the T–S spectrum. Filtering at $2f_{PEM} \pm f_m$, demodulation at $2f_{PEM}$ and subsequently at f_m yields the LD-(T–S) spectrum. In *ADMR* B_0 is always zero [77]

The whole experiment has to be repeated at various intensities of the microwave field $\bar{B}_1$, and R extrapolated for $|\bar{B}_1| \to 0$. A typical result is shown in Fig. 10, where the T–S and LD-(T–S) signals measured at the long-wavelength absorbance band of a bacterial photosynthetic reaction center are plotted against each other. The onset of saturation is clearly visible. Of course, for $|\bar{B}_1| \to \infty$ the LD-(T–S) intensity should become zero, as the change in triplet concentration then approaches an isotropic distribution (almost all triplet states are affected by the $\bar{B}_1$ field, regardless of orientation), whereas the T–S intensity levels off to a constant value. Since our highest $\bar{B}_1$ field is still comparatively low, we could not attain this limit, and the curve of Fig. 10 just approaches a plateau. At low values of $|\bar{B}_1|$ the T–S intensity is proportional to $|\bar{B}_1|$, so that the slope of the curve of Fig. 10 at low T–S intensity yields the ratio R. Note that this procedure is more accurate than just determining R at very low $\bar{B}_1$ field, because the signal-to-noise ratio then becomes proportionally low.

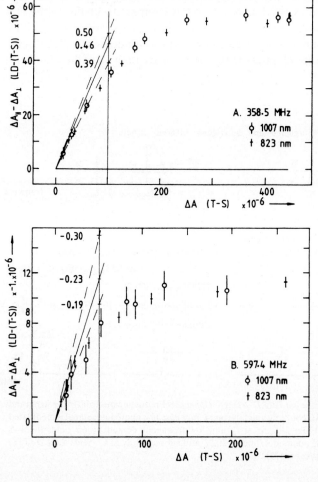

Fig. 10. LD-(T–S) versus T–S signal of *Rps. viridis* at 823 and 1007 nm at 1.2 K scaled at the point of maximum applied microwave power. (A) $|D|-|E|$ transition, (B) $|D|+|E|$ transition. The slope at low values of ΔA yields the value R of Eq. (1), from which α can be evaluated. For 1007 nm $\alpha_x = 72° \pm 5°$, $\alpha_y = 15° \pm 5°$ [66]

The *shape* of the LD-(T–S) spectrum is *not* dependent on the intensity of $\bar{B}_1$, so for the study of the orientation dependence of the influence of the triplet state on neighbouring pigments, expressed as band shifts, bleachings and appearing bands in the LD-(T–S) spectrum, one works best at comparatively high $\bar{B}_1$ amplitudes.

3 Applications of ODMR in Photosynthesis

As related in Section 2.1 the zero field splitting parameters and the sub-level populating probabilities and decay rates are fingerprints of molecular structure. They are also sensitive to environmental effects, which allows one to use the triplet state as a probe of secondary and tertiary structure, e.g. of pigment environment in photosynthetic structure. In this section the application of ODMR in photosynthesis will be briefly discussed, with emphasis on ADMR.

The use of ODMR in photosynthesis research has proved to be a particularly fruitful field of its application in biology. By far the most important triplet state in photosynthetic membranes is that generated on the primary electron donor by radical recombination under conditions in which forward electron transport is blocked by the prereduction of one of the acceptors (or its physical deletion):

$$D\ A_1A_2^- \xrightarrow{h\nu} D^*A_1A_2^- \xrightarrow{\leq 3\ ps} D^+A_1^-A_2^- \longrightarrow\ ^3D\ A_1A_2^-. \qquad (2)$$

Reaction (2) takes place in the so-called reaction center (RC), a specialized pigment-protein complex. In the reaction a photon first generates the excited singlet state of the primary donor D^2, D^*, either directly or by energy transfer from a so-called antenna pigment-protein complex. In less than 3 ps (established in photosynthetic bacteria, probable in the plant photosystems), an electron is transferred from D^* to the first acceptor A_1, creating a radical pair consisting of the cation D^+ and the anion A_1^-. From A_1^- the electron cannot travel further down the acceptor chain, because A_2 is prereduced (it cannot normally accept two electrons) or deleted from the RC. The radical pair $D^+A_1^-$ is not stable. It lives about 20 to 50 ns, and then recombines to either the singlet excited or ground state of D, or to its triplet state, ^3D. The latter reaction is almost 100% effective at cryogenic temperatures. Bear in mind that the charge separation reaction proceeds from a singlet state, $^*D$, which has zero spin angular momentum. The law of conservation of angular momentum does not allow conversion to a triplet state (which has a spin of $S = 1$) without some exchange of angular momentum from the electronic orbital motion. This cannot take place in the short time (3 ps) available. The radical pair, however, lives long enough to allow the conversion $^1[D^+A_1^-] \to\ ^3[D^+A_1^-]$, mostly because these states are almost isoenergetic (the exchange energy in the radical pair is very small), and the conversion may be driven by hyperfine interactions between

---

[2] The name of the primary donor varies with organism and with photosystem. We will use the generic label D, but for specified organisms it is convenient to use the label P with a number that designates the peak of the most red absorbance band of D. Thus, P860 in certain photosynthetic bacteria, P700 in photosystem I of plants, P680 in photosystem II, etc.

electron and nuclear spins. The triplet radical pair then recombines to ^3D (again because of conservation of angular momentum). The details of the radical pair singlet-triplet interconversion will not concern us here; for an extensive review see e.g. [13].

3.1 FDMR Spectroscopy

The primary donor triplet state was discovered by Dutton et al. [14] in RC of a photosynthetic bacterium by high field EPR. The first paper employing ODMR was by Clarke et al. [15], who presented the results of an FDMR experiment on whole cells of the photosynthetic bacterium *Rhodobacter sphaeroides* R-26. Shortly thereafter, a number of papers from the groups of Clarke, Schaafsma (Wageningen, The Netherlands), Wolf (Stuttgart, FRG) and from Leiden (The Netherlands) showed that with FDMR accurate values of |D| and |E| of the reaction center triplet and of triplets on plant antenna pigments and (bacterio) chlorophyll precursors could be obtained. Small, but well-observable differences in the ZFS parameters of the reaction center triplet, for example, showed that the primary electron donor was subtly different in closely related species [16].

The values of the k_i of the reaction center triplet that were obtained by FDMR were at first somewhat contentious, but the discrepancies were later proved to be due to the use of different methods, the saturation method being in this case much less accurate than the pulse method [4,17].

The earlier work up to 1982, all with FDMR, on plant photosynthetic material and on bacterial photosystems has been reviewed in [17] and [18] respectively, and will not be discussed here. Later applications of FDMR included the first experiments on isolated bacterial reaction centers (difficult because of the low fluorescence yield) [19,20], FDMR of triplet states in antenna complexes of several photosynthetic bacteria [21-25] and of plant photosystem I [26-28].

Apart from structural information, mainly from the ZFS parameters, a number of questions were raised and partly answered on singlet energy transfer within the antenna protein and between the antenna pigments and the reaction center. We will refrain from a further discussion of this work, and instead focus on the application of absorbance-detected magnetic resonance (ADMR), a variant of ODMR, which, after a first rather unpromising experiment on a pigment solution [29], has been developed in Leiden to a high degree of sophistication (reviewed recently in [11,30]).

3.2 ADMR Spectroscopy of Reaction Centers

The great advantage of the ADMR variant above fluorescence or phosphorescence detection is that it can always be applied, regardless of quantum yields of emission, provided the lifetime of the triplet state is not too short (this holds for all cw ODMR techniques) and a sufficient optical density can be attained (OD~0.7 gives maximal signal). Both conditions hold for the photosynthetic triplets, and in the first application of ADMR to isolated bacterial reaction centers [31] it was shown that the sensitivity of ADMR was several orders higher than that of FDMR on the same material.

The high sensitivity of the ADMR method opened the way to studies of numerous isolated reaction centers, pigment solutions, etc., both of bacterial and plant origin. Accurate values of $|D|$, $|E|$ and the k_i were determined (Tables 1 and 2). From the tables it is readily seen that the values of $|D|$ for the primary donor triplet in bacterial reaction centers are lower by 20–30% compared to that of the isolated bacteriochlorophyll (BC) pigment. At first, this was attributed to delocalization of the triplet state over the two BC making up the primary donor, and attempts were made to derive the structure of the BC dimer from the in vivo and in vitro values of $|D|$, $|E|$ and the k_i [32,33] in which an angle of 48° between the two BC planes was derived [32]. The more accurate values of k_i for BC obtained by ADMR [34], however, showed that the dimer structure was compatible with a more or less parallel dimer. The recently resolved X-ray structure of crystals of RC of the photosynthetic bacteria *Rps. viridis* and *Rb. sphaeroides* R-26 (Sect. 3.3 and Fig. 15) has fully sustained this result. From recent EPR data on the reaction center triplet state in single crystals [35] and from ADMR spectroscopy (Sect. 3.3), it was

Table 1. Representative zero field splitting parameters $(cm^{-1} \times 10^4)$ of triplet states of (bacterio)chlorophylls in vivo and in vitro

Pigment	$|D|$	$|E|$	Reference
Bacteriochlorophyll			
Rhodobacter (Rb.) sphaeroides			
Strain R-26, cells	187.2 ±0.2	31.2 ±0.2	72
Reaction centers	188.0 ±0.4	32.0 ±0.4	31
Strain 2.4.1	185.9 ±0.6	32.4 ±0.3	73
Rhodospirillum (R.) rubrum			
Strain S1, cells	187.8 ±0.6	34.3 ±0.3	72
Strain FRI, cells	187.9 ±0.6	34.3 ±0.3	73
Rhodobacter (Rb.) capsulatus			
Strain ATC 23872, cells	184.2 ±0.6	30.3 ±0.3	73
Chromatium (C.) vinosum			
Strain D, cells	177.4 ±0.6	33.7 ±0.3	73
Chloroflexus (Cfl.) aurantiacus			
Reaction centers	197.7 ±0.7	47.3 ±0.7	42
Prosthecochloris (P.) aestuarii			
Reaction centers	208.3 ±0.7	36.7 ±0.7	74
Rhodopseudomonas (Rps.) viridis[a]			
Cells	156.2 ±0.7	37.8 ±0.7	49
Reaction centers	160.3 ±0.7	39.7 ±0.7	49
BChl *a* in methyltetrahydrofuran	230.2 ±2.0	58.0 ±2.0	34
BChl *b* in methyltetrahydrofuran	221.0 ±2.0	57.0 ±2.0	34
Chlorophyll			
Photosystem I particles	281.7 ±0.7	38.3 ±0.7	45
Photosystem II particles	285.5 ±0.7	38.8 ±0.7	46
Chl *a* in methytetrahydrofuran	281.0 ±6.0	39.0 ±3.0	75

[a] Contains BChl *b* instead of BChl *a*.

Table 2. Triplet sublevel decay rates in s^{-1}

Species	k_x	k_y	k_z	Reference
Rb. sphaeroides				
Strain R-26, cells	9000 ± 1000	8000 ± 1000	1400 ± 200	72
R. rubrum				
Strain S1, cells	8000 ± 700	7200 ± 700	1350 ± 150	72
Cfl. aurantiacus				
Reaction centers	$12\,660 \pm 750$	$14\,290 \pm 800$	1690 ± 50	42
P. aestuarii				
Reaction centers	6790 ± 500	3920 ± 300	1275 ± 100	74
Rps. viridis				
Cells	$< 16\,000$	$> 16\,000$	< 2600	49
Reaction centers	$13\,700 \pm 900$	$16\,100 \pm 1300$	2420 ± 90	49
BChl *a* in MTHF	$11\,950 \pm 700$	$15\,900 \pm 1300$	1635 ± 50	34
BChl *b* in MTHF	$12\,400 \pm 900$	$14\,900 \pm 1300$	1300 ± 30	34
PS I particles	990 ± 100	1010 ± 100	93 ± 5	45
PS II particles	930 ± 40	1088 ± 50	110 ± 5	46

concluded that, at least in *Rps. viridis*, the triplet state is largely localized on one of the dimer BC, viz. D_A. The deviation of the $|D|$ value between the in vivo and in vitro BC triplet was ascribed to admixture of charge transfer (CT) states of the form $^3[D_A^+ \cdot D_B^-]$ to the monomeric ^3D state. Ca. 23% percent admixture lowers $|D|$ of *Rps. viridis* by the required amount [35]. A postulated CT state $^3[D^+ B_A^-]$ is rendered unlikely by the recent observation [36] that the values of $|D|$ and $|E|$ are virtually independent of temperature in the range 4.2–75 K.

In plant RC the values of $|D|$ and $|E|$ are practically the same as those of monomeric chlorophyll (Chl) in solution (Table 1). This can be explained by either one of three possibilities: (1) the primary donor of both plant photosystems I and II is a monomeric Chl *a* molecule, (2) the primary donor is a plane-parallel sandwich (Chl *a*) dimer with a fully delocalized triplet state (strong exciton coupling), (3) the primary donor is a dimer on which the triplet state is fully localized on one of the monomeric constituents and does not have CT admixture. In the latter case the term dimer in the sense of two *interacting* molecules obviously applies only to the singlet and possibly oxidized states of the primary donor. At present it is difficult to choose between the three possible explanations. It has recently become clear that the reaction center of photosystem II is quite homologous to that of the purple bacteria, both with respect to protein structure and the prosthetic groups. Since in the reaction center of these bacteria the primary donor is certainly dimeric, one is inclined to take the donor P680 in PS II also dimeric. Arguments in favour of this derive from the observation of a narrow EPR line of P680$^+$ (line width about $1/\sqrt{2}$ times that of monomeric Chl *a*$^+$, [37,38]), the interpretation of the optical singlet absorbance difference spectrum P680$^+$–P680 [39] and of the triplet-minus-singlet absorbance difference spectrum (Sect. 3.2.1). For photosystem I the latter spectrum also provided strong evidence that its primary donor, P700, is a Chl *a* dimer.

Remarkably, the orientation of $^3P680$ in the photosynthetic membrane is different from that found for bacteria [40]. Thus, the P680 dimer might consist of two non-parallel Chl a's, with the triplet state localized on a Chl a that is oriented more or less perpendicular to the postulated C_2 axis of the reaction center (case 3). A way to check this would be the study of triplet orientation by LD-ADMR. Preliminary results show indeed LD-(T–S) spectra that could be interpreted to result from a triplet localized on a non-parallel dimer (E.J. Lous and K. Satoh, unpublished results).

The values of the k_i of the two plant photosystems obtained by ADMR are practically identical to those of monomeric Chl a (Table 2). Again, this points to a triplet state localized on a non- or weakly interacting dimer or monomer. The line widths of the $|D| \pm |E|$ transitions are quite similar to those of the bacterial RC; the lines are approximately Gaussian and are presumably inhomogeneously broadened as a result of slight, frozen-in conformational differences.

3.2.1 T–S Spectra of Bacterial Reaction Centers

ADMR was first used to record T–S spectra in 1982 by Den Blanken et al. [8]. It was immediately clear that this technique permitted the recording of low temperature T–S spectra far more accurately than was possible with conventional flash techniques. For the first time the complicated structure in the 800 nm region for $Rb.$ $sphaeroides$ and the 830 nm region for $Rps.$ $viridis$ could be accurately measured (Figs. 11, 12). In these regions, the accessory BC's absorb. The long-wavelength absorption bands of the D_A and D_B BC's of the primary donor are shifted to the red, compared to the in vitro monomeric BC absorption, partly as a result of the strong excitonic coupling between D_A and D_B. If a (partly) localized triplet state is formed on D, then the excitonic coupling is greatly diminished. For 100% localization one BC is in the triplet state and absorbs only little in the red and near-infrared region, and one is in the singlet state, which is more or less unperturbed compared to the in vitro state, because of the absence of strong excitonic coupling. The result is that ^3D will then mostly absorb in the 800 or 830 nm region (for $Rb.$ $sphaeroides$ and $Rps.$ $viridis$ respectively), which absorption shows up as a strong positive band in the T–S spectrum [8]. The two smaller features at the long- and short-wavelength side of the positive band were attributed to band shifts of the two accessory BC's, B_A and B_B, induced by the change $D \rightarrow {}^3D$ [8].

The bleaching at 890 nm ($Rb.$ $sphaeroides$) or 990 nm ($Rps.$ $viridis$) is due to the disappearance of the red-shifted long-wavelength band of D. As stated above, exciton coupling is much weaker in the triplet state of ^3D, irrespective of its degree of localization. Therefore, in a localized ^3D state, part of the BC absorption is bleached and part shifts back to the 'normal' wavelength of BC absorption in a protein matrix.

With minor differences the above qualitative picture holds for all purple bacteria investigated [41]. The T–S spectra of green photosynthetic bacteria are more complex and have only been tentatively interpreted [42], with the exception of the green gliding bacterium $Chloroflexus$ $aurantiacus$, whose RC is very similar to that of purple bacteria [43,44].

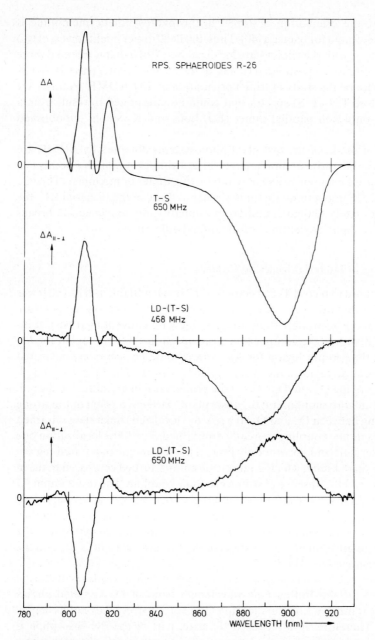

Fig. 11. T–S spectrum and LD-(T–S) spectra at 1.2 K for the two *ADMR* transitions of *Rb. sphaeroides* R-26 [48]

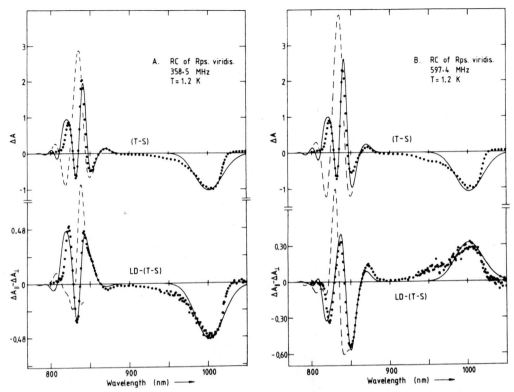

Fig. 12. T–S spectra and LD-(T–S) spectra at 1.2 of *Rps. viridis* for the two *ADMR* transitions at the |D| − |E| (**A**) and |D| + |E| (**B**). *Dots* Measured spectra; *drawn line* simulation assuming a triplet state of D localized on the D_A monomer; *dashed line* simulation assuming ^3D to be localized on D_B [66]

3.2.2 T–S Spectra of Plant Reaction Centers

The T–S spectra of reaction centers of the plant photosystems are shown in Fig. 13. Both are characterized by a strong bleaching of donor bands at 703 en 682 nm for photosystems I and II respectively [45,46]. At the blue side of this bleaching a small positive band appears, at a wavelength close to that of the absorption of Chl *a* in vitro. A similar appearing band was observed in a solution of dimeric Chl *a*, but not in one of monomeric Chl *a* (Fig. 13, bottom). The appearing band for (Chl *a*)$_2$ was attributed to a ^{3}Chl *a* state localized on one half of the dimer [47]. Following this assignment we attributed the appearing band in the in vivo T–S spectra to the absorption of one of the monomeric Chl *a* molecules of the primary donor dimer on which the triplet state is localized on the other Chl *a* molecule. Thus, in that interpretation the small appearing band in the T–S spectra of the plant photosystems is analogous to the strong appearing band in the T–S spectra of the purple bacteria, and the plant T–S spectra support the notion that in both plant photosystems the primary donor is a dimeric Chl *a* complex.

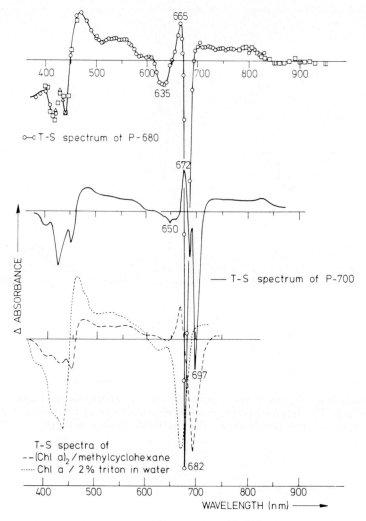

Fig. 13. T–S spectra at 1.2 K of particles of photosystem I (P700), photosystem II (P680), and of Chl *a* monomer and dimer (at -78 °C) [76]

3.2.3 Linear Dichroic T–S Spectroscopy

The information content of any absorbance spectrum is considerably enhanced when it is compared with the linear dichroism spectrum. Usually, such a LD spectrum is recorded for a sample that is oriented by stretching (e.g. RC in polyvinylalcohol) or by uni- or biaxial compression (e.g. RC in a gel or in photosynthetic membranes that are dried on a flat surface), and of which the absorbance spectrum is measured with light polarized parallel ($A_{//}$) and perpendicular ($A_\perp$) to the axis of orientation. The LD spectrum is then $A_{//} - A_\perp$ as a function of wavelength, and provides a wealth of information on the orientation of the various transition moments (tm) with respect to the axis of orientation and with respect to each other.

With known orientation of the tm in the molecule this in turn gives us the orientation of the various pigment molecules in the RC. An alternative is not to orientate the RC themselves, but to produce an oriented distribution of, e.g. photo-oxidized primary donors, by exciting the immobilized unoriented sample with a beam of linearly polarized light of a wavelength corresponding to a particular absorption band, e.g. that of D. D complexes that are oriented such that their long-wavelength tm is perpendicular to the $\bar{E}$ vector of the light are not photo-oxidized, and therefore not bleached. The (axial) distribution of bleached centers is interrogated with light polarized parallel and perpendicular to the $\bar{E}$ vector of the exciting light; the resulting difference spectrum is a *photoselection* spectrum.

As explained in Section 2.4 LD-(T–S) spectroscopy is very similar to that of photoselection, and has the advantage that the dichroic spectrum can be recorded with respect to two axes of reference that are perpendicular to each other. This reduces considerably the ambiguity of unraveling the orientation of the various tm in the T–S spectrum. Complex T–S spectra such as those of the photosynthetic RC, where many overlapping features are present as appearing bands, bleachings, and band shifts to the red and to the blue, can only be interpreted with certainty if the corresponding LD-(T–S) spectra are available. Often, a consistent interpretation is possible without a full spectral simulation. This is illustrated in Fig. 11, taken from Ref. 48.

The T–S spectrum of *Rb. sphaeroides* R-26 (curve a in Fig. 12) shows the bleaching of the long-wavelength band of the primary donor P860 (which at cryogenic temperatures absorbs at about 890 nm) and pronounced features in the BC absorption region between 790 and 830 nm. The strong peak at 807 nm *appears* because of ^{3}D formation. It is too strong and too narrow for a triplet-triplet transition, so it must be due to either an accessory BC or to D itself. The peak is much like one seen earlier in the *Rps. viridis* T–S spectrum [8], where it was flanked to the high and low energy side by features both resembling a BC shift. Thus, the strong central peak was attributed in [8] to a BC of D, which had become uncoupled from its companion BC in D and whose absorbance consequently had shifted from 990 nm to the region where, in general, BC in a protein matrix absorb. The aspect of the 800 nm region of *Rb. sphaeroides* R-26 is quite similar to the corresponding region in the T–S spectrum of *Rps. viridis* (Fig. 12). Hence we provisionally interpret it in the same way, which interpretation can now be checked by considering the LD-(T–S) spectra.

As explained before, the two LD-(T–S) spectra are recorded with microwaves corresponding to two mutually perpendicular microwave transition moments $\bar{\mu}_x^{mw}$ and $\bar{\mu}_y^{mw}$. From Eq. (1) it then follows that, if the optical transition moment of the 890 nm band, $\bar{\mu}_y^{opt}$, is about parallel to one of the $\bar{\mu}^{mw}$'s, say $\bar{\mu}_y^{mw}$ ($\alpha_y = 0$), then the 890 nm band in the LD-(T–S) spectrum should have an intensity ratio of $\frac{1}{2}$ ($\alpha_y = 0$)/$-\frac{1}{3}$ ($\alpha_x = 90°$) $= -\frac{3}{2}$ for the 468 MHz to the 650 MHz spectrum. (The 650 MHz and 468 MHz transitions are $\bar{x}$- and $\bar{y}$-polarized respectively, for D > 0, E > 0.) Note that the sign of the 890 nm band in the T–S spectrum is negative. Note also that the sign of the bands in the LD-(T–S) spectrum are determined by Eq. (1) and can be positive or negative, regardless of whether it is a bleaching or an appearing band. Experimentally, the ratio is –1.4, so we may take to a good approximation $\bar{\mu}^{opt} // \bar{\mu}_y^{mw}$.

In fact, the 890 nm band peaks at somewhat different wavelengths in the spectra of Fig. 12. This is due to site-selection: The microwave frequencies are set at a particular frequency in the $|D|\pm|E|$ ADMR transitions. For T–S spectra recorded at microwave frequencies that are slightly different but still within the ADMR transition, slightly different peak positions of the 890 nm band result. This is attributed to a correspondence between the exact values of $|D|$ and $|E|$ and the peak wavelength of the long-wavelength absorption of D, which are different for slightly different configurations of D (which probably are in a dynamic equilibrium at 300 K but are frozen in at 1.2 K [49]).

At the low energy flank of the '890' nm band in the T–S spectrum, a pronounced shoulder, which has about the same relative intensity in both the LD-(T–S) spectra, is visible. The transition moment of this shoulder has, therefore, an orientation that is parallel to the main band. Hence, it is unlikely that it is due to a charge transfer transition, which is expected to have a different orientation. (For a discussion of this point see Refs. 48 and 50.)

Turning now to the 800 nm region, we see that the 807 nm band always has a sign opposite to that of the 890 nm band. This indicates that its transition moment must, to a good approximation, be parallel to that of the 890 nm. Its intensity ratio for the LD-(T–S) spectra at 468 and 650 MHz is about –1.0, however, which is incompatible with $\alpha_y \sim 0$. In fact, inspection of Eq. (1) shows that the intensity ratio is always smaller than $-\frac{3}{2}$ or larger than $-\frac{2}{3}$ for $\bar{\mu}^{opt}$, $\bar{\mu}_x^{mw}$ and $\bar{\mu}_y^{mw}$ lying in one plane. The problem is, of course, that the 807 nm band is not a single band, but contains important contributions from the BC shifts. Inspection of Fig. 11 shows that the accessory BC absorptions are oriented differently from the 807 nm band, so that the intensity of the mixed 807 nm band cannot be used to calculate the precise value of α. This can only be done with confidence by a spectral simulation in which the contributions of all bands and their orientations are taken into account. This will be illustrated with recent simulations of the T–S and LD-(T–S) spectra of *Rps. viridis*, to which we now turn.

3.3 Spectral Simulations with Exciton Theory

The full power of T–S and LD-(T–S) spectroscopy is realized when the spectra can be compared with spectra simulated with exciton theory. It is beyond the scope of this chapter to explain that theory in any detail; several excellent reviews are available [51–53]. The gist of the theory is shown in Fig. 14. Two non-interacting identical molecules in identical environment have identical energies of the ground state and the singlet excited state. Allowing the two molecules to interact electrostatically via the dipole-dipole coupling between their electronic transition moments causes a shift and a splitting of the excited state, which is now composed of two levels with (apart from the shift) energies given by the original energy, plus or minus the interaction energy. The corresponding wavefunctions are the symmetric (+) and antisymmetric (–) combination of the original wavefunctions. The original absorption band of the two non-interacting molecules is now split, and the intensity of each component depends on the geometry of the dimer: For a parallel

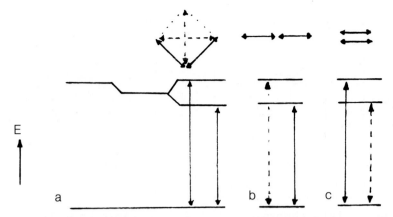

Fig. 14a-c. Two exciton coupled transition moments making an angle of **a** 90°, **b** 180° (parallel head-to-tail), **c** 0° (parallel side-by-side). Allowed transitions are indicated by *drawn arrows;* forbidden by *dashed arrows.* Note that for an angle of 90°, the two transitions are both allowed, and perpendicularly polarized

head-to-tail configuration and positive interaction energy, the lower energy component is allowed with twice the monomer intensity, and the higher energy transition is forbidden (zero intensity). For a parallel side-by-side configuration of the two tm's, the situation is reversed, whereas for non-parallel tm's the relative intensities depend on the angle θ between the tm's, according to the relations $|M(\pm)|^2 = |M|^2 (1 \pm \cos \alpha)$ and $|M(+)|^2/|M(-)|^2 = \cos^2(\alpha/2)$.

The theory of two excitonically interacting pigments is readily extended to n pigments, each pigment interacting differently with the other n–1 pigments. Since the exciton coupling between any two pigments is (for distances large compared to the molecular dimensions) a dipole-dipole interaction between their tm vectors, it depends on the angle between the tm's and that between each tm and the connecting distance vector of the two pigments, on the dipole strengths of the tm's (in Debye²), as well as on the inverse cube of the distance. The band splittings and band intensities of the coupled system are functions of the exciton coupling, but also of the ground state energies of the non-interacting pigments, which need not be the same because of differences in molecular structure (e.g. BC versus bacteriopheophytin, Φ) or because of environmental effects ('site splitting'). This can all be taken into account and, given a full set of n tm vectors (orientation, strength, distance, unperturbed ground state and excited state energies), the absorption spectrum of the excitonically coupled spectrum is readily obtained. Generally the resulting bands are mixtures of the original uncoupled bands, with tm's that are vectorial combinations of the original tm's.

Bacterial RC and also the RC of plant photosystem II contain six pigments (2 (B)Chl of the primary donor, 2 accessory (B)Chl and 2 (bacterio)pheophytins, Φ). This is a manageable exciton problem with respect to number, but obviously one has to limit parameter space with respect to angles of tm orientation, distances, etc. Fortunately, we are greatly aided by the crystal structure of the RC of *Rps. viridis*

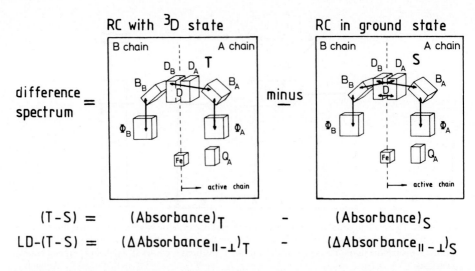

$$(T-S) = (Absorbance)_T - (Absorbance)_S$$
$$LD-(T-S) = (\Delta Absorbance_{\|-\perp})_T - (\Delta Absorbance_{\|-\perp})_S$$

Fig. 15. Exciton coupling between the pigments in the reaction center of *Rps. viridis* (schematized from [54]). In the *left panel* the triplet state is localized on the D_A monomer of D, and several exciton couplings are missing that are present in the *right panel*, which represents the singlet ground state of the reaction center (After [77])

that has recently become available [1,54] (Fig. 15). Exciton calculations based on this structure yielded a satisfactory simulation of the RC absorbance spectrum and of its LD measured on flat crystals [55,56], with the proviso that the 'unperturbed' energies of the primary donor BC in the near-infrared had to be taken considerably red-shifted with respect to the absorption band of BC in in vitro solution. (The unperturbed energies of the accessory BC were only slightly red-shifted.) There are many reasons why such a red shift could occur, such as interactions with the protein matrix, charge transfer effects, etc. If due to the latter, one should really take these into account from the start and incorporate them into the exciton calculation, but the difficulty is that they are hard to calculate exactly. In recent papers [57,58], advanced quantum chemical calculations are performed using the atomic coordinates of the *Rps. viridis* structure, and spectral simulations are done. One conclusion is that the long-wavelength band of P has considerable charge transfer character. This conclusion is supported by measurements of the effect of an electric field on the spectrum (Stark effect) [59–61] and by hole burning spectroscopy [62–63]. One may provisionally take this charge transfer character into account by allowing a pre-exciton red shift of the D_A and D_B absorption.

The above-discussed spectral simulation based on exciton theory has been successfully applied to the T–S and LD-(T–S) spectra [64–66]. We illustrate this in Figs. 12 and 15, which show the ADMR-monitored T–S and LD-(T–S) spectra of *Rps. viridis* for two microwave transitions and their simulations based on the coupling scheme of Fig. 15. Band shapes were assumed to be Gaussian (which for the long-wavelength band of P is only a first approximation). Nevertheless, it is seen that the simulation is quite good.

The spectral fits allow one to draw quite an array of conclusions, several of which are really assumptions that are vindicated by the success of the fit:

1. The Q_y transition moment of monomeric BC in vitro is oriented very close to the N_I-N_{III} axis;
2. The triplet is localized on the D_A component of D;
3. The y spin axis of the localized triplet state is very close to the N_I-N_{III} axis at an angle of $10°$, and the x and y spin axes lie approximately in the plane of the BC macrocycle. (Sublevel ordering y, x, z for D,E $>$ 0.);
4. The tm of the long-wavelength band of D is oriented close to the y spin axis of the triplet state, making an angle of $24°$ with the N_I-N_{III} axis, and lies close to the plane of the macrocycle. This agrees with the simple exciton picture of the two coupled BC of D, which yields a D(−) exciton transition at an angle of about $17°$ with either BC;
5. ^{3}D has a triplet-triplet absorption at about 872 nm, which is oriented along the N_{II}-N_{IV} axis;
6. The appearing band at 830 nm is mostly due to the D_B pigment, with contributions of the adjacent accessory BC;
7. The 'shifts' at 820 and 845 nm are due to the accessory BC's, whose excitonic coupling to ^{3}D is mostly with D_B, whereas in the singlet state they couple to both D_A and D_B. The feature at 820 nm is a shift to higher energy of one of the accessory BC, which at 850/870 nm is composed of a negative band that is part of a band shift to higher energy of the other accessory BC, and a positive band at 870 nm that is due to triplet-triplet absorption of ^{3}D;
8. The two Φ pigments, Φ_A and Φ_B, contribute only small difference bands in the 790 nm region. In other words, they are only weakly coupled to D and hardly sense the difference between D and ^{3}D.

Conclusions 2–5 are summarized in Fig. 16. Conclusions 6–8 fully bear out the original tentative interpretation of the T–S spectrum of *Rps. viridis* [8]. In addition to the above conclusions, the calibrated T–S and LD-(T–S) spectra allowed us to calculate the position of the Q_y tm of all coupled BC's and Φ's in the x, y, z spin axes coordinate frame, and consequently all their mutual angles.

Work similar to that done on *Rps. viridis* is now in progress for RC of *Rb. sphaeroides* R-26, whose crystal structure has recently also been elucidated [67,68]. Preliminary results indicate that here the triplet state is only partly localized. (See also [65] for a calculation based on the uncalibrated T–S and LD-(T–S) spectra published in [48].)

The above results show that, once the crystal structure is known, one can simulate the optical difference spectra quite well. From this a number of important conclusions can be drawn, not regarding the spatial positions of the pigments themselves (they were given by the crystal structure), but regarding the position of their tm's and the excitonic coupling between them. Now that the (fairly simple) exciton treatment appears to be good enough to accurately simulate optical spectra, one might ask whether the reverse is also possible: predicting the crystal structure from a simulation of optical (difference) spectra. For completely unknown RC this obviously is a tall order, in view of the very large parameter space. For RC closely

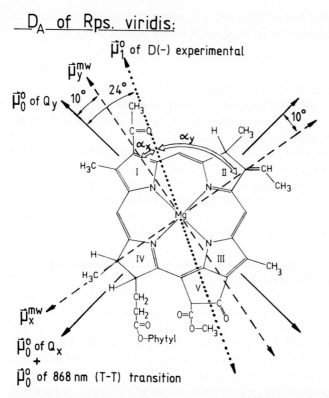

Fig. 16. Results of *ADMR*-monitored T–S and LD-(T–S) spectroscopy of *Rps. viridis*. The orientation of the microwave transition moments, the Q_y monomer and the Q_y dimer (D) transition moments and the T–T transition moment of ^{3}D have been determined in the molecular frame of D_A. For details see [66] (From [77])

related to those of *Rps. viridis* and *Rb. spaeroides* R-26, however, this can indeed be done, as was recently demonstrated by Scherer and Fischer [65,69], Vasmel [40] and H. Vasmel (unpublished simulations), who were able to simulate and predict remarkably well the T–S and LD-(T–S) spectra of *Cfl. aurantiacus* [44] and of a chemically modified RC of *Rb. sphaeroides* R-26 [69–71]. In the latter RC, the accessory B_B pigment is removed from the RC by borohydride treatment. Scherer and Fischer showed that moving the D dimer 2 Å towards the hole left by the B_B pigment sufficed to account for all the spectral difference between modified and native RC.

4 Conclusions and Prospects

These are as follows:

1. The triplet state is a versatile probe of structure and function in biology on a molecular level;
2. ODMR of triplet states is a powerful tool to study the structure of pigmented proteins, such as are found in photosynthetic membranes, energy transfer between these pigment-protein complexes, etc.;
3. ADMR permits the recording of isotropic and linear dichroic triplet-minus-singlet absorbance difference spectra of pigment-protein complexes with un-parallelled accuracy and sensitivity. Spectral simulation provides a wealth of information on the structure of, and coupling between, chromophores;
4. Applications of ODMR in photosynthesis will focus on the study of the structure of reaction centers and of antenna complexes, on energy transfer, and on relating the optical properties of reaction centers and antenna complexes to their structure and function via ADMR-monitored triplet absorbance difference spectroscopy;
5. ODMR is a physical technique, developed and applied by physicists and chemists. Its applications in biology show great promise for elucidating a number of important problems. Up to now, however, these applications have been carried out in only a handful of laboratories. Further growth and dissemination can only happen when physical laboratories acknowledge the importance of applying sophisticated physical techniques to biological problems, and biology departments acknowledge the necessity of investing in expensive physical techniques. Regarding both, there is much left to be desired.

Acknowledgements. Much of the work carried out in Leiden was performed by Drs. H.J. den Blanken and E.J. Lous, who, as graduate students, paired tremendous enthusiasm with great experimental and theoretical skill. I owe a special debt to Profs. J.H. van der Waals and J. Schmidt of the Leiden Centre for the Study of Excited States of Molecules, who provided equipment, laboratory space and stimulating interest. Mrs. Tineke Veldhuyzen assisted, with remarkable efficiency, in the preparation of the manuscript.

References

1. Clarke RH (ed) (1982) Triplet state ODMR spectroscopy. Wiley-Intersci, NY
2. Van Dorp WG, Schaafsma TJ, Soma M, Van der Waals JH (1973) Investigation of the lowest triplet state of free base porphin by microwave induced changes in its fluorescence. Chem Phys Lett 21:221–225
3. Van Dorp WG, Schoemaker WH, Soma M, Van der Waals JH (1975) The lowest triplet state of free base porphin. Determination of its kinetics of populating and depopulating from microwave-induced transients in the fluorescence intensity. Mol Phys 30:1701–1721
4. Hoff AJ, Cornelissen B (1982) Microwave power dependence of triplet state kinetics as measured with fluorescence detected magnetic resonance in zero field. An application to the reaction centre bacteriochlorophyll triplet in bacterial photosynthesis. Mol Phys 45:413–425

5. Chiha PA, Clarke RH (1978) Triplet-state intersystem crossing rates from optically detected magnetic resonance spectroscopy. J Magn Res 29:535–543

6. Beck J, Kaiser GH, Von Schütz JU, Wolf HC (1981) Optically excited triplet states in the bacteria *Rhodopseudomonas sphaeroides* 'wild-type' detected by magnetic resonance in zero-field. Biochim Biophys Acta 634:165–173

7. Beck J, Von Schütz JU, Wolf HC (1983) Optically detected magnetic resonance of porphyrin complexes in the bacterium *Rhodopseudomonas sphaeroides*. Z Naturforsch 38c:220–229

8. Den Blanken HJ, Hoff AJ (1982) High-resolution optical absorption-difference spectra of the triplet state of the primary donor in isolated reaction centers of the photosynthetic bacteria *Rhodopseudomonas sphaeroides* R-26 and *Rhodopseudomonas viridis* measured with optically detected magnetic resonance at 1.2 K. Biochim Biophys Acta 681:365–374

9. Vermeglio A, Breton J, Paillotin G, Cogdell R (1978) Orientation of chromophores in reaction centers of *Rhodopseudomonas sphaeroides*: a photoselection study. Biochim Biophys Acta 501:514–530

10. Hoff AJ (1985) Triplet-minus-singlet absorbance difference spectroscopy of photosynthetic reaction centers by absorbance detected magnetic resonance. In: Michel-Beyerle ME (ed) Antennas and reaction centers of photosynthetic bacteria. Structure, interaction and dynamics. Springer, Berlin Heidelberg New York Tokyo, pp 150–163

11. Meiburg RF (1985) Orientation of components and vectorial properties of photosynthetic reaction centers. Doctoral dissertation, Univ Leiden

12. Hardy WN, Whitehead LA (1981) Split-ring resonator for use in magnetic resonance from 200–2000 MHz. Rev Sci Instrum 57:213–216

13. Hoff AJ (1981) Magnetic field effects on photosynthetic reactions. Quart Rev Biophys 14:599–665

14. Dutton PL, Leigh JS, Seibert M (1972) Primary processes in photosynthesis: in situ ESR studies on the light induced oxidation and triplet state of reaction center bacteriochlorophyll. Biochem Biophys Res Commun 46:406–413

15. Clarke RH, Connors RE, Norris RF, Thurnauer MC (1975) Optically detected zero-field magnetic resonance studies of the photoexcited triplet state of the photosynthetic bacterium *Rhodospirillum rubrum*. J Am Chem Soc 97:7178–7179

16. Hoff AJ, Van der Waals JH (1976) Zero field resonance and spin alignment of the triplet state of chloroplasts at 2 K. Biochim Biophys Acta 423:615–620

17. Hoff AJ (1982) ODMR spectroscopy in photosynthesis. II. The reaction center triplet in bacterial photosynthesis. In: Clarke RH (ed) Triplet state ODMR spectroscopy. Wiley-Intersci, NY, pp 367–425

18. Schaafsma TJ (1982) ODMR spectroscopy in photosynthesis I. The chlorophyll triplet state in vitro and in vivo. In: Clarke RH (ed) Triplet state ODMR spectroscopy. Wiley-Intersci, NY, pp 292–365

19. Den Blanken HJ, Van der Zwet GP, Hoff AJ (1982) Study of the long-wavelength fluorescence band at 920 nm of isolated reaction centers of the photosynthetic bacterium *Rhodopseudomonas sphaeroides* R-26 with fluorescence detected magnetic resonance in zero field. Biochim Biophys Acta 681:375–382

20. Angerhofer A, Von Schütz JU, Wolf HC (1984) Fluorescence-ODMR of reaction centers of *Rhodopseudomonas viridis*. Z Naturforsch 39c:1085–1090

21. Beck J, Von Schütz JU, Wolf HC (1983) Fluorescence-ODMR of chlorophyll in photosynthetic bacteria. Reaction centers of *Rhodopseudomonas sphaeroides* R-26. Chem Phys Lett 94:141–146

22. Beck J, Von Schütz JU, Wolf HC (1983) Fluorescence-ODMR of chlorophyll in photosynthetic bacteria. LH-1 antenna complexes of *Rhodopeudomonas capsulata* Ala⁺ pho⁻. Chem Phys Lett 94:147–151

23. Angerhofer A, Von Schütz JU, Wolf HC (1985) Fluorescence-ODMR of light harvesting pigments of photosynthetic bacteria. Z. Naturforsch 40c:379–387

24. Angerhofer A, Von Schütz JU, Wolf HC (1985) Fluorescence-detected magnetic resonance of the antenna bacteriochlorophyll triplet states of purple photosynthetic bacteria. In: Michel-Beyerle ME (ed) Antennas and reaction centers of photosynthetic bacteria. Structure, Interaction and Dynamics. Springer, Berlin Heidelberg New York Tokyo, pp 78–80

25. Angerhofer A, Von Schütz JU, Wolf HC (1987) Optical excited triplet states in antenna complexes of the photosynthetic bacterium *Rhodopseudomonas capsulata* Ala⁺ detected by magnetic resonance in zero field. In: Biggins J (ed) Progress in photosynthesis research vol 1. Nijhoff, Dordrecht, pp 427–430

26. Searle GFW, Koehorst RBM, Schaafsma TJ, Möller BL, Von Wettstein D (1981) Fluorescence detected magnetic resonance (FDMR) spectroscopy of chlorophyll-proteins from barley. Carlsberg Commun 46:183–194

27. Schaafsma TJ, Searle GFW, Koehorst RBM (1982) Fluorescence-detected chlorophyll triplet states in chlorophyll-proteins. J Mol Struct 79:461–464

28. Van Wijk FG, Schaafsma TJ (1984) Low temperature magnetic resonance and optical spectroscopy of chromatophores and (crystalline) reaction centers of *Rhodopseudomonas viridis*. In: Sybesma C (ed) Advances in photosynthesis research vol II. Nijhoff/Dr Jung, Den Haag, pp 173–176

29. Clarke RH, Connors RE (1975) An investigation of the triplet state dynamics of zinc chlorophyll b by microwave-induced changes in the intensity of fluorescence and singlet-singlet absorbance. Chem Phys Lett 33:365–368

30. Hoff AJ (1986) Optically detected magnetic resonance (ODMR) of triplet states in vivo. In: Staehelin LA, Arntzen CJ (eds) Photosynthesis III. Photosynthetic Membranes and Light Harvesting Systems vol 19. Encycl Plant Physiol, New Ser, Springer, Berlin Heidelberg New York Tokyo, pp 400–421

31. Den Blanken HJ, Van der Zwet GP, Hoff AJ (1982) ESR in zero field of the photoinduced triplet state in isolated reaction centers of *Rhodopseudomonas sphaeroides* R-26 detected by the singlet ground state absorbance. Chem Phys Lett 85:335–338

32. Clarke RH, Connors RE, Frank HA, Hoch JC (1977) Investigation of the structure of the reaction center in photosynthetic systems by optical detection of triplet state magnetic resonance. Chem Phys Lett 45:523–528

33. Hägele W, S⁻hmid D, Wolf HC (1978) Triplet-state electron spin resonance of chlorophyll a and b molecules and complexes in PMMA and MTHF. II. Interpretation of the experimental results. Z Naturforsch 33a:94–97

34. Den Blanken HJ, Hoff AJ (1983) Sublevel decay kinetics of the triplet state of bacteriochlorophyll *a* and *b* in methyltetrahydrofuran at 1.2 K. Chem Phys Lett 96:343–347

35. Norris JR, Lin CP, Budil DE (1987) Magnetic resonance of ultrafast chemical reactions. Examples from photosynthesis. J Chem Soc Faraday Trans 1(83):13–27

36. Ulrich J, Angerhofer A, Von Schütz JU, Wolf HC (1987) Zero-field absorption ODMR of reaction centers of *Rhodopseudomonas sphaeroides* at temperatures between 4.2 and 75 K. Chem Phys Lett 140:416–420

37. Ghanotakis DF, Babcock GT (1983) Hydroxylamine as an inhibitor between Z and P680 in photosystem II. FEBS Lett 153:231–234

38. Hoff AJ (1979) Applications of ESR in photosynthesis. Phys Rep 54:75–200

39. Van Gorkom HJ (1976) Primary reactants in photosystem 2. Doctoral dissertation, Univ Leiden

40. Rutherford AW (1985) Orientation of EPR signals arising from components in photosystem II membranes. Biochim Biophys Acta 807:189–201

41. Hoff AJ, Vasmel H, Lous EJ, Amesz J (1988) Triplet-minus-singlet optical difference spectroscopy of some green photosynthetic bacteria. In: Olson JM, Ormerod JG, Amesz J, Stackebrandt E, Trüper HG (eds) Green Photosynthetic Bacteria. Plenum Press, NY, pp 119–126

42. Den Blanken HJ, Vasmel H, Jongenelis APJM, Hoff AJ, Amesz J (1983) The triplet state of the primary donor of the green photosynthetic bacterium *Chloroflexus auarantiacus*. FEBS Lett 161:185–189

43. Vasmel H, Meiburg RF, Amesz J, Hoff AJ (1987) Optical properties of the reaction center of *Chloroflexus aurantiacus* at low temperature. Analysis by exciton theory. In: Biggins J (ed) Progress in Photosynthesis Research vol 1. Nijhoff, Dordrecht, pp 403–406

44. Vasmel H (1986) The photosynthetic membranes of green bacteria. Doctoral dissertation, Univ Leiden

45. Den Blanken HJ, Hoff AJ (1983) High-resolution absorbance-difference spectra of the triplet state of the primary donor P-700 in photosystem I subchloroplast particles measured with absorbance-detected magnetic resonance at 1.2 K. Evidence that P-700 is a dimeric chlorophyll complex. Biochim Biophys Acta 724:52–61

46. Den Blanken HJ, Hoff AJ, Jongenelis APJM, Diner BA (1983) High-resolution triplet-minus-singlet absorbance difference spectrum of photosystem II particles. FEBS Lett 157:21–27

47. Periasamy N, Linschitz H (1979) Photodisaggregation of chlorophyll *a* and *b* dimers. J Am Chem Soc 101:1056–1057

48. Hoff AJ, Den Blanken HJ, Vasmel H, Meiburg RF (1985) Linear-dichroic triplet-minus-singlet absorbance difference spectra of reaction centers of the photosynthetic bacteria *Chromatium vinosum, Rhodopseudomonas sphaeroides* R-26 and *Rhodospirillum rubrum* S1. Biochim Biophys Acta 806:389–397

49. Den Blanken HJ, Hoff AJ (1983) Resolution enhancement of the triplet-singlet absorbance-difference spectrum and the triplet-ESR spectrum in zero field by the selection of sites. An application to photosynthetic reaction centers. Chem Phys Lett 98:255–262

50. Den Blanken HJ, Jongenelis APJM, Hoff AJ (1983) The triplet state of the primary donor of the photosynthetic bacterium *Rhodopseudomonas viridis*. Biochim Biophys Acta 725:472–482

51. Davidov AS (1981) Theory of molecular excitons. Plenum Press, NY

52. Kasha M (1963) Energy transfer mechanisms and the molecular exciton model for molecular aggregates. Radiat Res 20:55–71

53. Pearlstein RM (1982) Chlorophyll singlet excitons. In: Govindjee (ed) Photosynthesis. Vol 1: Energy conversion in plants and bacteria. Academic Press, Lond NY, pp 293–329

54. Michel H, Deisenhofer J (1986) X-ray diffraction studies on a crystalline bacterial photosynthetic reaction center: a progress report and conclusions on the structure of photosystem II reaction centers. In: Staehelin LA, Arntzen CJ (eds) Photosynthesis III. Photosynthetic Membranes and Light Harvesting Systems vol 19. Encycl Plant Physiol, New Ser, Springer, Berling Heidelberg New York Tokyo, pp 371–381

55. Zinth W, Knapp EW, Fischer SF, Kaiser W, Deisenhofer J, Michel H (1985) Correlation of structural and spectroscopic properties of a photosynthetic reaction center. Chem Phys Lett 119:1–14

56. Knapp EW, Fischer SF, Zinth W, Sander M, Kaiser W, Deisenhofer J, Michel H (1985) Analysis of optical spectra from single crystals of *Rhodopseudomonas viridis* reaction centers. Proc Natl Acad Sci USA 82:8463–8467

57. Warshel A, Parson WW (1987) Spectroscopic properties of photosynthetic reaction centers. I. Theory. J Am Chem Soc 109:6143–6152

58. Parson WW, Warshel A (1987) Spectroscopic properties of photosynthetic reaction centers II. Application of the theory to *Rhodopseudomonas viridis*. J Am Chem Soc 109:6152–6163

59. Lockhardt DJ, Boxer SG (1987) Magnitude and direction of the change in dipole moment associated with excitation of the primary electron donor in *Rhodopseudomonas sphaeroides* reaction centers. Biochemistry 26:664–668

60. Losche M, Feher G, Okamura MY (1987) The Stark effect in reaction centers from *Rhodobacter sphaeroides* R-26 and *Rhodopseudomonas viridis*. Proc Natl Acad Sci USA 84:7537–7541

61. Braun HP, Michel-Beyerle ME, Breton J, Buchanan S, Michel H (1987) Electric field effect on the absorption spectra of reaction centers of *Rhodobacter sphaeroides* and *Rhodopseudomonas viridis*. FEBS Lett 221:221–225

62. Meech SR, Hoff AJ, Wiersma DA (1985) Evidence for a very early intermediate in bacterial photosynthesis. A photon-echo and hole-burning study of the primary donor in *Rhodopseudomonas sphaeroides*. Chem Phys Lett 121:287–292

63. Boxer SG, Lockhardt DJ, Middendorf TR (1986) Photochemical hole-burning in photosynthetic reaction centers. Chem Phys Lett 123:476–482

64. Knapp EW, Scherer POJ, Fischer SF (1986) Model studies of low-temperature optical transitions of photosynthetic reaction centers. A-, LD-, CD-, ADMR- and LD-ADMR-spectra for *Rhodopseudomonas viridis*. Biochim Biophys Acta 852:295–305

65. Scherer POJ, Fischer SF (1987) Model studies on low-temperature optical transitions of photosynthetic reaction centers. II. *Rhodobacter sphaeroides* and *Chloroflexus aurantiacus*. Biochim Biophys Acta 891:157–164

66. Lous EJ, Hoff AJ (1987) Exciton interactions in reaction centers of the photosynthetic bacterium *Rhodopseudomonas viridis* probed by optical triplet-minus-singlet polarization spectroscopy at 1.2 K monitored through absorbance-detected magnetic resonance. Proc Natl Acad Sci USA 84:6147–6151

Chang CH, Tiede D, Tang J, Smith U, Norris J, Schiffer M (1986) Structure of *Rhodopseudomonas sphaeroides* R-26 reaction center. FEBS Lett 205:82–86

68. Allen JP, Feher G, Yeates TO, Komiya H, Rees DC (1986) Structure of the reaction center from *Rhodobacter sphaeroides* R-26: The cofactors. Proc Natl Acad Sci USA 84:5730–5734

69. Scherer POJ, Fischer SF (1987) Application of exciton theory to optical spectra of sodium borohydride treated reaction centres from *Rhodobacter sphaeroides* R-26. Chem Phys Lett 137:32–36

70. Beese D, Steiner R, Scheer H, Robert B, Lutz M, Angerhofer A (1988) Chemically modified photosynthetic bacterial reaction centers: circular dichroism, Raman resonance, low temperature absorption, fluorescence and ODMR spectra and polypeptide composition of borohydride treated reaction centers from *Rhodobacter sphaeroides* R-26. Photochem Photobiol 47:293–304

71. Angerhofer A, Beese D, Hoff AJ, Lous EJ, Scheer H (1988) Linear-dichroic triplet-minus-singlet absorbance difference spectra of borohydride-treated reaction centres of *Rhodobacter sphaeroides* R-26. Applications of molecular biology in bioenergetics of photosynthesis. Narosa Publ House, New Delhi, (in press)

72. Hoff AJ (1976) Kinetics of populating and depopulating of the components of the photoinduced triplet state of the photosynthetic bacteria *Rhodospirillum rubrum, Rhodopseudomonas sphaeroides* (wild type), and its mutant R-26 as measured by ESR in zero field. Biochim Biophys Acta 440:765–771

73. Hoff AJ, Gorter de Vries H (1978) Electron spin resonance in zero magnetic field of the reaction center triplet of photosynthetic bacteria. Biochim Biophys Acta 503:94–106

74. Vasmel H, Den Blanken HJ, Dijkman JT, Hoff AJ, Amesz J (1984) Triplet-minus-singlet absorbance difference spectra of reaction centers and antenna pigments of the green photosynthetic bacterium *Prosthecochloris aestuarii*. Biochim Biophys Acta 767:200–208

75. Kleibeuker JF, Schaafsma TJ (1974) Spin polarization in the lowest triplet state of chlorophyll. Chem Phys Lett 29:116–122

76. Hoff AJ (1986) Triplets: phosphorescence and magnetic resonance. In: Govindjee, Amesz J, Fork DC (eds) Light Emission by Plants and Bacteria. Academic Press Inc, Orlando, pp 225–265

77. Lous EJ (1988) Interactions between pigments in photosynthetic protein complexes. Doctoral dissertation, Univ Leiden

78. Hoff AJ, Govindjee, Romijn JC (1977) Electron spin resonance in zero magnetic field of triplet states of chloroplasts and subchloroplasts particles. FEBS Lett 73:191–196

Laser Physical Methods:
Laser Microprobe Mass Spectrometry

G. Heinrich[1]

1 Introduction

Soon after the advent of the laser, several mass spectroscopists realized the potentials of a laser mass spectrometer combination. Laser-induced mass spectrometry was attempted as early as 1966, but without very promising results. Since the Laser Microprobe Mass Analyzer (LAMMA[R]) became commercially available in 1978, its application has been increasing rapidly in different fields, due to various particular advantages. It soon became obvious that this method allows the identification of complex molecules through finger-prints of the fragment ions. Especially in organic mass spectrometry and in particle analysis, laser desorption mass spectrometry (LDMS) is becoming a unique technique for analyzing non-volatile and/or thermally labile organic compounds on a micron-size level. The application of lasers in mass spectrometry (MS) implied a breakthrough in the further development of microanalysis. Laser radiation can be concentrated to a very high power density during extremely short periods of time. The use of the laser offered, therefore, the means to vaporize, excite or ionize a very small volume of solid material, which opened new perspectives for localization in the microscopical level. The aim of this article is to give a description of the commercial instruments for laser mass spectrometry (LMS), to explain how analysis is carried out, to point out the advantages and disadvantages of this technique and to show, on the basis of different applications, in which fields the instrument can be applied. A review of the areas of application will be presented with special emphasis on applications in the life sciences.

2 Historical Survey

The following brief historical survey is based on the paper of Hillenkamp (1982) and Michiels et al. (1984). The first reports of laser application as an ion source for mass spectrometric investigations of solid surfaces appeared in the literature of the early sixties. In these investigations lasers of various types had been used, either for evaporation only, followed by an independent ionization, e.g. by classical electron impact, or for simultaneous evaporation and ionization. Honig and Woolston (1963) used a pulsed ruby laser in conjunction with a double-focusing mass spectrometer, leading to broad kinetic energy spreads and, therefore, to limited spatial resolution. Later research concentrated on reducing laser pulse lengths and spot diameters by focusing the laser beam. Fenner and Daly (1966) developed a laser microprobe mass spectrometer with a Q-switched ruby laser with a pulse length of 30 ns and a beam of 20 μm, with detection limits of 10^{-8}–10^{-10}g. Further advance came with the development of a time-of-flight (TOF) mass spectrometer equipped with a modified laser focusing system utilizing microscope optics to focus the laser beam to a spot size of about 0.5 μm (Hillenkamp et al. 1975). Very low detection limits, down to

10^{-20} g, and a high spatial resolution could be obtained. The introduction of a time-focusing ion reflector, developed by Mamyrin et al. (1973) into the TOF-MS, which allows a high ion transmission, results in an increased mass resolution (Kaufmann et al. 1979a,b). This instrument, the LAMMA 500^R, is built by Leybold-Heraeus, Cologne. The chosen arrangement of the microscope and MS in LAMMA 500 restricts the use of this instrument mostly to thin perforable samples, e.g. tissue sections or microparticles on carrier film supported by three millimeter grids for electron microscopy (transmission mode of operation). Non-perforable samples can be analyzed according to this principle, if incident laser light is used (reflection mode). This concept was realized in the LAMMA 1000 as the first commercial instrument with reflecting laser light. The development of a new sample chamber permits the analysis of large bulk samples by using reflection geometry, with laser focusing and ion extraction on the same side of the sample (Heinen et al. 1983, Vogt et al. 1983). LAMMA 500 analysis can be described as a "transmission" technique, as the laser beam has to penetrate the sample. LAMMA 1000 analysis represents a "remission" method; the laser beam is focused onto the sample, which may be of varying thickness, and laser irradiation comes from the same side as ion emission.

Another "Laser Ionization Mass Analyzer" (LIMA) using a second order energy focusing TOF MS also became commercially available. This instrument is able to perform analysis in either a transmission or a reflection mode and is equipped with a carousel sample carrier (Dingle and Griffiths 1981; Dingle et al. 1982; Evans et al. 1983).

Another approach, which was intensively followed by Eloy (1978), is represented by the laser probe mass spectrograph (LPMS), using a magnetic sector spectrometer to analyze bulk specimens including biological samples. This instrument has recently been modified substantially with respect to the laser probe system itself, and to the mode of ion detection. Conzemius and Svec (1978) described a "Scanning laser mass spectrometer milliprobe", combining a ND-YAG (1060 nm/1 mJ) laser with a double focusing MS of the Mattauch-Herzog type. It is equipped with both electrical and photoplate ion detection. Jansen and Witmer (1982) used a Q-switched laser with a Mattauch-Herzog mass spectrometer. The performance of the five mentioned types of milli- or microprobes are compared by Michiels et al. (1984).

3 Instrumentation

All three types of laser microprobe mass analyzers, which are commercially available, are based on essentially identical principles. They use short laser pulses for the ionization of a small volume of matter and TOF mass spectrometry. We will limit our description mostly to the design of the LAMMA-500.

3.1 LAMMA 500

The LAMMA instrument (Fig. 1A) is described in detail (Hillenkamp et al. 1975; Wechsung et al. 1978; Kaufmann et al. 1979a,b; Heinen et al. 1980, 1981; Vogt et al. 1981; Verbueken et al. 1988) and works as follows: A high-power-Q-switched Neodymium-Yttrium Aluminium Garnet (ND-YAG) laser with frequence quadrupling (265 nm, 15 ns) is focused onto the sample down to a diameter of less than 1 μm, by the objective of an optical microscope, which also serves for visualizing the sample to be analyzed. The laser is used to generate very short and intense light pulses of about 2 mJ in 15 ns at the UV-wavelength of 265 nm. A Helium-Neon-laser which continuously emits red light is aligned co-linear to the invisible UV-light of the high power laser. A red spot of the He-Ne-laser marks the area where the high power laser will ionize a small volume of the sample. The laser intensity can be attenuated to 2% of the initial value by a 25-step UV-absorbing filter system.

The laser-induced ions are accelerated to 3000 V and then collected through an ion optical "einzel-type" lens into the drift tube of a 1.8 m TOF MS, including a "time focusing" ion reflector for compensating the spread of initial ion energies. In this way a complete mass spectrum of positively or negatively charged ions is available with every laser shot.

The ion detection system consists of an open secondary electron multiplier with 17 copper-beryllium dynodes. The mass spectra are stored in a fast channel transient wave-form recorder of 8-bit resolution. The recorded spectrum is displayed on a fast cathode-ray-tube (CBT) screen. It can be plotted by a strip-chart recorder or

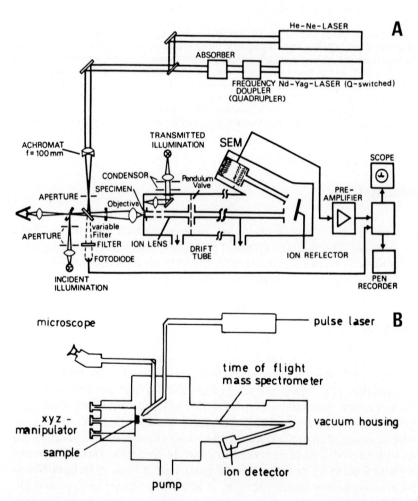

Fig. 1A,B. Schematic diagrams of **A** the LAMMA 500 (Kaufmann et al. 1979a) and **B** the LAMMA 1000 (Wink et al. 1984)

transferred into a computer for mass scale calibration, peak area calculation, spectra averaging and further statistical data treatment. More technical data and specifications of the apparatus are summarized by Verbueken et al. (1988). For sample exchange, the MS part and the sample chamber are separated by a pendulum valve. Only the sample chamber is vented with dry nitrogen, whereas the spectrometer part is continuously evacuated to 10^{-7}–10^{-6} mbar by an ion getter pump. After sample exchange, the sample chamber is evacuated by a turbomolecular pump within a few minutes.

3.2 LAMMA 1000

Figure 1B shows a schematic view of the instrument (Heinen et al. 1983; Vogt et al. 1983), which operates in the incident laser irradiation mode and opens the vast field of bulk samples to laser microprobe mass analysis. LAMMA 1000 is to LAMMA 500 what the scanning electron microscope is to the transmission electron microscope. Solid samples up to 200 mm can be subjected to the microarea analysis with a lateral resolution as small as 3 μm. A single laser pulse typically generates a crater of one μm in depth. Several shots at the same spot result in a rough depth profile (layer by layer analysis).

3.3 LIMA and LIMA-SIMS

The Laser Ionization Mass Analyzer (LIMA), Cambridge Mass Spectrometry Ltd., Cambridge, U.K., features a combination of the transmission- and reflection-type geometry and is able to perform analyses of thin sections or bulk samples in either a transmission or a reflection mode. A focused, Q-switched Nd:YAG laser is incorporated to ionize a microvolume of the specimen. Mass analysis is effected by means of the TOF technique.

LIMA SIMS is a combined laser ionization mass analyzer and a TOF/SIMS system. A pulsed ion/neutral gun is incorporated into the LIMA instrument to allow the operator to perform static SIMS and static imaging SIMS. In this particular instrument two ionization techniques, laser and ion beam, can be chosen for analysis.

4 Specimen Preparation

The LAMMA 1000 allows the analysis of bulk specimens as large as 200 mm in diameter. The LAMMA 500 sample chamber has been designed for the analysis of thin sections, mounted on 3 mm grids as used in electron microscopy. The LAMMA 500 requires, in most cases, samples that can be perforated or vaporized by laser impact. In laser desorption experiments thicker specimens, which cannot be perforated by the laser, may also be of advantage (Seydel and Lindner 1983).

Typical sample mounts include:

1. Particles with diameters between 0.2 and 20 μm, which can be attached to a thin polymer formvar or collodium film on a supporting grid. This can be done by bringing small particles into close contact with coated TEM grids so that the particles adhere sufficiently. Aerosols can be collected by aid of collector devices, with or without size classification (Kaufmann et al. 1980; Seiler et al. 1981; Wieser et al. 1980, 1981; Bruynseels and Van Grieken 1984, 1985; Bruynseels et al. 1985a,b);

2. Particles of a size between 0.1–1 mm, which can be inserted in a "sandwich grid". They have to be analyzed with laser light at grazing incidence;

3. Larger particles, e.g. fibres, which may be attached to the grid without any supporting film.

4. If the sample cannot be perforated easily, multiple laser shots can be used until complete perforation is achieved producing a mass spectrum from the final shot. Since preferential vaporization of some ions may lead to nonrepresentative analytical results, it is preferable to vaporize a sample region in the absence of the accelerating electric field. The evaporated material is recondensed and can be analyzed in the usual way (Kaufmann et al. 1980).

5. In the case of biological sample preparation for microanalysis, a basic requirement for the success of analytical studies is to minimize the movements and dissolution of specimen components during preparation. The usual processing methods for the fixation of tissues with subsequent dehydration and the usual embedding procedures for the electron microscopy are unsuitable for microanalysis. Many soluble materials will be extracted and the in vivo distribution of ions and other mobile components will be radically changed. Methods which combine shock-freezing, freeze-drying and dry plastic embedding or shock-freezing, followed by cryosection, may satisfy the requirements and may help to achieve a natural preservation of the native specimen structure. The latter is necessary, since the specimen analyzed with the LAMMA can later on be observed with the electron microscope (Edelmann 1981; Schröder 1981; Meyer zum Gottesberge 1988; Orsulakova et al. 1981).

Low temperature methods for the preservation of tissues are extensively described by Robards and Sleytr (1985).

Precipitation techniques to immobilize elements of interest prior to or during the wet chemical fixation can also be used for specimen preparation (Heinen and Schröder 1981; Lorch and Schäfer 1981a,b; De Nollin et al. 1984). One advantage of these techniques is that they are usually less time-consuming than the cryotechniques. At the same time they allow the analysis of larger areas of the specimen and can easily be combined further with various staining techniques. In many cases it is not certain if the results obtained represent the in vivo distribution of the elements under investigation. Another disadvantage of precipitation procedures is the fact that usually only one or a group of chemically similar substances can be analyzed, while other groups may be partially or fully removed and/or redistributed during the preparation.

Freeze-substitution can also serve as a method for preparing biological specimens for the localization of water-soluble substances. Material to be used for the localization of water-soluble components must be dry-sectioned, even excluding non-polar flotation media, to avoid loss of soluble material.

Soluble compounds or suspensions may also be investigated. Aliquots of about 1 μm of dilute solutions (10^{-6} mol/l) have been dipped onto formvar coated electron microscopical grids. The grids have been deep-frozen and rapidly transferred into a vacuum drying instrument, yielding thin, homogeneous substance films with occasional small crystallites on the film. Depending on the substance investigated, aqua tridest., aceton, methanol or acetic acid have been used as solvents.

Preparation methods for LAMMA analysis of dental hard tissue are described by Gabriel et al. (1981); for sections of soils by Henstra et al. (1981). Thin sections of soils can be analyzed by laser milling. Sections of weathered granite from Spain, 15 μm thick, are not directly perforable by multiple laser shots. But by starting at the edge of the section with a laser beam at grazing incidence, it was possible to produce a micro-channel; the beam was able to penetrate into the sample like a fretsaw.

In the course of the experiments the method of sample preparation and the homogeneity of the analyzed sample areas turned out to be the experimental parameter of prime importance.

5 Features of the Instrument

5.1 Advantages

All types of microprobes have their own particular merits and demerits. The advantages of the LAMMA are the following:

1. Analysis of inorganic ions in biological specimens is routinely possible. Main difficulties in this area are the sample preparation and quantification of results;
2. The lateral resolution with a diameter of < 1 μm, obtained with a 32 × objective, and of < 0.5 μm, obtained with the 100 × objective (Fig. 2A), permits the analysis of inorganic ions and organic compounds within the compartments of the cells, e.g. the content of different ions in vesicles, vacuoles, mitochondria, plastids and the nucleus. In favorable situations it would be possible to compare ER-rich regions with those poor in ER. Since analysis is performed with the aid of an optical microscope, in some cases it is necessary to use a TEM microscope to correlate the ultrastructure of the specimen with the analytical results of the laser shot (Fig. 2B). Since LAMMA analysis is destructive, TEM micrographs must be taken before and after LAMMA analysis of the regions of interest, in order to provide information about the fine structural details of the analyzed tissue. Fig. 2B shows a micrograph of secretory cells of the *Aloe transvalensis* septal nectary. Secreting cells are characterized by their highly branched wall protuberances. Blackenings are lead deposits resulting from glucose-6-phosphatase localization. It is not possible to analyze lead within the cells, although the isotopes of lead and those

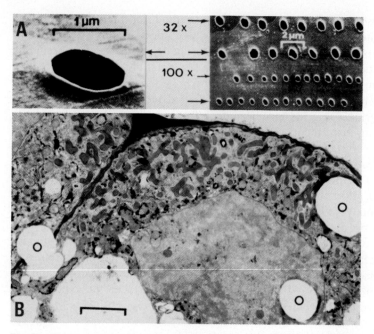

Fig. 2A. Electron micrograph of an epoxy polymer section with laser perforations obtained with a 32× or 100× objective respectively (Vogt et al. 1981); **B** TEM picture of secretory cells of the septal nectary of *Aloe transvalensis*. Note laser holes of different diameters, due to attenuating with the help of a filter set

of O_sO_4 do not interfere, but isotopes of osmic oxides, which can be found in the tissue, do so and therefore osmium fixation must be avoided;

3. As the analytical information is derived from mass spectrometry, the detection sensitivity is extremely high, so that all the elements of the periodic table can be detected;

4. From a single laser pulse, a whole mass spectrum of one polarity can be recorded;

5. Non-conductive samples can be analyzed as well, since the electrical conductivity is not a prerequisite for ionization with a laser;

6. One important application of LAMMA is the identification and localization of toxic trace elements, such as lead (Schmidt 1986; Lorch and Schäfer 1981a,b; Schmidt et al. 1983; Linton et al. 1985) or cadmium (Schmidt et al. 1986) and aluminium (Verbueken et al. 1985b);

7. Quasi-molecular ions are detected even from thermally labile and non-volatile molecules. The positive-negative ion operation is easy by push button selection;

8. The amount of energy deposited on a given sample volume to be ionized can be experimentally controlled by choosing an appropriate laser irradiance;

9. Another advantageous aspect of LAMMA analysis is the short time needed for performing one duty circle. Thus analytical and statistically significant in-

formation about several 100 individual particles, e.g. aerosol particle populations (Wieser et al. 1981) or organisms (Lindner and Seydel 1983, 1984; Seydel and Lindner 1981; Seydel et al. 1985; Heinrich et al. 1986b, 1987) can be obtained in a short period of time. A two-step procedure can be applied for the analysis of micrometer-sized particles, such as aerosols, as well as for surface and core characterization. A laser shot of low irradiance with a slightly defocused beam is placed at the particle surface. The particle remains intact, and selective evaporation of the surface material is obtained. A second laser shot of high irradiance and with a focused laser beam is aimed at the center of the particle, resulting in a complete destruction of the particle and analytical information about its core. This two-step procedure has proved to be very useful for the surface-center characterization of aerosols (Bruynseels and Van Grieken 1985) and asbestos fibers (De Waele and Adams 1985; De Waele et al. 1983);

10. With LAMMA it is possible to hold the sample outside the vacuum. In this case the ions are able to enter the vacuum through a hole which is so small that it has no significant effect on the vacuum (Holm et al. 1984; Heinen and Holm 1984). A thin polymeric carrier film or the sample itself serves as a vacuum seal to the mass spectrometer. Even after being perforated by the laser beam, the vacuum is hardly affected.

5.2 Disadvantages

These are as follows:

1. The LAMMA technique is of a destructive nature and, therefore, the spectrum of only one polarity can be recorded for the same part of the specimen;
2. The need to bring the specimen into the vacuum will create problems when the analysis of liquid or volatile particles is of interest. However, a suitable cold stage could help to overcome this difficulty;
3. With each laser-induced particle evaporation, parts of the supporting film or embedding material are also evaporated, revealing some organic background. This fact must be considered when trace elements are analyzed. However, since the spectra of the supporting foil or the embedding material are usually well-defined and highly reproducible, background subtraction can be applied;
4. As in most other microprobe techniques, quantitative analysis imposes many problems;
5. Since an optical microscope is used for sample observation, there is often no precise information in which compartment of the cell the analysis in persued. Therefore the sample must be viewed later on by TEM.

5.3 Detection Limits and Sensitivity

Since most of the physiologically important cations, such as Na, Mg and Ca, have low ionization potentials, the sensitivity of the LAMMA for these elements is extremely high. In the mass spectrum of positive ions some essential elements such as Zn, Cl or P, S and N, and toxic heavy metals, such as Hg, Cd and Pb, have high ionization energies and comparably poor detection limits. However, ions with high affinities for electron capture, such as F, P or Cl, can be detected in the spectrum of negative ions at rather good sensitivity (Kaufmann et al. 1979a).

Detection limits depend on the properties of the specimen as well as on operational conditions. The first ionization potential for the formation of positive ions and electron affinity for negative ions, the binding state of the elements with the surrounding matrix, the bond strength of organic molecules, the stability of the generated ions and the probability of recombination processes in the laser-induced microplasma have an impact on the sensitivity. In addition, instrumental parameters, such as laser wavelength, pulse duration, irradiance in the focus, as well as specific ion optical properties, also affect the analytical sensitivity (Verbueken et al. 1988).

Relative sensitivity factors in LAMMA-analysis have been published for trace metals in epoxy-resin reference material (Kaufmann et al. 1979a), for microspheres and fibres made from NBS reference glasses (Kaufmann et al. 1980), for NBS reference glass standards and SRM 1633 fly ash standards (Surkyn and Adams 1982) and for sucrose particles labelled with known amounts of inorganic metal salts (Wieser et al. 1982).

In Spurr's low viscosity medium, organometallic complexes of different metals were dissolved. Thin sections were cut by means of an ultramicrotome and mounted onto an electron microscopic grid (Kaufmann et al. 1979a). Detection limits in ppm_w for a number of metals in different matrices, listed in Table 1, demonstrate the high relative and absolute sensitivity of the instrument. A few thousand atoms of an element present in the ionized volume of the sample will suffice for detection. In the case of potassium, 10^{-20} g, corresponding to 150 potassium atoms, are needed. The absolute detection limits indicate that only a small amount of material is necessary

Table 1. Relative (ppm_w) and absolute (g) detection limits for elemental analysis (positive ions). Data provided by Leybold-Heraeus

Metal	Absolute	Relative
Li	2×10^{-20}	0.2
Na	2×10^{-20}	0.2
Mg	4×10^{-20}	0.4
Al	2×10^{-20}	0.2
K	1×10^{-20}	0.1
Ca	1×10^{-19}	1.0
Cu	2×10^{-18}	20.0
Rb	5×10^{-20}	0.5
Cs	3×10^{-20}	0.3
Sr	5×10^{-20}	0.5
Ag	1×10^{-19}	1.0
Ba	5×10^{-19}	0.5
Pb	1×10^{-19}	0.3
U	2×10^{-19}	2.0

to give a mass spectrum. The relative detection limits are not as impressive because of the small sample volume ($1-3 \times 10^{-13}$ cm^3) ionized by the laser.

The data given in Table 1 may serve for orientation purposes. Further data concerning detection limits of elements in the positive ion formation mode are listed in Michiels et al. (1984), Kaufmann (1986) and Verbueken et al. (1988).

5.4 Quantification

Quantitative analysis with LAMMA is difficult, particularly if one looks for absolute quantitation. For relative quantitation the situation is not so restricted, provided the instrument parameters are not changed between measurements and comparable samples are analyzed. A linear relationship between concentration and signal intensity (Fig. 3) together with reproducibility in consecutive LAMMA spectra constitute the basic prerequisite for quantitative analysis (Kaufmann et al. 1979a).

Standards for embedded specimens can be made by dissolving the organic complexes of the elements of interest in the same resin used for embedding the samples (Schröder 1981; Wieser et al. 1982). Dotation of the embedding media with small amounts of an internal standard often results in an inhomogeneously distribution within the resin material. Schröder (1981) obtained a repetitive pattern by vacuum deposition of the standard material onto the sections with the aid of another grid, which serves as a mask. A quartz thin film monitor should be used to control the amount of material deposited. This method allows the analysis of one spectrum with and one without the standard.

Another possibility for LAMMA calibration is the use of anionic surfactant films containing cations (Gijbels et al. 1982).

Homogeneously doped Chelex-100 ion chelating resin beads, co-embedded with the specimen itself, can also serve as a standard for calibration in biological thin

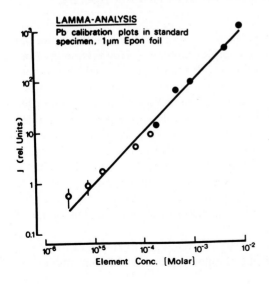

Fig. 3. Linearity between signal intensity and element concentration. Calibration plots for Pb in standard specimens (Kaufmann et al. 1979a)

sections (Verbueken et al. 1984). The standard is present in a similar chemical environment and in the same section as the analyte.

The LAMMA analysis of particulate matter implies another type of calibration, i.e. particle standards. Particles well-defined in shape and size, e.g. synthetic highly porous silica spheres, can be loaded by soaking with solutions of various salts.

6 Experimental Parameters of Different Groups

Almost every conceivable combination of suitable laser, mass spectrometer and sample stage has been used. Ruby lasers emitting at 347 and 694 nm, cwCO$_2$ lasers and CO$_2$ lasers with a 1060 nm wavelength and the 1064 nm wavelength of the Neodymium-YAG or Neodymium-glass laser in the far-infrared, as well as the frequency-doubled line at 532, the tripled line at 353 and the quadrupled line at 265 nm in the far-ultraviolet. The ruby line at 694 and the frequency doubled ruby line at 347 nm have also been applied by several groups. A N$_2$-laser is used at the Institute of Spectroscopy in Moscow.

The pulse time to irradiate the sample varies from 10^{-1} to 8×10^{-9} s.

TOF mass spectrometers have been used more often than other instruments, such as static or scanning quadrupoles, magnetic sector and douple focusing instruments.

Therefore, it is not easy to compare the results obtained by using different laser systems and sample preparations in more detail. Laser parameters of organic compounds used by the different groups in LDMS of organic compounds are summarized by Hillenkamp (1983) and van der Peijl (1984).

7 Ion Formation Mechanisms and Characteristics of Spectra

Mass spectrometry can provide a rapid and sensitive approach to the molecular weight and structural determination of complex molecules. However, many samples cannot be analyzed using classical mass spectrometry because of the low vapour pressure of the samples. In particular, nonvolatile and thermally labile samples require a special evaporation and ionization technique. Various methods have been introduced to produce stable molecular ions of nonvolatile samples. Among these are field desorption (FD), ^{252}Cf-Plasma Desorption (PD), Direct Chemical Ionization (CI) and Secondary Ion Mass Spectrometry (SIMS). The latter is also called Fast Atom Bombardment (FAB), if a neutral primary particle beam and/or a liquid matrix for dissolution of the sample is used. The common feature of all these techniques is the ionization from a liquid or solid state, hence they are also called desorption ionization techniques. The laser beam offers another possibility of producing ions of solid matter.

The laser intensity can easily be varied by an attenuating filter set and, therefore, laser microprobe mass analyzers might be operated in two different modes. Irradiances close to the threshold of ion formation, using moderate laser power at irradiances of $D < 10^8 W$ cm^2, will preferentially desorb ions from the sample surface, leading to parent ion formation (Hillenkamp 1982; Hillenkamp et al. 1987; Karas et al. 1985). Irradiances three- to tenfold above this threshold will lead to evaporation of the full target volume and yield information on the bulk composition. In particle analysis the nonthermal laser desorption (LD) as a soft ionization process is employed, if one aims at obtaining information about the target surface. Thermal evaporation and ionization (laser pyrolysis) is applied to get information about the element composition of the whole particle. The high-power ionization causes extensive fragmentation and molecular rearrangement. Molecular species originating under these conditions give some additional information about the chemical status of inorganic compounds, but limited information concerning organic constituents of the particle. A two-step procedure can be applied for the analysis of micrometer-sized particles, e.g. aerosols (Bruynseels and Van Grieken 1985). The laser desorption spectra show similarities to spectra obtained by other desorption techniques (e.g. FD–MS, SIMS, FA–BMS, PD–MS).

The mechanisms of laser-induced ionization and volatilization of solids is not well understood. In the region of direct laser impact, which is probably best characterized as a plasma, ionization must occur. An effective temperature of 7000 K has been noted in many experiments. It seems likely that only small atomic and molecular fragments are emitted from this region. The region immediately surrounding the zone of direct laser impact will not have the same temperature as the plasma, but ion emission can occur. The surface of this region is probably a liquid or highly compressed gas. A "cloud" is produced by expulsion of material into the vacuum. The absorption of the radiation in the "cloud" or plume may occur through inverse bremsstrahlung processes, which involve the absorption of photons by quasi-free electrons. Through collisions the energy of the electrons can be transferred to the neutrals and ions of the vapour plume, in which collisionally induced processes of excitation, de-excitation, dissociation, ionization and recombination may occur.

The general features of laser ionization are described by Hercules (1984) as follows:

1. The most important process for producing ions is loss or gain of protons $(M + BH \rightarrow MH^+ + B^-; MH + B \rightarrow M^- + BH^+)$.
2. In addition, spectra are generated by gain or loss of electrons $(M \rightarrow M^{+\cdot} + e^-; M + e^- \rightarrow M^{-\cdot})$. The loss of electrons corresponds to photoionization, and the gain of electrons corresponds to electron attachment. This process is not a major ion-producing mechanism in LMS;
3. A third important mechanism is the ionization of salts $(M^+ X^- \rightarrow M^+ + X^-)$, as a direct ionic dissociation of an organic salt molecule into the cation and anion. This process also occurs under purely thermal conditions, i.e. cations and anions of quaternary ammonium salts desorb from the surface through the actions of temperature alone.
4. A further type of reaction is ion attachment $(M + C^+ \rightarrow MC^+; M + A^- \rightarrow MA^-)$. High molecular weight saccharides can be volatilized by this process. The

dominant ionization process in organic LMS is alkali capture by polar organic molecules. Cationization by alkali-metal ions is quite frequent. Cationization by other metal ions, e.g. Mg, Cu, Ag etc., has also been observed (Schueler et al. 1981) as anionization by chlorine.

The laser-induced ion formation of organic solids has been discussed in detail by Hillenkamp (1982, 1983) and Hillenkamp et al. (1985, 1987).

In principle, four ion formation processes can be distinguished: Thermal evaporation of ions from the solid; thermal evaporation of neutral molecules from the solid followed by ionization in the gas phase; laser desorption; and ion formation in a laser generated plasma. Only the first two processes are called thermal.

Some more general features of spectra according to Hillenkamp (1982, 1983):

1. Polarity is not a prerequisite for desorption, in contrast to thermal evaporation where polar groups must be derivatized to achieve volability of the molecules;
2. Ions of both polarities are generated, usually at comparable abundances. For acidic compounds, specific ions such as $(M-H)^-$ are found mostly in the negative ion spectra. In the case of basic compound, ions such as $(M+H)^+$ and $(M+Alkali)^+$ are in the positive mode spectra;
3. Cluster ions such as $(2M+Na)^+$ are frequent.

   Lindner and Seydel (1984a, 1985), Seydel and Lindner (1983) and Seydel et al. (1984) used relatively thick samples which were not perforated by the laser. Mass spectra of high-molecular-weight compounds were obtained by irradiating the back surface of the sample at high laser power. The absorbed energy leads to electron excitation, the resulting electronic energy is converted to vibrational energy of the molecules and a shock front traverses the sample. When the shock front reaches the surface of the sample, intact molecules and fragments are released, e.g. quasi-molecular ion formation should then take place by attachment of alkali ions.

8 Efficiency of Different Microprobe Methods

An instrument for material analysis which can carry out microanalytical observations and measurements at high spatial resolution is called a microprobe. Up to now the Electron Probe X-ray Microanalysis, EPXM (Neumann 1979), and SIMS, the Secondary Ion Mass Spectrometry (Truchet 1975; Benninghoven 1983), have been mainly used for microanalysis. However electron microprobe instruments are restricted to purely elemental analysis. They exclude the possibility of giving information about the nature of the compounds and of carrying out in-depth analysis. Instruments equipped with a wavelength dispersive spectrometer (WDS) can only detect elements with atomic numbers >4, and for instruments with an energy dispersive spectrometer (EDS) the elemental coverage is >11. Since detection limits are in the 100 (WDS) or 700–1000 ppm (EDS) range, in many cases it is not sensitive enough to detect trace elements. The fact that isotopes cannot be discriminated is a further disadvantage. SIMS is less restricted, as there are two possibilities of sample analysis. The beam of primary ions can either be focused into a selected spot of interest or selected secondary ions can be taken for imaging the specimen. In the latter case, one obtains a pattern of the spatial distribution of an element. Both types of instruments have proved to be much more sensitive than EPXM. They are neither restricted to certain elements nor do they exclude the possibility of studying polyatomic ions. The spatial resolution lies in the range of 1–2 μm. The technical features of commercially available laser microprobes are summarized by Michiels et al. (1984) and Kaufmann (1986).

In the transmission-type LAMMA 500 the angle of laser beam incidence to sample surface is –90°, in the reflection type instrument LAMMA 1000 it is –45° and in the LIMA –90/90°. The angle of ion extraction in all three types is 90°. The sample consumption of a single laser shot is about 10^{-12}–10^{-13} g for both LAMMA types. The transmission of the TOF is 1–10%. Under practical conditions the mass range ends somewhere between 1000 and 2000 amu; the heaviest ions so far observed have m/z 5000.

9 LAMMA Applications, Inorganic Ions

The application of LAMMA has, in the meantime, been extended from biological and medical work to quite different fields such as general chemistry, mineralogy, geology, criminology, environmental research (aerosols, asbestos studies, heavy metals) and others. To give an impression of the obtainable results of LMS, some typical spectra of inorganic ions and of different classes of bio-organic compounds will be shown and discussed.

9.1 Inorganic Salts

An extensive knowledge of the mechanisms of ion formation in inorganic substances is a prerequisite for the specification of elements in the environment. Laser-induced mass spectra of inorganic salts ($CaSO_4$. $KMnO_4$, Nb_2O_5, Cu_2S) as well as organic compounds (barbital, atropine, stachyose) are presented and discussed by Heinen et al. (1980). Calcium salts were analyzed by Bruynseels and Van Grieken (1983). The LAMMA technique has also been used to study the fragmentation pattern of the nitrates of group I (Li, Na, K, Rb, and Cs) and of group II (Mg, Ca, Sr, Ba) of the periodic table by Landry and Dennemont (1984). For example, in the LAMMA spectra of potassium chloride (Fig. 4) the characteristic ion species are thought to be formed by decomposition of the parent molecules, followed by ion-molecule interactions as follows: $KCl + K^+ \rightarrow K_2Cl^+$; $K_2Cl^+ + KCl \rightarrow K_3Cl_2^+$; $K_3Cl_2^+ + KCl \rightarrow K_4Cl_3^+$ for the positive spectra and for the negative ions; $KCl + Cl^- \rightarrow KCl_2^-$; $KCl + KCl_2^- \rightarrow K_2Cl_3^-$ (Dennemont and Landry 1985).

LAMMA spectra of alkali halogenides can generally be interpreted in the following way: the positive ions consist of C^+ and cluster ions of the type $(C_{n+1}A_n)^+$,

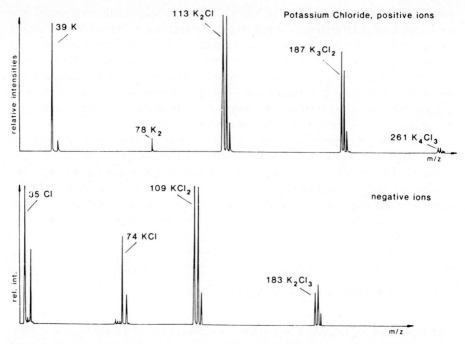

Fig. 4. LAMMA positive and negative spectra of potassium chloride (Dennemont and Landry 1985)

(C = cation, A = anion of the salt), whereas the negative ion spectrum shows A⁻ and cluster ions of the type $(C_nA_{n+1})^-$ (Heinen et al. 1980). Bruynseels and Van Grieken (1984) examined positive and negative LAMMA spectra of sodium sulfoxy salts with a different stoichiometric sulfur-to-oxygen ratio. The negative ion mode spectra of Na_2SO_4, Na_2SO_3 and $Na_2S_2O_3$ are dominated by SO_n^- ions with identical patterns which do not allow identification. This is also true for the $Na_2O_n^+$ clusters of the positive ion mode spectra. Only the positive spectrum of $Na_2S_2O_3$ is shown, since the spectra of Na_2SO_4 and Na_2SO_3 look quite similar (Fig. 5A). The ratio of the strong $Na_3SO_3^+$ to $Na_3SO_4^+$ mass peaks offers the possibility of unambiguous identification (Fig. 5B).

9.2 Aerosol Research

Microprobe techniques allow analysis of the organic and inorganic composition of aerosol particles and can help to provide information about their internal hete-rogeneity. Such information is useful for a detailed interpretation of the origin and formation mechanisms of an individual particle and of its behavior during trans-port. Particles of marine, continental or pollution-derived origin can suffer phy-sico-chemical interactions with the transport medium or with other particles.

The first LAMMA studies for single particle analysis of atmospheric aero-sols were published by Kaufmann et al. (1980) and Wieser et al. (1980). Since

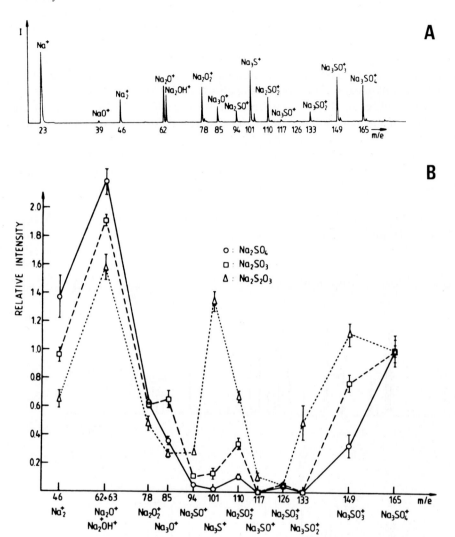

Fig. 5A. Typical LAMMA positive spectrum of Na₂S₂O₃; **B** cluster ion intensities as a function of the m/z for the sulfoxy compounds Na_2SO_4, Na_2SO_3, and $Na_2S_2O_3$. The intensities are the average of 6-fold measurements; they have been normalized to the m/z 165 mass peak; the *error bars* represent the standard deviation of the mean (Bruynseels and Van Grieken 1984)

that time the application of the LAMMA for detecting aerosol microstructures has been reported by several authors (Wieser et al. 1980, 1981;Seiler et al. 1981; Denoyer et al. 1982, 1983; Bruynseels and Van Grieken 1984; Bruynseels et al. 1985a,b; Mauney and Adams 1984; Mauney et al. 1984). Seiler et al. (1981) applied the LAMMA to study silicon-rich and sulfur-rich aerosol particles. Adams et al. (1981) and Surkyn et al. (1983) used the LAMMA for aerosol source identification.

Niessner et al. (1985) studied the behaviour of surface enriched polycyclic aromatic hydrocarbons (PAHs) under various conditions. Artificially generated NaCl particles were coated with PAHs, using a condensation technique, and exposed to reactive gases like ozone, bromine and nitrogen dioxide. They were then investigated by means of fluorometric analysis and LAMMA.

Bruynseels et al. (1985a) compared aerosol particles collected at a beach site with those of a heavily polluted industrialized area. Figure 6A gives an impression of the positive mode spectrum of a natural sea-salt particle from a beach location on the Atlantic Ocean. The major mass peaks of the positive spectra of these particles, which were formed by bubbles bursting in the ocean, are identified as Na- and K-chlorides (Na_n or K_nCl_{n-1})$^+$, the peak at m/z 165 (Na_3SO_4) is typical for alkali sulfates.

The negative LAMMA spectra are dominated by (Na or K_{n-1} Cl^{-1}), $MgCl_3^-$, ($MgCl_2$)OH^- and $CaCl_3^-$ (Fig. 6B).

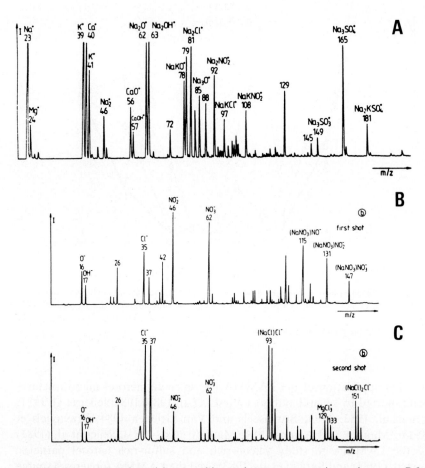

Fig. 6A-C. Natural sea-salt particles; **A** positive mode spectrum; negative mode spectra; **B** first and **C** second shot, indicating that the sea-salt particle is coated with a layer of $NaNO_3$ (Bruynseels et al. 1985a)

When applying the laser desorption mode, it becomes clear that some of these particles are coated with a layer of $NaNO_3$ devoid of chloride (Fig. 6C).

Aerosols sampled above the North Sea were described by Bruynseels et al. (1985b).

The LAMMA spectra of fly ash particles sampled at an incineration station (Dennemont and Landry 1985) are characterized by high lead peaks, which are often found in man-made polluted particles. About 30% of the particles collected near the ground at a site 10 km from Stuttgart contain lead (Wieser et al. 1980).

LAMMA was used for the characterization of oil and coal fly ash particles (Denoyer et al. 1982, 1983).

Coal mine dust particles (0.5–2 μm) were collected at the sites of 12 European pit coal mines (Kaufmann et al. 1986), and the data obtained by LAMMA analysis were correlated with the cyto- and histotoxicity data determined for each sample. The results of this study, basically confirming the role of quartz as a determinant of toxicity of coal mine dust particles, give strong evidence that there are more important toxic components related to some mineral and not to the organic constituents.

9.3 Fingerprint Analysis of Single Cells

Mycobacterium leprae. Single cell mass spectrometry has particular advantages in instances where only limited numbers of bacteria are available, as in Hansen's disease (leprosy), which is caused by the *Mycobacterium (M.) leprae*, which is not cultivable in vitro. Mass fingerprints of single cells can be used for the classification of closely related strains of *M. leprae* (Seydel and Heinen 1980). The intracellular Na/K ratio can serve as a criterion for the physiological state of a cell and can prove to be a sensitive indicator of its viability (Seydel and Lindner 1981). Changes in the intracellular Na/K ratio are induced by antibacterial agents (Lindner and Seydel 1983; Seydel et al. 1985). This enables the possibility of an in vitro drug screening and an in vitro therapy in leprosy and other diseases. A physiologically normal cell shows low intracellular sodium and high potassium contents, whereas the ratio of the treated cell has drastically changed.

Freshwater Organism, Iron and Manganese Mineralization. Envelopes, loricae and cell walls of a number of freshwater organisms were assayed for Fe and Mn mineralization by LMS and energy dispersive X-ray spectroscopy. Three types of mineralization could be distinguished: A small number of organisms exclusively showed iron or manganese incrustation, whereas most organisms investigated accumulated both Fe and Mn into their mineralized parts. All three types of mineralization may be found within the same water sample. It is therefore concluded that Fe and Mn mineralization does not solely depend upon the chemistry of the water (Heinrich et al. 1986b). Even in an iron-rich natural environment the iron bacterium *Leptothrix ochracea* occasionally accumulates manganese, while more often iron prevails.

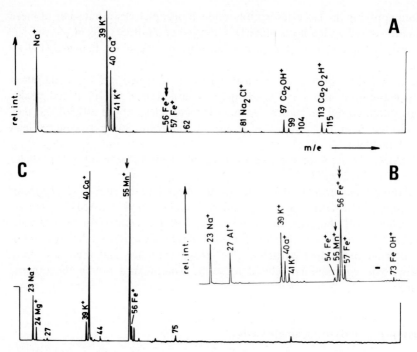

Fig. 7A. Water of the habitat, sampled in the close vicinity of **B** *Anthophysa steinii*, which possesses Fe incrustation; **C** the mineralized stalks of an *Anthophysa* collected in the river Rönne, showed considerably more Mn than Fe

Most organisms accumulate both metals; sometimes iron is the major component, other times it is manganese. An example of such an organism is *Anthophysa steinii* (Fig. 7A-C).

Tests of *Trachelomonas (T.) hispida* from a little pond accumulate only iron (Fig. 8), whereas *T. lefebre* from the same water sample shows all transitions from a predominating iron incrustation with traces of manganese to a predominating manganese incrustation with traces of iron only.

Negative mass spectra of *T. volvocina* and *T. volvocinopsis* tests are characterized by Cl⁻ peaks; those of *T. hispida* show PO_2- and PO_3- peaks; chlorid-peaks are insignificant (Heinrich et al. 1987).

9.4 Analysis of Plant Exudates

Trapping slimes of carnivorous plants, the slime of stigma papillae, nectar, latex and guttation sap show characteristic spectra. In the trapping slimes of most insectivorous plants, calcium is the most abundant ion, followed by magnesium (Heinrich 1984). This will be demonstrated for *Drosera rotundifolia* (Fig. 9A).

All trapping slimes contain the same cations as in the *Nepenthes* pitcher fluid, but in much lower concentrations. Divalent cations, such as magnesium and

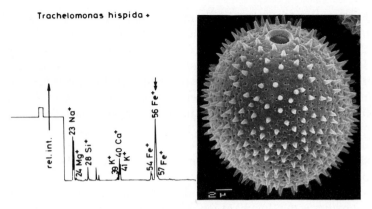

Fig. 8. *Trachelomonas hispida* with Fe incrustation

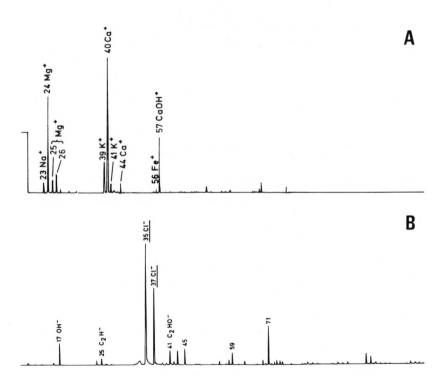

Fig. 9A. LAMMA positive spectrum of the trapping slime of *Drosera rotundifolia*, demonstrating high amounts of divalent cations; **B** in the negative spectrum of the *Nepenthes alata* slime Cl⁻ is dominant

calcium, prevail, whereas potassium and sodium occur in minor amounts. In the negative spectra, chloride predominates, as demonstrated by the trapping slimes of *Nepenthes* (Fig. 9B). As trapping slimes are extruded via dictyosome vesicles (Schnepf 1963, 1972), the exudate may represent the content of golgi vesicles.

In the nectar of most plants, potassium is the prevailing ion. In most nectars the other cations can be listed according to their decreasing amounts: Na, Ca, Mg, Al, Fe, Mn. As for the cation contents, the nectars are more similar to the latices than to the trapping mucilages of carnivorous plants. In general the magnesium content is much higher in latices than in nectars.

In the stigma slime of *Rhododendron*, calcium prevails, whereas in the nectar of the same plant potassium is abundant (Fig. 10).

The positive ion spectra of latices show that in certain families, such as Papaveraceae, potassium is the dominant ion. In other families, e.g. in Euphorbiaceae, there are species in which potassium is dominant and others in which magnesium and sometimes calcium prevails (Heinrich 1984). Latices containing rubber show carbohydrate cluster ions corresponding to C_n^- and C_nH^-.

In the positive mode spectra of the latex of fungi, potassium predominates. Negative spectra of fungi distinctly differ from those of most higher plants with regard to PO_2, PO_3 sequences (Fig. 11). In all latices of fungi examined, the mass peaks 63 (PO_2^-) and 79 (PO_3^-) were found, as well as the masses 26, 42 and 97, corresponding to the compounds CN^-, CNO^-, and $KSCN^-$. It is assumed that these peaks indicate cyanogenesis (Heinrich 1989).

9.5 Localization of Inorganic Ions and of Toxic Metals in Tissues

In microanalysis of biological tissue the preparation procedure must preserve finestructural details of the sample, and the elements to be analyzed must not get lost or be displaced from their original site. For the analysis of highly diffusable, watersoluble electrolytes, cryotechniques or precipitation methods must be employed.

LAMMA proved to be a useful instrument for element localization in single wood cells of poles treated with preservatives, such as CCB (Chronium-copper-

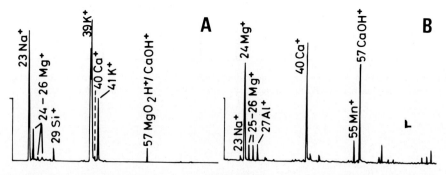

Fig. 10A. LAMMA positive spectra of the nectar; **B** the stigma slime of *Rhododendron laetum*

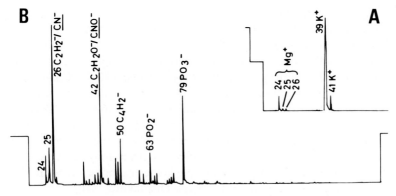

Fig. 11A. LAMMA positive and **B** negative spectra of latex of *Lactarius chrysorrheus*

boron) solutions (Klein and Bauch 1981). It also helps to investigate the distribution of elements in different cells of bark and wood of the fine roots of healthy and diseased conifers (Bauch and Schröder 1982; Bauch 1983; Stienen and Bauch 1988; Schröder et al. 1988). In the case of diseased trees growing on acidified soils in sections of freeze-dried and methacrylat embedded fine roots, a lack of the nutrient elements calcium and magnesium was demonstrated (Fig. 12).

In order to determine the primary causes for coniferous fine root damage and to study nutrient deficiencies in acidic soils, hydroponic cultures of young spruce trees in pH neutral, acidic and metal ion-amanded media were established (Stienen and Bauch 1988). Nutrient uptake by the fine roots was blocked by the following ions in order to decreasing severity: $Al^{3+} > H^+ > Fe^{3+} > Mn^{2+}$. Magnesium- and calcium-uptake, as shown by LAMMA and the X-ray microprobe, was greatly inhibited by these blockers. Al^{3+} and Fe^{3+} act predominantly in the root cortex, while Mn^{2+} is mobile throughout the whole plant.

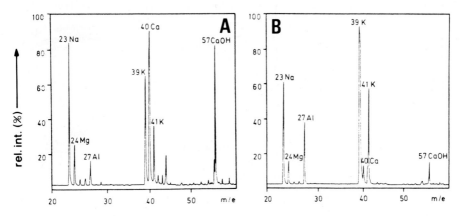

Fig. 12A. Positive mass spectrum of a cell wall in the primary xylem of a healthy fir, compared with that of **B** a diseased fir (Bauch and Schröder 1982)

Another example of LAMMA application in tissues is the qualitative analysis of the intracellular cation distribution in muscle and stria vascularis (inner ear) specimens. The distribution of cations in the lateral wall of the cochlear duct was shown not to be uniform. The K/Na ratio in the basal cells is higher (12:1) than in the spiral ligament and middle part of the stria vascularis (7:1). Changes of concentration ratios recorded after anoxia suggest that an energy dependent active transport mechanism must exist in the stria vascularis (Orsulakova et al. 1981).

Microanalytical investigations of the melanin granules of the pigment cells in the inner ear showed the affinity of divalent ions, such as Mg, Sr and Ba, for melanin (Fig. 13). It is assumed that melanin represents a physiologically important reservoir for essential trace elements and may play a key role in the enzymatically controlled processes of ionic pumps (Meyer zum Gottesberge-Orsulakova 1986).

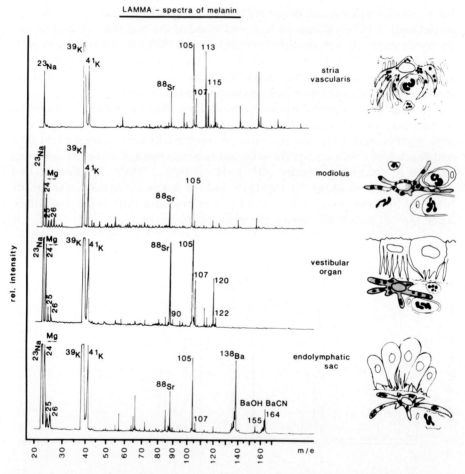

Fig. 13. LAMMA spectra of pigment granula in different parts of the inner ear. Note the differences in divalent ions Mg, Sr and Ba, and the absence of Ca (Meyer zum Gottesberge-Orsulakova 1986)

The LAMMA technique has also been applied extensively in vision research (Kaufmann et al. 1979a; Heinen et al. 1980; Schröder and Fain 1984; Fain and Schröder 1985). The calcium distribution in vertebrate and invertebrate retinas is of major interest in this field, as calcium was thought to play an important role in photoreception by acting as an internal neurotransmitter.

Using the LAMMA technique, Ca contents and Ca movements in "red" rod photoreceptors in the dark-adapted retina of the toad, *Bufo marinus*, were analyzed (Fain and Schröder 1985). Large amounts of Ca appear to be uniformly distributed within the part of the rod which contains the photopigments, whereas in the inner segment Ca levels were below the limit of detection of the LAMMA technique. In vertebrate photoreceptor pigments significant amounts of barium were found next to calcium, in contrast to the invertebrate species (Heinen et al. 1980). In the cow retina, barium is also present next to high amounts of calcium (Verbueken et al. 1985a). In the retina tissues of cats, frogs and man Ba appears to be associated with Ca and Mg, strongly bound to the pigment granula of the pigment epithelium and choroid, whereas only trace amounts of Ba were found in other cell layers of the retina (Kaufmann et al. 1979a).

Some LAMMA studies have been performed on muscle physiology as well. Measurement of cation distribution in muscle cells is highly relevant, since their function as chemicomechanical energy converters depends on transmembrane and subcellular movements of inorganic ions in a highly compartmentalized kinetic system. The Na^+/K^+ ratio as measured by LAMMA was shown to be a most sensitive criterion for the quality of the muscle preparation (Kaufmann et al. 1979a; Hirche et al. 1981).

The Ca^{2+} distribution in the vascular smooth muscle preparations of the rat and in the myocardial cells of the dog was visualized by oxalate-pyroantimonate precipitation and demonstrated to be in accordance with parallel LAMMA measurements (De Nollin et al. 1984).

Isolated single muscle fibres of the frog and small mammalian heart muscle specimens were investigated for their subcellular ionic distribution under various experimental conditions (Kaufmann et al. 1979a). Depending on membrane damages occurring during specimen preparation, the Na/K/Ca ratio can vary. Such damage induces an inward leakage current of Ca, which is rapidly taken up by the sarcoplasmic reticulum (SR) so that, finally, the Ca concentration in the SR exceeds that of K.

LAMMA has been used in studies on the accumulation of heavy metals by microorganisms, and was used to localize the distribution of lead in artificially contaminated cultures of the desmid *Phymatodocis nordstedtiana* after fixation in glutaraldehyde saturated with hydrogen sulfide (Lorch and Schäfer 1981a,b). LAMMA was able to provide information about the distribution of lead in the compartments of cells, with sensitivity superior to that of EDX analysis (Fig. 14). The subsequent examination by TEM clearly demonstrates the importance of a careful preservation of the ultrastructure. Large relative concentrations of lead were found in the nucleus as well as in parts of the chloroplast and the pyrenoid of the cell. In the cytoplasm only small amounts of lead could be demonstrated. Representative mass spectra are shown in Fig. 15.

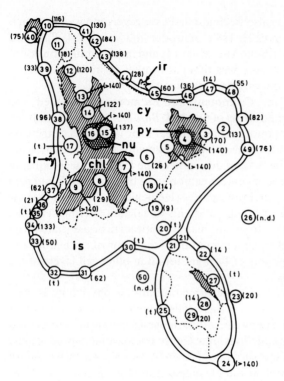

Fig. 14. Line drawing of a semi-thin section of *Phymatodocis* after LAM-MA analysis. Sample areas are designated by *numbers 1* to *50*. The relative lead concentration (calculated from peak height in the LAMMA spectra and diameter of the sample area) are given in parentheses; *n.d.* not detectable; *t* trace; *chl* chloroplast; *cy* cytoplasm; *ir* isthmus region; *is* intracellular space due to shrinkage of protoplast during preparation of samples; *nu* nucleus; *py* pyrenoid (Lorch and Schäfer 1981a)

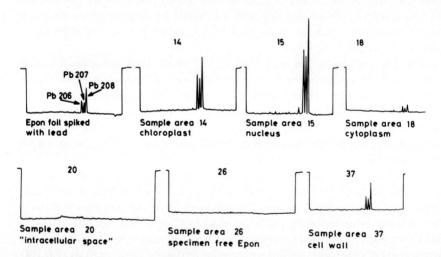

Fig. 15. LAMMA spectra of some of the sample areas and an Epon foil spiked with lead, showing only the lead peaks (Lorch and Schäfer 1981a)

The elemental composition of the halotolerant *Dunaliella salina* was studied by Bochem and Sprey (1979), and the uptake of uranium into the cells of this alga by Sprey and Bochem (1981). *Dunaliella* was shown to be resistant to high (> 1000 ppm) uranium concentrations in artificial media. Although the algal cells take up uranium, they show no accumulation of this metal and, therefore, this alga seems not to be suitable for the accumulation of uranium from seawater.

A bibliography for the years 1963–1982 has been compiled by Conzemius et al. (1983). Verbueken et al. (1985b, 1988) reported on the application of LAMMA in medicine, biology and environmental research.

10 LAMMA Applications Organic Compound Analysis

For the detection of organic compounds with LAMMA, the laser desorption mode (MD) is used. A systematic LAMMA analysis of the most important groups of biorganic molecules is a necessary prerequisite for the identification of molecules by a fingerprint analysis of spectra obtained from complex matrices (Schiller et al. 1981). Ion generation from large, nonvolatile and thermally labile organic molecules, for identification preferably via the parent molecular ion and/or structure analysis via specific fragments, poses special problems. The goal is to desorb those molecules by breaking their often many ionic or polar bonds to the substrat or surrounding matrix without exciting the intramolecular vibrational or rotational states, in order to avoid excessive fragmentation and unspecific secondary chemical reactions. This is the description from Hillenkamp et al. (1987) for what is usually called "soft" desorption.

10.1 Saccharides

Oligosaccharides are known to be thermally labile, — stachyose more than raffinose and sucrose —, and are therefore useful for studying the thermal stress exerted by ionization techniques. The appearance of a molecular ion signal of stachyose in a mass spectrum provides evidence for a thermally soft ionization. Lindner and Seydel (1985) found direct evidence for the existence of two different desorption modes, thermal and nonthermal, depending on laser power density and sample thickness. Figure 16 shows that the softest desorption occurs in spectrum a, obtained from a thick sample layer at highest laser power density and without perforating the sample. The spectrum comprises only the abundant quasi-molecular peak $(M + Na)^+$ of m/z 689 and one low-intensity fragment peak $C_{18}H_{32}O_{16}Na^+$ at m/z 527. In spectrum b, low laser power density and a thin sample layer result in a high level of fragmentation with peaks at m/z 527, m/z 365 ($C_{12}H_{22}O_{11}Na^+$) and m/z 203 ($C_6H_{12}O_6Na^+$), which stem from glycosidic bond ruptures. The quasi-molecular peak is reduced nearly to zero intensity. Spectrum c resembles the pattern of spectrum b in respect to the intensive fragments at m/z 203, 265, 527 and the NaI cluster peaks. Since the impact of a laser beam of $p \approx 10^{11}$ Wcm^{-2} leads to some thousand degrees at the sample surface, resulting in drastic fragmentation, the

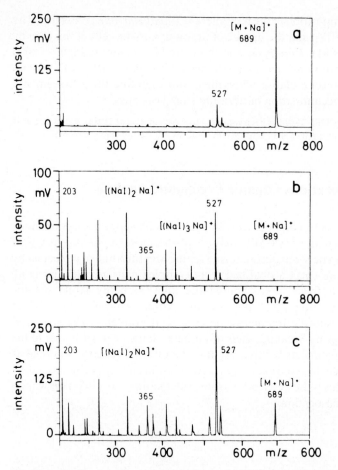

Fig. 16a-c. Positive ion LD mass spectra (average of 25 single spectra) of stachyose and NaI at a molar ratio 5:1, obtained for different sample thicknesses and laser power densities: **a** 20 $\mu m/10^{11}$ W cm^{-2}, no sample perforation; **b** 1 $\mu m/10^{8}$ W cm^{-2}, perforation; **c** 1 $\mu m/10^{11}$ W cm^{-2}, perforation (Lindner and Seydel 1985)

desorption process in Fig. 16a must be nonthermal. Therefore, Lindner and Seydel (1985) postulated a process triggered by a laser-driven shock wave. The shock-wave-driven desorption mode from organic solids is favourable, especially for molecular weight determination. An example for this application is given in Fig. 17. An unknown saccharide isolated from the bacterial species *Shigella flexneri* was mixed with stachyose and an octasaccharide for mass scale calibration and KI. The spectrum shows intense quasi-molecular ion peaks of the three sugars at m/z 705, 1340 and 1982, and two peaks each 104 mass units lower than the respective quasi-molecular ions which show alkali attachment. The unknown compound was shown to be a dodecasaccharide.

Structural studies on the lipid A component of enterobacterial lipopolysaccharides were performed by Seydel et al. (1984).

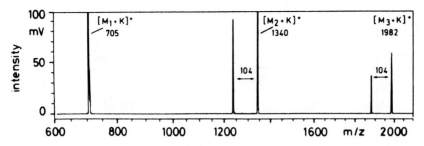

Fig. 17. Positive ion LD mass spectrum of a mixture of stachyose M_1, an octasaccharide M_2, and a dodecasaccharide M_3 and KI (sample thickness 20 µm, laser power density 10^{11} W cm^2, no perforation of sample) (Lindner and Seydel 1985)

10.2 Amino Acids, Oligopeptides and Alkaloids

Mass spectra of amino acids obtained by means of several ionization sources, such as electron impact (EI), FD, CI, SIMS, FAB and LAMMA show similarities, e.g. quasi-molecular $(M + H)^+$ ions (EI shows H^+·) and peaks corresponding to loss of formic acid from the quasi-molecular ion (Parker and Hercules 1985). Some papers describe results of systematic LAMMA investigations of biologically important amino acids (Schiller et al. 1981; Hercules et al. 1983; Karas et al. 1985; Parker and Hercules 1986; Hillenkamp et al. 1987). Amino acids are characterized by their different residues R, attached to the unsymmetrically substituted Cα-atom that also binds the acidic COOH$^-$ and the basic NH$_3^+$ group. The influence of the wavelength in LDMS of aliphatic and aromatic amino acids, as well as a number of dipeptides, is exemplified by Karas et al. (1985). Samples were prepared from a 10-mmol/l aqueous solution, and the analysis was performed with the LAMMA 1000. All threshold irradiances are given as a multiple of those from tryptophan (E_0^{Trp} at 266 nm), which was the lowest of all samples under all conditions. At 266 nm the threshold irradiances of tyrosine and phenylalanine were factors of 2 and 5 respectively, for all aliphatic AAs of 12 above that of Trp.

At their threshold irradiance, the mass spectra of all aromatic AAs, as well as the dipeptides containing at least one aromatic AA, showed intense $(M + H)^+$ ions. No alkaline or substrate ions are observed. With increasing irradiance the $(M + H)^+$ signals decrease and fragment ions, due to decarboxylations $(M + H-46)^+$ or the aromatic residue (R^+) become the base peaks. Above $5 \times E_0^{Trp}$, alkaline ions appear in the mass spectra and cationized molecules are observed. Positive ion spectra of Phenylalanine at 266 nm and threshold irradiance show only $(Phe + H)^+$ and $(Phe + H-46)^+$ ions. At only two times the threshold irradiance, strong alkali signals appear and fragmentation increases at the expense of the parent ion. At a wavelength of 355 nm, at which linear absorption is negligible, the threshold irradiance is a factor of 10 above that at 266, and no alkali-free spectrum can be generated (Fig. 18).

Dipeptids with strong absorption at 266 nm, such as Trp-Trp and Val-Trp, possess simple spectra without alkali signals. The spectrum of Trp-Trp, at low

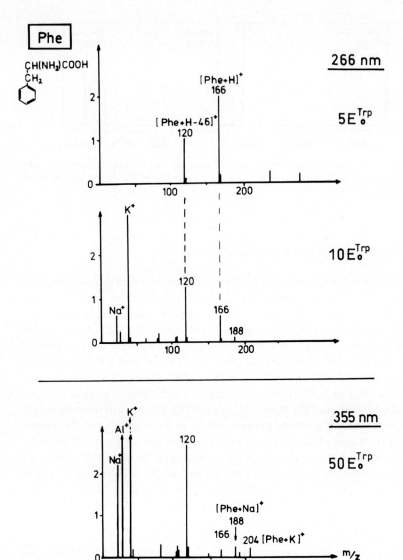

Fig. 18. Positive ion spectra of phenylalanine at 266 nm and threshold irradiance; two times threshold irradiance; and at 355 nm (Hillenkamp et al. 1985)

irradiance, shows $(M + H)^+$, and a fragment ion $(AA-45)^+$. Val-Pro with negligible absorption at 266 nm shows strong fragmentation, high alkali signals and ions such as $(M + Na)^+$ and $(M-H + 2Na)^+$. This observation can be taken as an indication of the involvement of very specific photochemical processes and that laser ion formation depends on the laser parameters, the irradiation geometries and sample properties.

In general, the relative intensities of the ion signals were found to depend on the laser irradiance. A higher degree of fragmentation is usually more apparent in the positive than in the negative spectra of acidic compounds. As an example for an acidic compound, the negative spectrum of nicotinic acid is shown (Fig. 19A,B).

Due to the amphoteric character of the compound, the positive spectrum shows strong signals of protonated molecules. CO_2 elimination from the acid molecule can be deduced from both the positive and the negative spectrum (Heinen et al. 1981). As an example for basic compounds the positive spectrum of atropine is given in Fig. 19C.

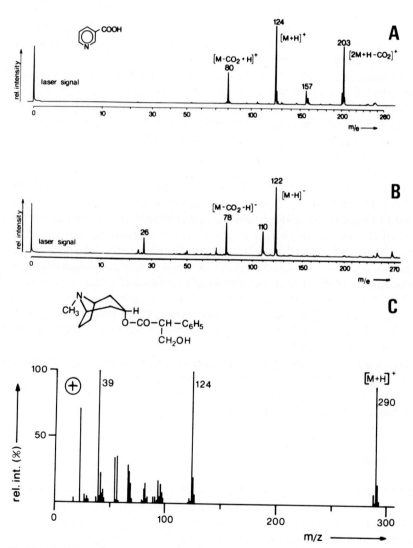

Fig. 19A. Positive and **B** negative spectrum of nicotinic acid; **C** LAMMA positive spectrum of atropine (Heinen et al. 1981)

All the organic compounds so far analyzed show a strong correlation between the chemical nature and the spectrum obtained. Except for alkali ions that often are present as impurities, atomic ions and a substantial contribution of non-specific organic ions appear at irradiances above those used for laser desorption.

10.3 Tracing of Organic Molecules in Plant Tissue

There are only a few examples demonstrating that LAMMA analysis is suitable for microanalytical characterization of organic molecules in plant cells and tissues. Microcrystals present in a cross section of the tropical microlichen *Laurera benguelensis* revealed the characteristic mass spectrum of lichexanthone with the $(M + H)^+$ parent molecule peak at 286 and some characteristic fragment ions at m/z 257 and 243 respectively (Mathey 1981).

The phytoalexin glyceollin could be detected at the cellular level in soybean cotyledons infected with an incompatible race of *Phytophthora megasperma* f. ssp. *glycinea*. 10 μm freeze microtome sections, freeze-dried on copper grids, were used. The glyceollin $(M-OH)^+$ peak at m/e = 321 was also present in the spectra of infected regions from a cotyledon, but absent in the spectra of uninfected tissue (Moesta et al. 1982).

Stem sections of *Lupinus polyphyllus* and *Cytisus scoparius* have been analyzed for distribution of the quinolizidine alkaloids sparteine and lupanine by LDMS (Wink et al. 1984). Both alkaloids could be recorded in the biological matrix and were found to be restricted to the epidermis and probably also to the one or two subepidermal cell layers. In *Cytisus scoparius* stems intensive ions at m/z 233 appeared in the peripheral cell layers but not in the inner part of the bark or the wood. This ion could not be detected in the material of *Lupinus polyphyllus* and other sparteine-free plants. In petiole sections of *Lupinus*, distinctive ions could be measured at m/z 247, in the epidermis and the neighbouring one or two subepidermal cell layers (Fig. 20). For this study the LAMMA 1000 was used, with the advantage that cross sections of varying dimensions can be analyzed directly without extensive sample preparation.

If the accumulation of an organic substance occurs, resulting in a rather high local concentration, LAMMA could be a valuable tool for plant biochemistry, especially for questions concerning spatial distribution, not only for inorganic but also for organic compounds in plant tissues.

11 Stable Isotopes as Markers

Because LAMMA can discriminate isotopes, stable isotopes can be used as a label for kinetic studies. In one and the same experiment different stable isotopes, e.g. ^{6}Li, ^{25}Mg, ^{41}K, ^{44}Ca and ^{54}Fe, could be applied as markers to study cellular or subcellular kinetics of ions or labeled inorganic and organic compounds. Highly enriched stable isotopes are available without the hazard of radiation damage or the need of precautions, as in radiotracer studies. Schröder (1981) examined the kinetics of Ca

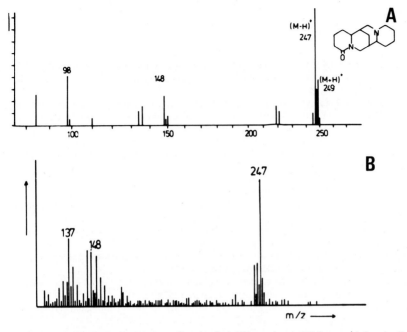

Fig. 20. LAMMA spectra of **A** the authentic alkaloid lupanine and **B** from peripheral cells of the stem of *Lupinus polyphyllus* (Wink et al. 1984)

uptake and accumulation by the DP granules within crayfish photoreceptor cells, using the ^{44}Ca stable isotope. ^{26}Mg, ^{41}K and ^{44}Ca in soil solutions were used to monitor the exchange, uptake and transport of these elements in the fine root system of spruce plants (Schröder et al. 1988).

With the aid of deuterium labelling studies, Parker and Hercules (1986) were able to demonstrate that intermolecular proton transfer is the major mechanism contributing to the production of quasi-molecular ions $(M+H)^+$ and $(M-H)^-$, in the LMS of amino acids. Isotope labelling was used to study cluster ion formation under laser bombardment (Musselmann et al. 1988). The potential for such isotope labelling studies represents a major advantage of LAMMA. With the exception of elements having no stable isotopes, such as ^{9}Be, ^{19}F, ^{23}Na, ^{27}Al, ^{31}P, ^{55}Mn, ^{59}Co, ^{75}As, ^{127}J, ^{133}Cs, ^{197}Au, there are many possibilities of using stable isotopes as a tracer. In this way the use of radioactive isotopes, such as ^{45}Ca, ^{64}Cu, ^{65}Zn, ^{86}Rb and ^{90}Sr can be avoided, and from which the handler should be protec' ,d by suitable high density shielding material such as lead.

Acknowledgements. I would like to thank Dr. H.J. Heinen of Leybold, Cologne, Prof. Dr. Ulrich Seydel, Dr. Buko Lindner, Borstel, and Dr. Walter H. Schröder, KFA Jülich, for their help in the LAMMA analysis. I am also very grateful for permission to use figures from various works. The financial support of the Fonds zur Förderung der wissenschaftlichen Forschung (Heinrich, P 6160) is also kindly acknowledged.

References

Adams F, Bloch P, Natusch DFS, Surkyn P (1981) Microscopical analysis for source identification in air chemistry and air pollution. In: Anagnostopoulos A (ed) Proc Int Conf Environ Poll, Thessaloniki, Greece, pp 20–24

Bauch J (1983) Biological alterations in the stem and root of fir and spruce due to pollution influence. In: Ulrich B, Pankrath J (eds) Effects of accumulation of air pollutants in forest ecosystems. Reidel Publ Co, pp 377–386

Bauch J, Schröder W (1982) Zellulärer Nachweis einiger Elemente in den Feinwurzeln gesunder und erkrankter Tannen (*Abies alba* Mill.) und Fichten (*Picea abies* [L.] Karst.) Forstwiss Centralbl 101:285–294

Benninghoven A (1983) Secondary ion mass spectrometry of organic compounds. In: Benninghoven A (ed) Proc Sec Int Conf Münster. Springer Ser Chem Phys 25:64–89

Bochem HP, Sprey B (1979) Laser microprobe analysis of inclusions in *Dunaliella salina*. Z Pfanzen-physiol 95:179–182

Bruynseels FJ, Van Grieken RE (1983) Molecular ion distributions in laser microprobe mass spectrometry of calcium oxide and calcium salts. Spectrochim Acta 38B:853–858

Bruynseels FJ, Van Grieken RE (1984) Laser microprobe mass spectrometric identification of sulfur species in single micrometer-size particles. Anal Chem 56:871–873

Bruynseels FJ, Van Grieken RE (1985) Direct detection of sulfate and nitrate layers on sampled marine aerosols by laser microprobe mass analysis. Atmos Environ 19:1969–1970

Bruynseels F, Storms H, Tavares T, Van Grieken R (1985a) Characterization of individual particle types in coastal air by laser microprobe mass analysis. Int J Environ Anal Chem 23:1–14

Bruynseels F, Storms H, Van Grieken R (1985b) Chemical characterization of airborne particulate matter above the North Sea. In: Lekkas TD (ed) Int Conf Heavy Metals Environ 1:189–191

Conzemius RJ, Svec HJ (1978) Scanning laser mass spectrometer milliprobe. Anal Chem 50:1854–1860

Conzemius RJ, Simons DS, Shankai Z, Byrd GD (1983) Laser mass spectrometry of solids: A bibliography 1963–1982. In: Gooley R (ed) Microbeam analysis – 1983. San Francisco Press, San Francisco, pp 301–328

Dennemont J, Landry JC (1985) Mass spectrometric identification of inorganic substances by laser microprobe mass analysis. In: Armstrong JT (ed) Microbeam analysis – 1985. San Francisco Press, San Francisco, pp 305–309

De Nollin S, Jacob W, Hentsens R (1984) Laser microprobe mass analysis (LAMMA) as a technique to quantitate Ca^{2+} ions. Dev Cardiovascular Med 40:52–58

Denoyer E, Mauney T, Natusch DFS, Adams F (1982) Laser microprobe mass analysis of coal and oil fly ash particles. In: Heinrich KFJ (ed) Microbeam analysis – 1982. San Francisco Press, San Francisco, pp 191–196

Denoyer E, Natusch DFS, Surkyn P, Adams FC (1983) Laser microprobe mass analysis (LAMMA) as a tool for particle characterization: a study of coal fly ash. Environ Sci Technol 17:457–462

De Waele JK, Adams FC (1985) Applications of laser microprobe mass analysis for characterization of asbestos. Scanning Electron Microsc III:935–946

De Waele JK, Vansant EF, Van Espen P, Adams FC (1983) Laser microprobe mass analysis of asbestos fiber surfaces for organic compounds. Anal Chem 55:671–677

Dingle T, Griffiths B (1981) A laser ion mass analyser (LIMA) for bulk samples with high spatial resolution and PPM hydrogen sensitivity. J Phys E: Sci Instrum 14:513

Dingle T, Griffiths BW, Ruckmann JC, Evans CA Jr (1982) The performance of a laser-induced ion mass analyzer system for bulk samples. In: Heinrich KFJ (ed) Microbeam analysis – 1982. San Francisco Press, San Francisco, pp 365–368

Edelmann L (1981) Selective accumulation of Li^+, Na^+, K^+, Rb^+ and Cs^+ at protein sites of freeze-dried embedded muscle detected by LAMMA. Fresenius Z Anal Chem 308:218–220

Eloy JF (1978) Chemical analysis of biological materials with the laser probe mass spectrograph. Microscop Acta Suppl 2:307–317

Evans CA Jr, Griffiths BW, Dingle T, Southan MJ, Ninham AJ (1983) Microanalysis of bulk samples by laser-induced ion mass analysis. In: Gooley R (ed) Microbeam analysis – 1983. San Francisco Press, San Francisco, pp 101–105

Fain GL, Schröder WH (1985) Calcium content and calcium exchange in dark-adapted toad rods. J Physiol 368:641–665

Fenner NC, Daly NR (1966) Laser used for mass analysis. Rev Sci Instrum 37:1068–1070

Gabriel E, Kato Y, Rech FJ (1981) Preparation methods and LAMMA analysis of dental hard tissue with special respect to fluorine. Fresenius Z Anal Chem 308:234–238

Gijbels R, Verlodt P, Tavernier S (1982) Anionic surfactant films as standards for quantitative laser microprobe mass analysis. In: Heinrich KFJ (ed) Microbeam analysis – 1982. San Francisco Press, San Francisco, pp 378–382

Heinen HJ, Holm R (1984) Recent development with the laser microprobe mass analyzer (LAMMA). Scanning Electron Microsc III: 1129–1138

Heinen HJ, Schröder W (1981) The technique of LAMMA micromass spectrometry. Biochem Soc Trans 9/6:591–593

Heinen HJ, Hillenkamp F, Kaufmann R, Schröder W, Wechsung R (1980) LAMMA – A new laser microprobe mass analyzer for biomedicine and material analysis. In: Frigerio A, McCamish M (eds) Recent developments in mass spectrometry in biochemistry and medicine, vol 6. Elsevier Sci Publ, NY, pp 435–459

Heinen HJ, Vogt H, Wechsung R (1981) Laser desorption mass spectrometry with LAMMA. Anal Chem 308:290–296

Heinen HJ, Meier S, Vogt H, Wechsung R (1983) LAMMA 1000, a new laser microprobe mass analyzer for bulk samples. Int J Mass Spectrom Ion Phys 47:19–22

Heinrich G (1984) LAMMA-Ionenspektren der Fangschleime carnivorer Pflanzen. Biochem Physiol Pflanz 179:129–143

Heinrich G, Schultze W, Schröder W (1986a) LAMMA Ionenspektren der Milchsäfte höherer Pflanzen. Biochem Physiol Pflanz 181:227–239

Heinrich G, Kies L, Schröder W (1986b) Eisen- und Mangan-Inkrustierung in Scheiden, Gehäusen und Zellwänden einiger Bakterien, Algen und Pilze des Süßwassers. Biochem Physiol Pflanz 181:481–496

Heinrich G, Kies L, Schröder W (1987) Untersuchungen über die Mineralisierung der Gehäuse von Trachelomonas-Arten (Euglenophyceae) mit Hilfe des Laser Mikrosonden Massenanalysators (LAMMA). Phyton (Austria) 26:219–225

Heinrich G (1989) LAMMA ion spectra of the latices of fungi. J Plant Physiol 133:770–772

Henstra S, Bisdom EBA, Jongerius A, Heinen HJ, Meier S (1981) Microchemical analysis on thin sections of soils with the laser microprobe mass analyzer (LAMMA). Fresenius Z Anal Chem 308:280–282

Hercules DM (1984) Solid state mass spectrometry using a laser microprobe. In: Vorhees KJ (ed) Anal pyrolysis Chapter I. Butterworths, Lond, pp 1–41

Hercules DM, Parker CD, Balasanmugam K, Viswanadham SK (1983) Laser mass spectrometry of organic compounds. In: Benninghoven A (ed) Ion formation from organic solids. Proc Sec Int Conf Münster, Springer Ser Chem Phys 25:222–228

Hillenkamp F (1982) Laser desorption technique of nonvolatile organic substances. Int J Mass Spectrom Ion Phys 45:305–313

Hillenkamp F (1983) Laser induced ion formation from organic solids. In: Benninghoven A (ed) Ion formation from organic solids. Proc Sec Int Conf Münster, Springer Ser Chem Phys 25:190–205

Hillenkamp F, Kaufmann R, Nitsche R, Unsold E (1975) A high-sensitivity laser microprobe mass analyzer. Appl Phys 8:341–348

Hillenkamp F, Karas M, Rosmarinowsky J (1985) Processes of laser-induced ion formation in mass spectrometry. In: Lyon PA (ed) Desorption mass spectrometry. ACS Symp Ser 291, Anal Chem Soc, Washington, pp 69–82

Hillenkamp F, Bahr U, Karas M, Spengler B (1987) Mechanisms of laser ion formation for mass spectrometric analysis. Scanning Microsc Suppl 1:33–39

Hirche H, Heinrichs J, Schaefer HE, Schramm M (1981) Preparation and analysis of heart and skeletal muscle specimens with LAMMA (laser microprobe mass analysator). Fresenius Z Anal Chem 308:224–228

Holm R, Kämpf G, Kirchner D (1984) Laser microprobe mass analysis of condensed matter under atmospheric conditions. Anal Chem 56:690–692

Honig RE, Woolston JR (1963) Laser induced emission of electrons, ions and neutral atoms from solid surfaces. Appl Phys Lett 2:138–139

Jansen JAJ, Witmer AW (1982) Quantitative inorganic analysis by Q-switched laser mass spectroscopy. Spectrochim Acta 378:483–491

Karas M, Bachmann D, Hillenkamp F (1985) Influence of the wavelength in high-irradiance ultraviolet laser desorption mass spectrometry of organic molecules. Anal Chem 57:2935–2939

Kaufmann R (1986) Laser microprobe mass spectroscopy (LAMMA) of particulate matter. In: Spurny KR (ed) Physical and chemical characterization of individual airborne particles. Horwood, Ltd Publ Chichister, pp 227–244

Kaufmann R, Wieser R (1980) Laser microprobe mass analysis (LAMMA) in particle analysis. In: Heinrich KFJ (ed) Characterization of particles. NBS Spec Pub 533, Washington, pp 199–223

Kaufmann R, Hillenkamp F, Wechsung R (1979a) The laser microprobe mass analyzer (LAMMA): a new instrument for biomedical microprobe analysis. Med Progr Technol 6:109–121

Kaufmann R, Hillenkamp F, Wechsung R, Heinen HJ, Schürmann H (1979b) Laser microprobe mass analysis: Achievements and aspects. Scanning Electron Microsc II:279–290

Kaufmann R, Wieser P, Wurster R (1980) Application of the laser microprobe mass analyzer LAMMA in aerosol research. Scanning Electron Microsc II:607–622

Kaufmann R, Barths G, Bruch J, Schmitz G (1986) The characterization of coal mine dust particles by LAMMA. In: Adams F, Van Vaeck L (eds) Third Int Laser Microprobe Mass Spectrometry Workshop. Antwerpen, p 121

Klein P, Bauch J (1981) Studies concerning the distribution of inorganic wood preservatives in cell wall layers based on LAMMA. Fresenius Z Anal Chem 308:283–286

Landry JCL, Dennemont J (1984) Speciation of elements in the environment by laser microprobe mass analyser (LAMMA). Arch Sci Genève 37:105–122

Lindner B, Seydel U (1983) Mass spectrometric analysis of drug-induced changes in Na^+- and K^+-contents of single bacterial cells. J Gen Microbiol 129:51–53

Lindner B, Seydel U (1984a) Laser desorption mass spectrometry of complex biomolecules at high laser power density. In: Benninghoven A (ed) Secondary ion mass spectrometry. Springer Ser Chem Phys 36:370–373

Lindner B, Seydel U (1984b) Results in taxonomy and physiological state of bacteria derived from laser-induced single cell mass analysis. J Phys Colloque (Paris) 45 (C2):565–568

Lindner B, Seydel U (1985) Laser desorption mass spectrometry of nonvolatiles under shock wave conditions. Anal Chem 57:895–899

Linton RW, Bryan SR, Schmidt PF, Griffis DP (1985) Comparison of laser and ion microprobe detection sensitivity for lead in biological microanalysis. Anal Chem 57:440–443

Lorch DW, Schäfer H (1981a) Laser microprobe analysis of the intracellular distribution of lead in artificially exposed cultures of *Phymatodocis nordstedtiana* (Chlorophyta). Z Pflanzenphysiol 101:183–188

Lorch DW, Schäfer H (1981b) Localization of lead in cells of *Phymatodocis nordstedtiana* (Chlorophyta) with the laser microprobe analyzer (LAMMA 500). Fresenius Z Anal Chem 308:246–248

Mamyrin BA, Korataev VI, Schmikk DV, Zagulin VA (1973) The mass reflectron, a new ion magnetic time-of-flight mass spectrometer with high resolution. Sov Phys JETP 37:45–48

Mathey A (1981) LAMMA: New perspectives for lichenology. Anal Chem 308:249–252

Mauney T, Adams F (1984) Laser microprobe mass spectrometry of environmental soot particles. Sci Environ 36:215–244

Mauney T, Adams F, Sine MR (1984) Laser microprobe mass spectrometry of environmental soot particles. Sci Environ 36:215–224

Meyer zum Gottesberge (1988) Physiology and pathophysiology of inner ear melanin. Pigment Cell Res 1:238–249

Meyer zum Gottesberge-Orsulakova A (1986) Melanin in the inner ear: Micromorphological and microanalytical investigations. Acta Histochem Suppl XXXII:245–253

Michiels E, Van Vaeck L, Gijbels R (1984) The use of the lasermicroprobe mass analyzer for particle characterization and as a molecular microprobe. Scanning Electron Microsc 111:1111–1128

Moesta P, Seydel U, Lindner B, Grisebach H (1982) Detection of Glyceollin on the cellular level in infected soybean by laser microprobe mass analysis. Z Naturforsch 37c:748–751

Musselmann IH, Linton RW, Simons DS (1988) Cluster ion formation under laser bombardment. Studies of recombination using isotope labeling. Anal Chem 60:110–114

Neumann D (1979) Die Röntgenstrahlmikroanalyse und ihre Anwendung in der Pflanzenphysiologie. Biol Rundsch 17:171–182

Niessner R, Klockow D, Bruynseels F, Van Grieken R (1985) Investigation of heterogeneous reactions of PAH's on particle surfaces using laser microprobe mass analysis. Int J Environ Anal Chem 22:281–295

Orsulakova A, Kaufmann R, Morgenstern C, D'Haese M (1981) Cation distribution of the cochlea wall (stria vascularis). Fresenius Z Anal Chem 308:221–223

Parker CD, Hercules DM (1985) Laser mass spectra of simple aliphatic and aromatic amino acids. Anal Chem 57:698–704

Parker CD, Hercules DM (1986) Intermolecular proton transfer reactions in the laser mass spectrometry of organic acids. Anal Chem 58:25–30

Robards AW, Sleytr UB (1985) Low temperature methods in biological electron microscopy. In: Glauert AM (ed) Practical methods in electron microscopy, vol 10. Elsevier, Amst NY, pp 1–551

Schiller Ch, Kupka KG, Hillenkamp F (1981) Investigation of the biologically relevant amino acids and some small peptides by laser induced mass spectrometry. Fresenius Z Anal Chem 308:304–308

Schmidt PF (1986) Detection of lead in human tissue by means of a laser-microprobe-mass-analyser (LAMMA). Trace Elem Med 3:43–44

Schmidt PF, Lehmann RR, Ilsemann K, Wilhelm AH (1983) Distribution patterns of lead in the aortic wall determined by LAMMA. Artery 12:277–285

Schmidt PF, Barckhaus R, Kleimeier W (1986) Laser microprobe mass analyser (LAMMA) investigations on the localization of cadmium in renal cortex of rats after long-term exposure to cadmium. Trace Elem Med 3:19–24

Schnepf E (1963) Zur Cytologie und Physiologie pflanzlicher Drüsen. 1. Über den Fangschleim der Insektivoren. Flora 153:1–22

Schnepf E (1972) Über die Wirkung von Hemmstoffen der Proteinsynthese auf die Sekretion des Kohlenhydrat-Fangschleimes von Drosophyllum lusitanicum. Planta 103:334–339

Schröder WH (1981) Quantitative LAMMA analysis of biological specimens. I Standards II. Isotope labelling. Fresenius Z Anal Chem 308:212–217

Schröder WH, Fain GL (1984) Light-dependent calcium release from photoreceptors measured by laser micro-mass analysis. Nature (Lond) 309:268–270

Schröder WH, Bauch J, Endeward R (1988) Microbeam analysis of Ca exchange and uptake in the fine roots of spruce: influence of pH and aluminum. Trees 2:96–103

Schueler B, Feigl P, Krueger FR, Hillenkamp F (1981) Cationization of organic molecules under pulsed laser induced ion generation. Org Mass Spetr 16:502–506

Seiler H, Haas U, Rentschler I, Schreiber H, Wieser P, Wurster R (1981) Einige Untersuchungen an atmosphärischen Aerosolpartikeln. Optik 58:145–157

Seydel U, Heinen HJ (1980) First results in fingerprinting of single mycobacterial cells with LAMMA. In: Frigerio A, McCamish M (eds) Recent developments in mass spectrometry in biochemistry and medicine. Elsevier Sci Publ, Amst 6:489–496

Seydel U, Lindner B (1981) Qualitative and quantitative investigations on Mycobacteria with LAMMA. Fresenius Z Anal Chem 308:253–257

Seydel U, Lindner B (1983) Mass spectrometry of organic compounds (< 2000 amu) and tracing of organic molecules in plant tissue with LAMMA. In: Benninghoven A (ed) Ion formation from organic solids. Proc Sec Int Conf Münster, Springer Ser Chem Phys 25:240–244

Seydel U, Lindner B, Wollenweber HW, Rietschel ET (1984) Structural studies on the lipid A component of enterobacterial lipopolysaccharides by laser desorption mass spectrometry. Eur J Biochem 145:505–509

Seydel U, Lindner B, Dhople AM (1985) Results from cation and mass fingerprint analysis of single cells and from ATP measurements of M. leprae for drug sensitivity testing: A comparison. Int J Leprosy 53:365–372

Sprey B, Bochem HP (1981) Uptake of uranium into the alga Dunaliella detected by EDAX and LAMMA. Fresenius Z Anal Chem 308:239–245

Stienen H, Bauch J (1988) Element content in tissue of spruce seedlings from hydroponic cultures simulating acidification and deacidification. Plant Soil 106:231–238

Surkyn P, Adams F (1982) Laser microprobe mass analysis of glass microparticles. J Trace Microprobe Techn 1:79–114

Surkyn P, De Waele J, Adams F (1983) Laser microprobe mass analysis for source identification of air particulate matter. Int J Environ Anal Chem 13:257–274

Truchet M (1975) Application de la microanalyse par emission ionique secondaire aux coupes his-
    tologiques: Localisation des principeaux isotopes de divers eléments. J Microsc 24:1–22
Van der Peijl GJQ (1984) Desorption and ionization processes in laser mass spectrometry. PhD Thesis,
    Univ Amsterdam
Verbueken AH, Van Grieken RE, Paulus GJ, De Bruijn WV (1984) Embedded ion exchange beads as
    standards for laser microprobe mass analysis of biological specimens. Anal Chem 56:1362–1370
Verbueken AH, Bruynseels FJ, Van Grieken RE (1985a) Laser microprobe mass analysis: A review of
    applications in the life sciences. Biomed Mass Spectrom 12:438–463
Verbueken AH, Van de Vyver FL, Van Grieken RE, De Broe ME (1985b) Microanalysis in biology and
    medicine: ultrastructural localization of aluminum. Clin Nephrol Suppl 1 24:558–577
Verbueken AH, Bruynseels FJ, Van Grieken R, Adams F (1988) Laser Microprobe mass spectrometry.
    In: Winevordner JD (ed) Chemical analysis. A series of monographs on analytical chemistry and its
    application, vol 95. Wiley, NY, pp 173–256
Vogt H, Heinen HJ, Meier S, Wechsung R (1981) LAMMA 500 – principle and technical description of
    the instrument. Anal Chem 308:195–200
Vogt H, Heinen HJ, Meier S (1983) LAMMA – a new approach for the application of lasers in the
    analysis of bulk material. Laser Optoelektronik 1:23–29
Wechsung R, Hillenkamp F, Kaufmann F, Nitsche R, Vogt H (1978) LAMMA – a new laser microprobe
    mass analyzer. Scanning Electron Microsc I:611–620
Wieser P, Wurster R, Seiler H (1980) Identification of airborne particles by laser-induced mass
    spectroscopy. Atmos Environ 14:485–494
Wieser P, Wurster R, Hass RU (1981) Application of LAMMA in aerosol research. Fresenius Z Anal
    Chem 308:260–269
Wieser P, Wurster R, Seiler H (1982) Laser microprobe mass analysis of doped epoxy resin standards.
    Scanning Electron Microsc IV:1435–1441
Wink M, Heinen HJ, Vogt H, Schiebel HM (1984) Cellular localization of quinolizidine alkaloids by laser
    desorption mass spectrometry (LAMMA 1000). Plant Cell Rep 3:230–233

Fast Atom Bombardment Mass Spectrometry[1]

D. RHODES

1 Introduction to Fast Atom Bombardment Mass Spectrometry

Traditionally, a major constraint in the application of mass spectrometry to analytes of biological origin has been the need to transfer the analyte to the gas phase for ionization by the conventional techniques of electron ionization (EI) and chemical ionization (CI) (Busch and Cooks 1982). This requirement for volatilization has limited the range of applications of the combined technique of gas chromatography-mass spectrometry (GC-MS) (Busch and Glish 1984). Whereas derivatization of the analyte(s) can overcome these specific limitations of volatility in GC-MS applications, derivatization is often restricted to analytes of relatively low molecular weight, and each derivatization scheme may not be appropriate for all compounds within any given analyte class. For nonvolatile or thermally fragile samples, heating the sample to vaporize it often leads to thermal degradation (Busch and Cooks 1982). These widely recognized limitations have spurred many recent developments in liquid chromatography-mass spectrometry (LC-MS), specifically, the development of new interfaces for transferring LC effluent into a mass spectrometer, and alternative ionization techniques which overcome the need for volatility of the analyte (Busch and Cooks 1982; Busch and Glish 1984). Of the latter techniques, molecular secondary ion mass spectrometry (SIMS) (Benninghoven and Sichtermann 1977) and fast atom bombardment mass spectrometry (FAB-MS) [liquid SIMS] (Barber et al. 1981) have emerged as powerful new research tools in recent years. These methods are now in routine use in a wide range of biological applications (Rinehart 1982; Busch and Glish 1984).

By bombarding a solid surface with ions (primary) and directing the secondary ions coming off the surface into a mass spectrometer, analysis of the surface composition of the solid is accomplished. This technique (SIMS) has obvious applications in depth profile analysis of solid samples (Busch and Cooks 1982). SIMS entails bombardment of the solid (condensed) phase with particles of several keV kinetic energy; typically Ar^+ and Xe^+ ions are used (Busch and Glish 1984). Fast atom bombardment (FAB) is similar to SIMS except that the analyte is dissolved in a liquid matrix (glycerol being the most common matrix employed) on the tip of a direct insertion probe, and the sample is bombarded with neutral particles (typically Ar or Xe). Secondary ions are sputtered from the sample surface into the gas phase. These secondary ions are then extracted from the ion source and mass-analyzed in either the positive or negative ion mode (Rinehart 1982; Busch

[1]Purdue University Agricultural Experiment Station Journal Article #11,933.

and Glish 1984). Typical sample sizes are about 1 nmol/μl glycerol. Spectra can be obtained for 15 to 20 min (Busch and Glish 1984). Of these two related ionization techniques, FAB-MS in particular has made possible the analysis of a wide range of fragile compounds of biological interest which were previously difficult, if not impossible to analyze by conventional ionization techniques (EI and CI) (Busch and Cooks 1982; Rinehart 1982).

Desorption ionization mass spectrometry is a general term describing those methods which sample directly from the condensed phase, whether this phase be solid (SIMS) or liquid (FAB-MS) (Busch et al. 1982; Busch and Cooks 1982). Desorption ionization mass spectrometry is particularly suited to analysis of preformed ionic species (Busch et al. 1982; Busch and Glish 1984). Although derivatization is traditionally employed to impart specific physical and chemical properties which enhance volatility, stability and/or ionization efficiency in conventional GC-MS applications, derivatization can also be employed to enhance sensitivity and selectivity in desorption ionization mass spectrometry (Busch and Cooks 1982; Busch and Glish 1984). As noted by Busch et al. (1982); "In conjunction with the development of desorption ionization mass spectrometry, a distinctive set of derivatization conditions is emerging which endow the analyte with the physical and chemical properties of nonvolatility and ionic character — characteristics which contrast with those traditionally sought in mass spectrometry derivatization. Succinctly stated, these chemical reactions are designed to produce preformed ions at the surface; energy input into the sample is then required only to effect a change of state; that is, to effect desorption without any requirement for concurrent ionization."

This review will consider some of the present and future potential applications of FAB-MS in plant sciences. Where possible, derivatization techniques to optimize selectivity and sensitivity for different classes of analytes will be discussed. Several of the illustrative examples will be drawn from the author's research applications of FAB-MS. Initially I will consider FAB-MS applications in the quantification and stable isotope analysis of relatively simple nitrogenous compounds of plant origin (quaternary ammonium compounds and amino acids) and subsequently discuss the multitude of uses of FAB-MS in the structural analysis of more complex biomolecules including oligosaccharides, polypeptides and glycoproteins.

2 Quantification and Stable Isotope Analysis of Quaternary Ammonium Compounds

Quaternary ammonium salts represent a useful starting point for considerations of FAB-MS applications in plant sciences. Firstly, quaternary ammonium salts have been the subject of extensive studies concerning ionization and fragmentation using a variety of desorption ionization mass spectrometry methods (Day et al. 1979; Ryan et al. 1980; Unger et al. 1981; Cotter et al. 1982; Sano et al. 1982; Davis et al. 1983), including FAB-MS (Busch and Cooks 1982; Rhodes et al. 1987). Secondly, this class of compounds effectively illustrates the potential of "reversed derivatization" to optimize selectivity and sensitivity of FAB-MS in routine analyte quantification and stable isotope analysis in plant extracts (Rhodes et al. 1987).

2.1 Betaines

Betaines occur widely in nature and often represent major nitrogenous solutes of higher plant tissues (Wyn Jones and Storey 1981). These compounds are zwitter-ionic, containing a permanently positively charged quaternary ammonium group and a carboxyl group; properties which render these compounds difficult to derivatize to impart the volatility required for conventional GC and GC-MS analytical applications, except as the pyrolytic product, trimethylamine (Hitz and Hanson 1980). LC methods for betaine separation and quantification employ "reversed derivatization" of the compounds to their p-bromophenacyl, phenacyl, p-nitrobenzyl or methyl esters, using UV absorbance as detection method (Gorham 1986). Reversed derivatization of betaines to betaine esters also proves to be an extremely valuable approach in FAB-MS applications for the routine quantification of these compounds (Rhodes et al. 1987).

Esterification of betaines to their n-propyl or n-butyl esters imparts increased mass, a permanent positive charge and polarized hydrophobic character which increases surface activity in a glycerol matrix (Rhodes et al. 1987); properties which have been found to be desired in desorption ionization mass spectrometry (Busch et al. 1982; Rinehart 1982; Busch and Glish 1984). Reversed derivatization further eliminates H^+, Na^+ and K^+ adduct ion formation (i.e., cation pairing with the betaine carboxylic anion) in the condensed phase which can represent a serious problem in quantifying betaines as the underivatized species in plant extracts. In the underivatized form, positive ion FAB-MS detection of betaines requires proton-ation (or adduct ion formation with Na^+ or K^+) in the condensed phase, a reversible process, to observe the positively charged adduct ion(s) in the sputtered gas phase. Detection of betaine esters, however, simply requires a change of state from the condensed to the gas phase at the surface of the glycerol matrix (Rhodes et al. 1987).

The increased signal intensity of the molecular cations of glycinebetaine esters (relative to the protonated glycerol dimer (m/z 185)) as the length of the carbon chain of the ester group is increased (Fig. 1), strongly suggests that hydrophobicity of the ester group may alter the surface activity of the betaine ester in glycerol, thereby increasing sensitivity and selectivity of detection by FAB-MS in positive ion mode (Rhodes et al. 1987). As the n-propyl ester, the lower limit of detection of glycinebetaine corresponds to 0.05 nmol/μl glycerol (Rhodes et al. 1987). Although the molecular cation of n-butyl betaine gave the strongest signal per nmol of all esters tested (Fig. 1), it should be noted that glycerol can contribute to the signal at m/z 183 (Field 1982), which can interfere with reliable quantification of low nmol amounts of d_9-glycinebetaine n-butyl ester at this same mass (Rhodes et al. 1987). Nevertheless, d_9-glycinebetaine represents a useful internal standard for routine quantitative applications (Rhodes et al. 1987).

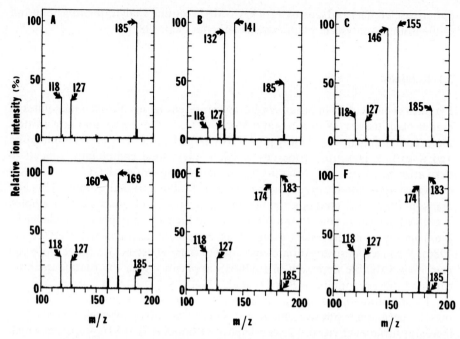

Fig. 1. FAB mass spectra (positive ion, glycerol) of mixtures of d_0- and d_9-glycinebetaine (at 22.5 nmol/μl and 21.5 nmol/μl glycerol, respectively) in the underivatized form (**A**) [(M + H)$^+$ = 118 and 127, respectively], or as methyl (**B**) [M$^+$ = 132 and 141, respectively], ethyl (**C**) [M$^+$ = 146 and 155, respectively], *n*-propyl (**D**) [M$^+$ = 160 and 169, respectively], *n*-butyl (**E**) [M$^+$ = 174 and 183, respectively], and isobutyl (**F**) [M$^+$ = 174 and 183, respectively] esters. Esters undergo fragmentation to yield protonated free acids [(M + H)$^+$ = 118 and 127, respectively]. The ion of mass 185 corresponds to the protonated glycerol dimer, which becomes a progressively smaller component of the spectrum as the length of the carbon chain of the betaine esters is increased (**A**→**F**). This data has been summarized in part by Rhodes et al. (1987)

The results of Fig. 1 illustrate the relative responses of d_0- and d_9-glycinebetaine esters in approximately equimolar mixtures using positive ion FAB. Note that betaine esters undergo unimolecular decomposition in the gas phase to yield fragment ions of mass 118 and 127 (for d_0- and d_9-glycinebetaine esters, respectively) corresponding to the protonated free acids. Straight chain esters tend to fragment to a lesser extent than branched-chain structures, and the d_9-glycinebetaine ester consistently tends to fragment to a lesser extent than the unlabeled species (Fig. 1) (Rhodes et al. 1987). Equimolar mixtures of d_0- and d_9-glycinebetaine *n*-butyl esters yield ion clusters in the mass range 383–401 in positive ion FAB, which are consistent with the following betaine ester dimers with Cl$^-$ as counter ion: [174$^+$ + 174$^+$ + 35$^-$]$^+$: 2×[174$^+$ + 183$^+$ + 35$^-$]$^+$: [183$^+$ + 183$^+$ + 35$^-$]$^+$ (not shown). These species thus have a single net positive charge. The chloride ion originates from acetyl chloride used in ester preparation (reaction of betaine with alcohol : acetyl chloride (5:1 v/v) at 120°C for 20 min) (Rhodes et al. 1987). The molecular ions, dimers with Cl$^-$ as counter ion, and the free acid fragmentation products (of variable ion intensity in comparison to the molecular cations), are highly diagnostic of the

specific betaine(s) present in any given sample. For example, γ-butyrobetaine, as the
n-propyl ester, exhibits a much lower degree of elimination to the protonated free
acid in comparison to either the n-propyl ester of glycinebetaine and the n-propyl
esters of d_6-stachydrine or homostachydrine (Rhodes et al. 1987). Betonicine
(hydroxyproline betaine) as the n-propyl ester yields a molecular cation of
$M^+ = 202$, and a protonated free acid fragment product at m/z 160 (Rhodes et al.
1987). Thus, care should be taken in interpreting the ions of mass 160 in n-propanol
esterified betaine fractions which could result from either the molecular cation of
glycinebetaine or the free acid fragment product of betonicine n-propyl ester.

Where both glycinebetaine and hydroxyproline betaine are encountered in the
same plant extracts, these species can be readily resolved by utilizing n-butyl esters
(d_0-glycinebetaine n-butyl ester = m/z 174; betonicine n-butyl ester = m/z 216;
betonicine protonated free acid fragment product = m/z 160). Certain alfalfa
genotypes that we have evaluated (Rhodes, Rich, Hendershot, Volenec and Wood;
unpublished) appear to contain several betaines including stachydrine (m/z 200),
homostachydrine (m/z 214), a hydroxyproline betaine (either betonicine and/or
turicine; m/z 216), and relatively low levels of glycinebetaine (m/z 174; Fig. 2B).
Moreover, certain alfalfa genotypes appear to lack the hydroxyproline betaine (Fig.
2A), but not the other betaines. Note the diagnostic ions at m/z 144, 158 and 160
(corresponding to the free acid fragment products of stachydrine, homostachydrine

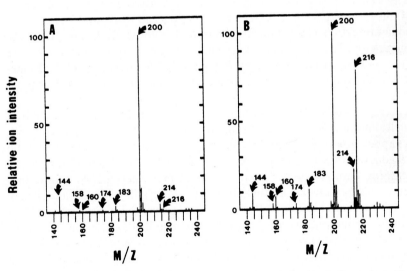

Fig. 2A,B. FAB-MS spectra (positive ion, glycerol) of the betaine fractions (n-butanol esterified) isolated
from immature leaf tissue (0.5 gram fresh weight per sample) of two alfalfa genotypes: 'SSER' (**A**) and
'9-2' (**B**). Samples received an internal standard of 442 nmol d_9-glycinebetaine ($M^+ = 183$ as the n-butyl
ester). The spectra reveal abundant quantities of prolinebetaine (stachydrine) ($M^+ = 200$ as the n-butyl
ester, $M^+ -56 = 144$ as the free acid fragmentation product) in both genotypes, a hydroxyproline betaine
($M^+ = 216$ as the n-butyl ester, $M^+ -56 = 160$ as the free acid fragmentation product) in genotype '9-2'
alone, and relatively low levels of pipecolic acid betaine (homostachydrine) ($M^+ = 214$ as the n-butyl
ester, $M^+ -56 = 158$ as the free acid fragmentation product) and glycinebetaine ($M^+ = 174$ as the n-butyl
ester) in both genotypes (Rhodes et al., unpublished results)

and hydroxyproline betaine respectively) occurring in approximately the same ratios as the molecular cations. The hydroxyproline betaine-deficient genotype 'SSER' exhibits a correspondingly lower intensity of the ion at m/z 160 in comparison to the hydroxyproline betaine-positive genotype '9–2' illustrated (Fig. 2). Here d_9-glycinebetaine (m/z 183 as the *n*-butyl ester) was used as internal standard. These examples not only illustrate the potential of FAB-MS in resolving complex mixtures of quaternary ammonium compounds in plant extracts, but also convey the potential of this methodology in identifying unique sources of genotypic variation for betaines. The striking differences in betaine composition among alfalfa genotypes may conform to a general pattern of genetic variability for N-methylated amino acids and their derivatives among species (Naidu et al. 1987), and among cultivars of domesticated crops (Hanson et al. 1983; Rhodes et al. 1987; Rhodes and Rich 1988).

This methodology also obviously lends itself to stable isotope analysis of betaines. Figure 3 shows typical mass spectra of the betaine fractions (*n*-butyl esters) of leaf discs of four hybrids of *Zea mays* (sweet corn) differing in glycinebetaine content (Rhodes et al. 1987) following incubation for 5 h with 1.6 mM d_3-betaine aldehyde chloride. The d_0-glycinebetaine-(m/z 174)-deficient genotype (hybrid '2708'; Fig. 3D) does not exhibit an impaired ability to oxidize exogenously supplied d_3-betaine aldehyde to d_3-glycinebetaine (m/z 177) in comparison to the d_0-glycinebetaine-positive genotypes (Fig. 3A-C). The mass spectra clearly reveal simultaneous deficiency of the molecular cation of d_0-glycinebetaine *n*-butyl ester (m/z 174), its protonated free acid fragment product (m/z 118) and deficiency of dimers containing d_0-glycinebetaine *n*-butyl ester in hybrid '2708' (Fig. 3D, inset). Only dimers (with Cl⁻) of the d_9- and d_3-glycinebetaine ester species are observed in the positive ion FAB spectrum of this genotype (Fig. 3D, inset). Again, d_9-glycinebetaine (m/z 183) was used as internal standard for quantification (Fig. 3; Rhodes et al. 1987).

Glycinebetaine deficiency in certain maize inbreds appears to be caused by a single, nuclear homozygous recessive gene (Rhodes and Rich 1988). Heterozygous genotypes appear to exhibit approximately half the betaine : total amino acid ratio of homozygous dominant genotypes (Rhodes and Rich 1988). Routine application of FAB-MS for quantification of betaine levels in maize germplasm in this laboratory has revealed a wide range of betaine levels among public inbred lines (Brunk et al. 1989). Sweet corn hybrid '1720' (Fig. 3C) in fact proved to be comprised of a 1:1 mixture of betaine-positive and betaine-deficient individuals, accounting for its "apparent" intermediate glycinebetaine level in comparison to the relatively high glycinebetaine levels of hybrids 'Spirit' and 'Silver Queen' and the betaine-deficient hybrid '2708' (Fig. 3). This phenomenon has now been traced to segregation of a single, nuclear encoded, recessive gene determining betaine deficiency in the female inbred parent of hybrid '1720' (Rhodes et al. 1989b). The male parent of this hybrid is homozygous recessive for this allele (Rhodes et al. 1989b). Our studies indicate that in glycinebetaine-positive individuals (heterozygous for the allele determining betaine accumulation capacity) glycinebetaine accumulation is salt stress-inducible. Treatment with 150 mM NaCl for 3 weeks leads to a four- to fivefold accumulation of glycinebetaine particularly in young expanding leaves. Salinity

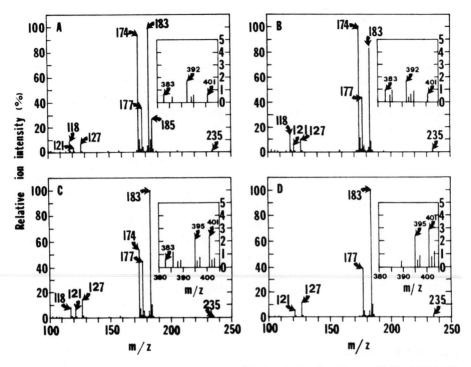

Fig. 3A-D. FAB mass spectra (positive ion, glycerol) of the betaine fractions (Dowex-50-H$^+$ 6 M NH$_4$OH eluants derivatized with n-butanol: acetyl chloride (5:1 v/v, 120 C, 20 min)) of leaf discs of four sweet corn hybrids (**A** 'Silver Queen'; **B** 'Spirit'; **C** '1720'; **D** '2708') following 5 h incubation with 1.6 mM d$_3$-betaine aldehyde chloride. All samples received an internal standard of 1000 nmol d$_9$-glycinebetaine [M$^+$ = 183 as the n-butyl ester; free acid fragment product (M$^+$–56) = 127]. Endogenous unlabeled glycinebetaine [M$^+$ = 174 as the n-butyl ester; free acid fragment product (M$^+$–56) = 118] is detectable in hybrids **A** 'Silver Queen', **B** 'Spirit' and **C** '1720', but not in **D** the betaine-deficient hybrid '2708'. However, note that all hybrids exhibit equal potential to oxidize exogenously supplied d$_3$-betaine aldehyde to d$_3$-glycinebetaine [M$^+$ = 177 as the n-butyl ester; free acid fragment product (M$^+$–56) = 121]. A small amount of d$_3$-betaine aldehyde as the di-n-butyl acetal (M$^+$ = 235) survived the isolation process. Note the occurrence of betaine ester dimers with Cl$^-$ in the mass range 383 to 401 (*insets*): [Md$_0^+$ + Md$_0^+$ + Cl$^-$]$^+$ = 383; [Md$_0^+$ + Md$_3^+$ + Cl$^-$]$^+$ = 386; [Md$_3^+$ + Md$_3^+$ + Cl$^-$]$^+$ = 389; [Md$_0^+$ + Md$_9^+$ + Cl$^-$]$^+$ = 392; [Md$_3^+$ + Md$_9^+$ + Cl$^-$]$^+$ = 395; [Md$_9^+$ + Md$_9^+$ + Cl$^-$]$^+$ = 401. These data have been summarized in part by Rhodes et al. (1987)

stress does not induce glycinebetaine accumulation in any organ of the betaine-deficient individuals of hybrid '1720' (Rhodes et al. 1989b). Tassel tissue of this hybrid is characterized by relatively low levels of glycinebetaine and the accumulation of a second betaine, nicotinic acid betaine (trigonelline) (Rhodes et al. 1989b). Glycinebetaine-deficient individuals of hybrid '1720' exhibit "normal" trigonelline levels, as evidenced by the relative ion intensities at m/z 194 versus m/z 183 (d$_9$-glycinebetaine internal standard) of n-butanol esterified betaine fractions of tassel tissue of salinized glycinebetaine-deficient and glycinebetaine-positive individuals of this hybrid (Rhodes et al. 1989b). Authentic trigonelline n-butyl ester yields a molecular ion of $M^+ = 194$ and a free acid fragment product at m/z 138 (not shown).

Where d_9-labeled precursors of glycinebetaine are supplied to plant tissues, and d_9-labeled glycinebetaine is sought, γ-butyrobetaine can be substituted as internal standard. The kinetics of conversion of d_9-choline to d_9-betaine aldehyde and d_9-glycinebetaine in spinach leaf discs using this internal standard are described by Rhodes et al. (1987). FAB-MS clearly resolved endogenous unlabeled glycinebetaine (m/z 160 as the n-propyl ester), newly synthesized d_9-glycinebetaine (m/z 169 as the n-propyl ester), newly synthesized d_9-betaine aldehyde (m/z 213 as the di-n-propyl acetal derivative), the protonated glycerol dimer (m/z 185), and the internal standard, γ-butyrobetaine (m/z 188 as the n-propyl ester; Rhodes et al. 1987). In principle, this methodology should be applicable to a wide range of stable isotope tracer studies on the betaine biosynthetic pathways of higher plants (including deuterium, ^{15}N, ^{13}C and/or ^{18}O tracer studies). FAB-MS deuterium and ^{18}O-labeling studies of the reaction mechanism of the choline monooxygenase of spinach have recently been reported (Lerma et al. 1988; Brouquisse et al. 1989). Hitherto, tracer studies on the glycinebetaine biosynthetic pathway(s) in higher plants have been largely restricted to ^{14}C-labeled intermediates (see e.g., Coughlan and Wyn Jones 1982; Hanson and Rhodes 1983).

2.2 Betaine Aldehyde

As noted above, betaine aldehyde yields di-alcohol acetal derivatives under the same esterification scheme used to derivatize betaines for FAB-MS analysis (Rhodes et al. 1987). The results of Fig. 4A illustrate the dramatic increase in signal intensity of betaine aldehyde as the length of the carbon chain of the alcohol used in derivatization is increased. The limit of detection of the di-n-butyl acetal derivative of betaine aldehyde is as low as 5 pmol/μl of glycerol (Rhodes et al. 1987). The exceptionally strong signals from betaine aldehyde dialcohol acetal derivatives (an order of magnitude more intense than the signals from the corresponding glycinebetaine esters) is most likely a function of the high surface activity of the betaine aldehyde derivatives in the glycerol matrix. Note again, the dimer with Cl⁻ as counter ion in the mass spectrum of d_3-betaine aldehyde di-n-butyl acetal; $M^+ = 235$; $(M^+ + M^+ + Cl^-)^+ = 505$ (Fig. 4B). FAB-MS has facilitated not only routine quantification of this intermediate of glycinebetaine synthesis in plant extracts (Lerma et al. 1988), but also in determination of pmol quantities of betaine aldehyde in isolated chloroplasts (Hanson et al. 1985). These derivatives of betaine aldehyde offer striking examples of the distinctive set of derivatization conditions which are emerging that endow the analyte with the physical and chemical properties of nonvolatility, ionic character, and surface activity desired in desorption mass spectrometry applications (Busch et al. 1982).

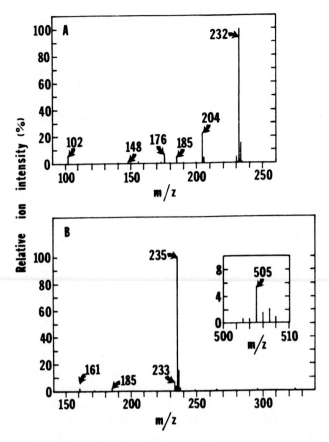

Fig. 4A,B. FAB mass spectrum (positive ion, glycerol) of an equimolar mixture of the isobutyl ($M^+ = 232$), n-propyl ($M^+ = 204$), ethyl ($M^+ = 176$) and methyl ($M^+ = 148$) dialcohol acetal derivatives, and underivatized ($M^+ = 102$) betaine aldehyde, each at 2 nmol/μl glycerol (**A**). FAB mass spectrum of the di-n-butyl acetal derivative of d_3-betaine aldehyde ($M^+ = 235$) at 5 nmol/μl glycerol (**B**). The ion at m/z 505 (**B**, *inset*) corresponds to the d_3-betaine aldehyde derivative dimer with Cl⁻ : $(M^+ + M^+ + Cl^-)^+$. These data have been summarized in part by Rhodes et al. (1987)

2.3 Choline

As noted earlier, choline is the immediate precursor of betaine aldehyde en route to glycinebetaine (Coughlan and Wyn Jones 1982; Hanson and Rhodes 1983; Hanson et al. 1985; Lerma et al. 1988). A useful derivatization scheme for choline to enhance sensitivity to detection by FAB-MS entails reaction of this quaternary ammonium compound with heptafluorobutyric anhydride (120°C, 10 min). The resulting O-heptafluorobutyryl derivative of choline yields a molecular cation at m/z 300 (not shown). The lower limit of detection of this derivative is approximately 50 pmol/μl glycerol (Rhodes, unpublished observations). Quantification of d_0-choline in plant extracts can be accomplished by using either d_9-choline or d_3-choline as internal standards (m/z 309 or 303 respectively, as their O-heptafluorobutyryl derivatives).

This methodology has found recent applications in ^{18}O tracer studies on the choline monooxygenase reaction mechanism of spinach (Lerma et al. 1988), in which $d_3,^{18}O$-choline (m/z 305 as the O-heptafluorobutyryl derivative) was supplied to leaf discs and $d_3,^{18}O$-labeled products sought by FAB-MS, recognizing sources of ^{18}O dilution by ion exchange chromatography, derivatization, and the rapid exchange of H_2O with the oxygen moiety of betaine aldehyde, which exists in equilibrium with the hydrate (Lerma et al. 1988). In studies of the latter equilibrium in vitro it was found necessary to incubate d_3-betaine aldehyde in $H_2^{18}O$ and to rapidly trap the ^{18}O incorporated into betaine aldehyde by chemically converting betaine aldehyde to choline with $NaBH_4$, subsequently monitoring the ^{18}O abundance of choline by FAB-MS (Lerma et al. 1988).

3 Quantification and Stable Isotope Analysis of Amino Acids

In determining quaternary ammonium compounds by FAB-MS in plant extracts, it is necessary to remove free amino acids prior to esterification by relatively simple ion exchange chromatography steps (Rhodes et al. 1987). Initially crude aqueous extracts can be passed over Dowex-1-OH⁻, and quaternary ammonium compounds recovered in the aqueous wash (and further purified by Dowex-50-H⁺ ion exchange chromatography). Amino acids retained on the Dowex-1-OH⁻ column can be eluted with 2.5 N HCl (Rhodes and Rich 1988). This removal of free amino acids is essential, because, in the derivatization process, amino acids also yield esters which can considerably complicate the FAB mass spectra; for example, valine yields an n-butyl ester, which, when protonated, gives a positive ion at m/z 174 which is identical to that of glycinebetaine n-butyl ester (Rhodes et al. 1987). Di-carboxylic amino acids tend to give much more intense signals than mono-carboxylic amino acids in equimolar mixtures, presumably as a function of the different surface activities of their esters in the glycerol matrix. This lends itself to sensitive stable isotope tracer studies of the labeling of aspartate and glutamate, or asparagine and glutamine using FAB-MS, as an alternative to conventional EI and/or CI GC-MS methods (cf. Rhodes et al. 1986). Asparagine and glutamine undergo acid hydrolysis to aspartate and glutamate, respectively, during esterification (Rhodes et al. 1986; 1989a), thus independent stable isotope ratio analyses of these compounds must be undertaken only following the separation of the acidic amino acids from the amides using Dowex-1-acetate ion exchange chromatography.

The results of Fig. 5 show positive ion FAB spectra of n-butyl derivatives of aspartate and glutamate isolated from ^{14}N-grown and ^{15}N-grown fronds of *Lemna minor*, employing ^{14}N-α-aminoadipic acid as the internal standard. Note the expected shift of 1 amu of the protonated molecular cations of the esters of aspartate and glutamate isolated from the ^{15}N-grown fronds, relative to ^{14}N-grown fronds. Simultaneous quantification and ^{15}N isotope abundance determination of aspartate and glutamate becomes feasible by FAB-MS without loss of sensitivity, without time-consuming GC separations of the component amino acids, and with only a single derivatization step (esterification with n-butanol), as opposed to the two steps

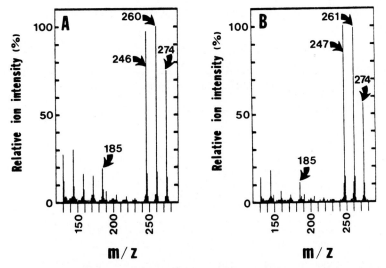

Fig. 5A,B. FAB-MS spectra (positive ion, glycerol) of the acidic amino acid fractions (*n*-butanol esterified) isolated from ^{14}N- (**A**) and ^{15}N-grown (**B**) fronds of *Lemna minor* (growth conditions and amino acid purification steps as described in Rhodes et al. (1989b)). Samples received an internal standard of 250 nmol α-aminoadipic acid ($M + H^+ = 274$ as the di-*n*-butyl ester) per 0.3 gram fresh weight sample. The spectra reveal aspartate and glutamate di-*n*-butyl esters ([M + H]$^+$ = 246 and 260, respectively) in **A**, and the expected isotopic shift of + 1 amu of both aspartate and glutamate (but not the internal standard) in **B** due to ^{15}N substitution (Rhodes et al., unpublished results)

required for derivatization to N-heptafluorobutyryl isobutyl esters and subsequent GC quantification, and ^{15}N abundance determination by conventional EI or CI GC-MS (Rhodes et al. 1986; 1989a). However, conventional GC-MS may remain the method of choice when amide-^{15}N abundance of asparagine and glutamine is sought (Rhodes et al. 1989a).

This application of FAB-MS in stable isotope tracer studies of amino acid metabolism in plants is further illustrated in Table 1, which compares the relative ion intensities of the molecular $(M + H)^+$ ion clusters of N-heptafluorobutyryl isobutyl derivatives of aspartate and glutamate as determined by CI GC-MS, with the corresponding molecular ion clusters of *n*-butyl esters of aspartate and glutamate as determined by FAB-MS. The glutamate and aspartate samples were isolated from *Lemna minor* during a brief exposure to a $^{13}CO_2$ atmosphere. The isotopic shifts due to in vivo synthesis of $^{13}C_1$- to $^{13}C_4$-labeled aspartate species and $^{13}C_1$- to $^{13}C_5$-labeled glutamate species appear similar in the two methods (Table 1) (Rhodes et al., unpublished results). FAB-MS tends to give a generally higher background of interfering ions from the glycerol matrix in comparison to CI GC-MS, especially for aspartate which was present at two- to threefold lower levels than glutamate in these samples. However, multiple ^{13}C-labeled amino acid species appear to be slightly more abundant in the FAB spectra in comparison to the CI spectra for identical samples, even after subtraction of the contribution of glycerol chemical noise in the FAB spectra. This could imply that multiple ^{13}C-labeled

Table 1. Comparison of relative ion intensities of the molecular ion $(M+H)^+$ clusters of aspartate and glutamate[a]

	$M+H^+$ $^{13}C_0$	$M+H^+$ $^{13}C_1$	$M+H^+$ $^{13}C_2$	$M+H^+$ $^{13}C_3$	$M+H^+$ $^{13}C_4$	$M+H^+$ $^{13}C_5$	$M+H^+$ $^{13}C_6$
Time (min) after exposure to $^{13}CO_2$	Aspartate CI GC/MS molecular ion intensities						
	m/z 442	m/z 443	m/z 444	m/z 445	m/z 446	m/z 447	
0	100	12.81	1.37	0.11	ND	ND	
10	100	12.30	0.92	0.05	ND	ND	
30	100	21.94	7.21	3.33	1.01	0.17	
75	100	30.08	16.41	12.03	7.07	0.53	
Time (min) after exposure to $^{13}CO_2$	Aspartate FAB-MS molecular ion intensities						
	m/z 246	m/z 247	m/z 248	m/z 249	m/z 250	m/z 251	
0	100	14.99	3.91	1.07	1.77	0.96	
10	100	14.62	3.28	0.87	1.10	0.80	
30	100	24.13	11.69	6.42	6.04	2.27	
75	100	32.87	22.31	18.19	13.36	2.37	
Time (min) after exposure to $^{13}CO_2$	Glutamate CI GC/MS molecular ion intensities						
	m/z 456	m/z 457	m/z 458	m/z 459	m/z 460	m/z 461	m/z 462
0	100	15.33	1.70	0.14	0.02	ND	ND
10	100	15.57	1.71	0.11	ND[b]	ND	ND
30	100	16.25	2.32	0.35	0.08	0.01	ND
75	100	20.81	7.03	2.74	1.26	0.75	0.14
Time (min) after exposure to $^{13}CO_2$	Glutamate FAB-MS molecular ion intensities						
	m/z 260	m/z 261	m/z 262	m/z 263	m/z 264	m/z 265	m/z 266
0	100	15.50	3.08	0.60	0.47	0.28	0.30
10	100	15.43	2.69	0.48	0.40	0.25	0.27
30	100	17.24	4.52	1.33	0.94	0.53	0.47
75	100	20.65	8.97	3.95	2.39	1.41	0.30

[a] Isolated from *Lemna minor* grown in a carbon/nitrogen stat device during a brief exposure to a $^{13}CO_2$ atmosphere, using CI GC/MS of N-heptafluorobutyryl isobutyl derivatives and FAB-MS (positive ion, glycerol) of *n*-butyl derivatives (Rhodes et al., unpublished results).
[b] ND = not detectable. Relative ion intensities are normalized on the most abundant ion ($^{13}C_0$ 100%) in each cluster. Fronds were supplied with $^{13}CO_2$ (99%) for the first 53 min (CO_2 concentration increased from 340 ppm to about 1100 ppm during this pulse phase) and were chased with $^{12}CO_2$ beginning at 53 min. $^{13}CO_2$ abundance of the gas phase was determined (at the following specified times) to be: 9% (20 min), 69% (31 min), 88% (54 min), 60% (72 min).

amino acid esters give more stable protonated molecular cations than unlabeled amino acid esters in FAB-MS [cf. differential fragmentation of d_9- and d_0-betaine esters (Rhodes et al. 1987)], and/or that CI GC-MS may systematically underestimate ^{13}C abundance. Systematic isotope ratio errors in GC-MS have been previously documented (Matthews and Hayes 1976). Regardless of this as yet unresolved minor discrepancy between the two methods, it is clear that glutamate and aspartate become labeled with ^{13}C to remarkably different extents, and this presumably reflects both differential turnover of aspartate and glutamate pools and/or differential rates of labeling of their precursor pools. It would be instructive to deduce the precise positions of ^{13}C in partially labeled species; e.g., among aspartate molecules containing two ^{13}C atoms, what are the relative frequencies of $^{13}C1 + {}^{13}C2$, $^{13}C1 + {}^{13}C3$, $^{13}C1 + {}^{13}C4$, $^{13}C2 + {}^{13}C3$, $^{13}C2 + {}^{13}C4$ and $^{13}C3 + {}^{13}C4$ labeled species? FAB-MS in conjunction with tandem mass spectrometry (MS/MS) methods may be required to address these questions (R.G. Cooks; personal communication). FAB-MS is expected to be particularly useful when combined with surface-induced dissociation (SID) in the second MS stage (Mabud et al. 1985). Long-lived signal and stable ion emissions generally associated with FAB-MS (Busch and Glish 1984; Rhodes et al. 1987), as opposed to the transient sample signals often associated with conventional GC-MS, could facilitate MS/MS applications without time-consuming and repetitive GC separations in the first stage (Busch and Glish 1984). The MS/MS spectrum is less susceptible to interfering ions from the matrix (Busch and Glish 1984). Such FAB tandem MS applications may open up unique opportunities to explore fluxes and compartmentation of amino acid metabolism in plants, i.e., the turnover of discrete amino acid pools drawing ^{13}C-label from discrete organic acid pools. The potential of these methods for analysis of compartmentation of amino acid metabolism in *Lemna minor* is currently being explored using dual isotopes ($^{15}NH_4^+$ and $^{13}CO_2$) in pulse-chase, steady-state labeling experiments (Rhodes et al., work in progress).

Whereas esterification with *n*-butanol can be used to derivatize the carboxyl groups of amino acids and to selectively enhance FAB signals for di-carboxylic amino acids, alternative reagents are available for derivatization of the amino groups for routine FAB applications in amino acid analysis. For example, dansyl amino acids have been analyzed by FAB, with sensitivity equivalent to that of fluorescence analysis (Beckner and Caprioli 1983). Although this specific application of FAB was in the N-terminal sequencing of polypeptides (see also Sect. 5), in principle this methodology could be applied to routine free amino acid analysis. Using dansyl aminobutyric acid as internal standard, dansyl amino acids can be determined quantitatively down to a level of 0.1 nmol (Beckner and Caprioli 1983). However, Beckner and Caprioli (1983) caution that in the use of this method it is best to use the non-saturated portion of the standard curve and to employ quantities of internal standard comparable to those of the analyte(s).

In the underivatized form, amino acids yield strong $(M + H)^+$ and $(M-H)^-$ signals in both SIMS (Benninghoven and Sichtermann 1977) and FAB-MS (Kulik and Heerma 1988). Mixtures of amino acids (e.g., resulting from protein hydrolysis) can thus be readily qualitatively analyzed. However, the $(M + H)^+$ ion intensities depend upon both the concentration and the surface activity of each amino acid;

because of the different surface activities the results should be interpreted quantitatively with caution (Kulik and Heerma 1988).

The FAB-MS method clearly holds promise as a tool to complement the existing battery of methods which have been applied to the characterization of the diverse array of non-protein amino acids and their derivatives in the plant kingdom (see e.g., Rosenthal 1982). An acidic compound isolated from the seed of the legume *Peltophorum africanum* was identified using FAB-MS as *trans*-4-hydroxypipecolic acid-4-sulphate; apparently the first naturally occurring sulphate ester of a non-protein amino acid to be described (Evans et al. 1985).

The above examples are intended to illustrate some of the advantages of FAB-MS over other analytical methodologies available for specific quantification and stable isotope ratio analysis of relatively simple nitrogenous solutes. This initial emphasis is not intended to imply that FAB-MS is restricted to such uses and such specific compounds, but rather has been designed to convey the following major points, using common plant metabolites as examples: (1) derivatization methods can be chosen to optimize selectivity and sensitivity of FAB; (2) FAB becomes a powerful quantitative tool when coupled with the use of stable isotope labeled or homolog internal standards of discrete mass; and (3) the method should be applicable to a wide range of stable isotope tracer studies in plant metabolism wherever FAB (and associated derivatization techniques) can be employed to produce molecular and/or fragment and/or adduct ions of interest.

Aside from these applications of FAB-MS as a quantitative analytical tool and in stable isotope tracer studies of metabolism, FAB-MS has found tremendous utility in the structural characterization of complex biological molecules. The remainder of this article will focus on these exceedingly important and versatile applications of FAB-MS.

4 Analysis of Oligosaccharides

As noted by Rinehart (1982), in the space of 1.5 years since the first announcement of the technique, FAB-MS had already been successfully applied to oligosaccharide-containing compounds such as glycosphingolipids. Glycosphingolipids have the general formula $R-CH(OR')-CH(NHCOR'')-CH_2OH$, where R and R'' are long-chain saturated alkyl groups and R' is an oligosaccharide substituent (Rinehart 1982). The intact underivatized glycosphingolipids yield intense molecular and fragment ions in both positive and negative ion FAB (Rinehart 1982). Briefly stated, the major cleavages occur within the oligosaccharide unit at (C–O) bonds (thereby identifying the sequence of oligosaccharide units) and at the amide (N–CO) bond (thereby specifying the R and R'' groups; Rinehart 1982).

Forsberg et al. (1982) employed positive ion FAB-MS to a detailed analysis of the structure of the 6-O-methylglucose polysaccharide of *Mycobacterium smegmatis*. Analysis of the underivatized methylglucose polysaccharide demonstrated that it contained 20 hexose units $[(M + Na)^+ = 3538]$. Analyses of amylase-glucoamylase-digested underivatized polysaccharide gave a result in agreement with 1 1 6-O-methylglucose units (in an uninterrupted sequence) and 5 glucose units in this fragment [(M + Na)$^+$ = 2876]. The FAB fragmentation patterns revealed the following: a) cleavage occurs at each glycosidic residue, the most abundant ions arising from charge retention on the non-reducing end fragment; b) cationization, particularly of the higher mass sequence ions, is a favored process; c) cleavage can occur on either side of the glycosidic oxygen to result in the characteristic pattern of signals 16 to 18 mass units apart for each sequence ion above m/z 900; and d) branch points are characterized by the absence of the latter pattern (Forsberg et al. 1982).

This methology has been extended to the structural characterization of glucans and plant cell wall oligosaccharides (see e.g., Dell et al. 1983a). Dell et al. (1983a) have analyzed a range of carbohydrates (including glycosphingolipids) in their native form or as acetyl, permethyl, methoxime, or pentafluorobenzyloxime derivatives, and present a useful comparison of derivatization techniques to optimize sensitivity, ionization and fragmentation of complex carbohydrates. High sensitivity for permethylated samples was achieved in positive ion mode by NH$_4^+$ cationization (Dell et al. 1983a). Spectra are reported up to m/z 6620 for permethylated glucans (Dell et al. 1983a). The β-D-(1→2)-linked D-glucans secreted by *Rhizobia* and *Agrobacteria* are shown (using both positive and negative ion FAB-MS) to be unbranched, cyclic molecules varying in size from 17 to at least 24 glucose residues (Dell et al. 1983b; Miller et al. 1986). Miller et al. (1986) further demonstrate that the regulation of the biosynthesis of these cyclic oligosaccharides in *Agrobacterium tumefaciens* parallels the osmotic regulation of membrane derived oligosaccharide biosynthesis of *Escherichia coli*.

The structure of the repeating oligosaccharide unit of the pneumococcal capsular polysaccharide type 18C, has also been investigated by FAB-MS (Phillips et al. 1983). The latter oligosaccharide was isolated from an aqueous hydrofluoric acid hydrolysate of the polysaccharide and was shown by positive ion FAB-MS to have a molecular weight of 928 [(M + H)$^+$ = 929] and to contain an O-acetyl group and a glycerol residue (Phillips et al. 1983). Spellman et al. (1983) have employed positive ion FAB-MS to elucidate the structure of a complex heptasaccharide isolated from a plant cell wall pectic polysaccharide rhamnogalacturonan II. This heptasaccharide was released by selective acid hydrolysis of the glycosidic linkages of apoiosyl residues, and FAB was performed on the per-(O)-(trideuteriomethyl)ated heptasaccharide alditol, revealing an (M + H)$^+$ ion at m/z 1349 [(M + Na)$^+$ = 1371] and its sugar sequence from the fragment ions.

Further additional discussions of FAB fragmentation patterns are found for peracetylated oligosaccharides (Naik et al. 1985; Tsai et al. 1986), oligogalacturonides (Jin and West 1984; Doner et al. 1988), and their di-pentafluorobenzyl-oxime derivatives (Nothnagel et al. 1983). The latter article illustrates the use of FAB-MS to characterize a dodeca-α-1,4-D-galacturonide as an elicitor of phytoalexin accumulation in *Glycine max* cell walls. The molecular ion (M–H)$^-$ of the di-pentafluorobenzyl-oxime derivative of the 12-residue oligogalacturonide was found to be at m/z 2519.

Recently, Dell et al. (1988) have reviewed the current status of carbohydrate analysis using FAB-MS and have reported on two new protocols which are being employed to optimize sensitivity and fragmentation. The first involves incorporating a moiety containing a permanent positive charge into the reducing end of the carbohydrate before permethylation or peracetylation. This introduction of a permanent positive charge is achieved by reaction with trimethyl *p*-aminophenyl ammonium chloride (TMAPA), and greatly enhances sensitivity. The second employs a strategy of FAB-MS analyses during chemical hydrolysis; this facilitates sequence analysis of highly branched oligosaccharides (Dell et al. 1988).

Applications of FAB-MS in sugar sequence analysis of glycosides and saponins will be discussed briefly in Section 8, but in the present context it is noteworthy that Rose et al. (1983) recommend the use of polyethylene glycol 200 for FAB-MS of complex plant glycosides such as digitonin. This matrix gave more abundant ions indicative of molecular weight and sugar sequence especially in the negative ion mode. Thus, as in other applications of FAB, not only chemical derivatization and cationization can be used to optimize sensitivity and fragmentation behavior, but the matrix can also be selected to meet specific needs.

5 Analysis of Polypeptides

As early as 1982 FAB-MS had been used to detect biologically active peptides to mass 5700 (Rinehart 1982). The technique has been especially valuable in determining the sequences of amino acids in polypeptides (Rinehart 1982). The latter review article illustrates examples of FAB sequencing of the polypeptide zervamicin IC $(M+H)^+ = 1839$, and the cyclic peptides didemnin A and B (didemnin A $(M+H)^+ = 949.6202$ as determined by high-resolution FAB). Generally, FAB fragment ions of peptides are observed in both the positive ion and negative ion modes, the molecular ion(s) provides the molecular weight of the peptide, ions of low masses are characteristic of individual amino acids, and ions at intermediate masses provide details of the sequence of the amino acids from both the NH_2- and COOH-termini for linear peptides. The work with zervamicin illustrated clean, regular fragmentation at the (CO-N) bonds of the polypeptide, with charge retention on the protected NH_2-terminal fragment (Rinehart 1982). Molecular weight information can usually be obtained on 0.1–1.0 nmol of sample, whereas sequence information may require 2–20 nmol (Busch and Glish 1984).

The molecular weight of the peptide may often exceed the mass-range limitation of the mass spectrometer, and chemical methods in combination with FAB must then be adopted (Busch and Glish 1984). These approaches have included the combination of Edman degradation, carboxypeptidase digestion and FAB (Bradley et al. 1982), dansylation of the N-terminus, acid hydrolysis and FAB of the dansyl amino acid (Beckner and Caprioli 1983), real-time enzymatic degradation of proteins as monitored by FAB within the glycerol matrix (Smith and Caprioli 1983), and FAB-MS analysis of carboxymethylated cyanogen bromide fragments of polypeptides, separated by gel filtration, ion exchange chromatography and HPLC (Nakano et al. 1986). Acetylation with 1:1 v/v acetic anhydride : d_6-acetic anhydride is commonly employed, leading to the production of equivalent N-acetyl derivatives whose acetyl groups contain either 3 hydrogen or 3 deuterium atoms generating a series of doublets 3 mass units apart in the FAB mass spectrum corresponding to N-terminal ions (Busch and Glish 1984). Stepwise degradations and derivatizations are combined in an overall strategy of FAB mapping (Morris et al. 1983; Busch and Glish 1984).

Gibson and Biemann (1984) have employed FAB-MS to confirm and correct regions from the amino acid sequences of 3 large proteins, glutaminyl- and glycyl-tRNA synthetase from *Escherichia coli* and methionyl-tRNA from yeast, whose primary structures were deduced from the base sequences of their corresponding genes. This strategy is based on comparison of the molecular weight of tryptic peptides predicted from all three reading frames of the gene sequences with those determined by FAB-MS. This method can be used to assess the correctness of the base sequence, identify errors that lead to frame shifts, premature stop codons, incorrect amino acids, or the presence of posttranslational modifications, using 5 to 20 nmol of sample (Gibson and Biemann 1984).

A novel posttranslationally modified residue (γ-N-methylasparagine) was detected using FAB-MS in the beta subunit of *Anabaena variabilis* allophycocyanin at position 71 in the protein (Klotz et al. 1986 and refs. cited therein). Recently, Mulligan et al. (1988) have employed FAB-MS to determine that the actual amino-terminal residue of the large subunit of ribulose bisphosphate carboxylase/oxygenase is proline at position 3 of the DNA-deduced sequence and that this proline is blocked with an N-acetyl moiety. Thus, posttranslational processing of the chloroplast-encoded large subunit of the enzyme must occur to remove Met-1 and Ser-2 and to acylate the amino terminus (Mulligan et al. 1988).

Whereas glycerol (and thioglycerol) are often employed as matrices for peptide sequencing, Meili and Seibl (1984) note that these matrices can occasionally yield an erroneous analytical answer if glycerol reacts with the sample upon bombardment. 3-NOBA (*m*-nitrobenzyl alcohol) has been recommended as an alternative to glycerol or thioglycerol (Meili and Seibl 1984). The cyclic peptide cyclosporin A (mol. wt. 1201) gave a barely visible protonated molecular ion cluster in glycerol with little or no relevant fragment information, while in 3-NOBA a prominent signal from the protonated molecular ion was observed, with a multiplicity of relevant fragments (Meili and Seibl 1984).

The applications of MS/MS in combination with FAB for protein sequencing have been noted by Busch and Glish (1984) and discussed extensively by Biemann and Scoble (1987) and Biemann (1988). Not only does the MS/MS spectrum provide confirmatory sequence data, but it is less susceptible to interfering ions from the matrix or other peptides in the sample (Busch and Glish 1984). Collision-induced dissociation of ions resulting from fast atom bombardment ionization is recognized as a powerful combination which can differentiate structural isomers (Kingston et al. 1985), and can be used to fragment the protonated molecular ion $(M+H)^+$ of peptides if FAB-MS alone produces insufficient fragmentation required for sequence interpretation (Biemann and Scoble 1987). The $(M+H)^+$ ion generated by FAB-MS in the first stage is collided with a neutral atom (e.g., helium) and the product ions are then mass-analyzed in a second mass spectrometer (Biemann and Scoble 1987). These techniques have been particularly useful in the structural characterization of peptides with blocked NH_2-termini, phosphorylated peptides, and in the sequencing of complex mixtures of peptides of synthetic and natural origin (Biemann and Scoble 1987; Biemann 1988). For thorough and up-to-date discussions of FAB-MS/MS fragmentation patterns of polypeptides, and algorithms for sequencing peptides using FAB and FAB tandem MS, the reader is referred to Biemann (1988) and Siegel and Bauman (1988).

Recent applications of FAB in combination with LC/MS, allowing a continuous flow of liquid to dynamically renew the surface of a FAB target (flow FAB), and the direct analysis of aqueous reactions (e.g., tryptic hydrolysis of peptides) using constant flow reaction monitoring and flow injection analysis, are described by Games et al. (1988) and Caprioli (1988). These newly emerging techniques of combined LC-FAB-tandem MS and the capability of constant flow reaction monitoring, provide extremely powerful and sensitive tools for polypeptide sequencing. Potentially, sensitivity could be further enhanced by using a pulsed fast atom beam, rather than a continuous fast atom beam (Tecklenburg et al. 1989). Tecklenburg et al. (1989) have compared ionization efficiency of continuous Xe FAB and pulsed Xe FAB with small dinitrophenyl peptides using tandem mass spectrometry (nitrobenzyl alcohol as matrix). Sample lifetimes and the yield of total $(M+H)^+$ ions were greatly enhanced using pulsed FAB; collision-induced dissociation signals were increased by a factor of 15-fold (Tecklenburg et al. 1989). Pulsed ionization appears to offer higher gains of sensitivity than can be achieved by flow FAB (Tecklenburg et al. 1989).

One of the first applications of liquid SIMS tandem mass spectrometry in sequencing of homologous peptides of plant origin is illustrated below. Steffens et al. (1986) employed cesium ion bombardment (5% acetic acid : thioglycerol (1:1 v/v) as matrix) triple quadropole mass spectrometry to determine the structure of non-protein metal-binding polypeptides (phytochelatins) derived from cadmium-resistant *Lycopersicon esculentum* cells. $(M+H)^+$ peptides were selected in the first quadropole and subjected to collisionally activated dissociation (CAD) in the second quadropole, with daughter ions scanned in a third quadropole at 200 amu/s. The peptide composed of 3:3:1 (Glu:Cys)$_3$:Gly yielded a mass spectrum with an $(M+H)^+$ ion of 772, and losses establishing it as a heptapeptide with the above sequence. Similar procedures confirmed the sequence of the nonapeptide (Glu:Cys)$_4$:Gly. Treatment of the peptides with acetic anhydride in pyridine allowed free carboxylates to undergo cyclization to the corresponding oxazalones, which possess an active proton at the alpha-carbon. Mass spectral peaks consistent with additions of 1 amu at each residue bearing Glu and at the carboxyl-terminal residue (Gly) were found after deuterium exchange of both peptides, establishing the structures as (γ-Glu-Cys)$_n$-Gly, where n = 3 or 4 (Steffens et al. 1986).

The above peptides are of course similar to glutathione, where n = 1. Glutathione and homoglutathione conjugates with acetochlor in maize and soybean respectively, have been identified as initial metabolites of this herbicide using FAB-MS (Breaux 1986). Breaux et al. (1988) have further described a FAB method for identifying biologically important thiol peptides (including glutathione and homoglutathione) using ^{14}C-labeled N-*p*-bromophenylmaleimide (BPM) to selectively derivatize thiols. The isotopic label facilitates isolation and purification of trace quantities of the maleimide thiol adducts by HPLC, and the use of the Br atom in the thiol derivative greatly enhances the detectability of the parent and fragment ions containing the BPM moiety (Breaux et al. 1988). The FAB method has also been applied in the study of the formation of styrene-glutathione conjugates by peroxidases (Stock et al. 1986), and the characterization of glutathione conjugates in combination with tandem MS (Haroldson et al. 1988).

6 Analysis of Glycoproteins

The successful application of FAB to glycoprotein characterization in part relies upon developments in FAB analysis of complex oligosaccharides, which were outlined in Section 4 above, and which have been discussed extensively by Dell et al. (1988). FAB-MS has been employed in the analysis of acetylated neutral and phosphorylated oligosaccharides from yeast glycoproteins and to their acetolysis products (Tsai et al. 1986). Acetylation by trifluoroacetic anhydride/glacial acetic acid is particularly convenient. Although this increases sample molecular weight and the complexity of the spectra, it enhances sensitivity, is applicable to samples containing salt, is especially useful for analysis of phosphorylated derivatives, and can be performed on a small amount of sample. Acetolysis by acetic anhydride/ glacial acetic acid/H_2SO_4 is performed on the acetylated oligosaccharides, and the acetylated fragments are recovered by solvent extraction prior to FAB-MS. The methodology allows molecular weight determinations and sequence analysis by acetolysis to be carried out on a few micrograms of isolated oligosaccharide in a few hours (Tsai et al. 1986). Similar protocols were employed by Naik et al. (1985) to determine the type of oligosaccharide chain present in glycoproteins; the procedure is based on acetolysis of the intact glycoconjugate, extraction of the peracetylated carbohydrate fragments, and analysis by FAB. The molecular ions define the composition of the oligosaccharide with respect to hexose, aminohexose, and sialic acid content. High mannose oligosaccharides yield a series of perace-tylated hexose oligomers, whereas complex-type oligosaccharides afford a series of N-acetyl-lactosamine containing species. The method was tested on three glyco-proteins of known structure; fetuin, ribonuclease B and erythrocyte Band 3, and on a glycoprotein of unknown structure — alpha-galactosidase I, an enzyme lectin from *Vicia faba*, which was found to contain high mannose carbohydrate chains (Naik et al. 1985).

Couso et al. (1987) report the use of FAB-MS to identify a novel intersecting N-acetylglucosamine residue in the high mannose oligosaccharides of *Dictyostelium discoideum* glycoproteins. Haavik et al. (1987) have employed FAB-MS to examine N-glycosidic linkages of a glycoprotein allergen of Timothy pollen. The acylated pronase-digested glycoprotein was found to contain N-acetylgluco-samine-asparagine linkages (Haavik et al. 1987).

Reddy et al. (1988) have recently applied FAB-MS to determine both the location of glycosylated sequons in yeast external invertase, and to correct the sequence of the SUC2 gene; a proline residue was found in place of alanine at position 390, which could result from a single base change in the triplet specifying the latter amino acid. To determine the location of the glycosylated sequons, external invertase was deglycosylated with endo.-beta.-acetylglucosaminidase H, and its component peptides analyzed by FAB-MS (Reddy et al. 1988).

7 Nucleotide Analysis

Barber et al. (1981) first illustrated the potential of FAB-MS in the analysis of nucleotides with the positive ion spectrum of ATP sodium salt and the negative ion spectrum of NAD. FAB-MS continues to find novel applications in nucleotide analysis, including the identification of cyclic-GMP in plants (Newton et al. 1984), resolution of isomeric cyclic nucleotides (Kingston et al. 1985), the analysis of ^{18}O isotope position in adenosine-5'-O-(1-thiotriphosphate) (Connolly et al. 1984), and the identification of a covalent cross-link adduct between mitomycin C and DNA (deoxyguanosine residues; Tomasz et al. 1987). Moreover, the technique has been employed successfully in oligonucleotide sequencing (Grotjahn et al. 1982; Panico et al. 1983). Typically only the negative ion FAB spectrum is used, as the positive ion spectrum is exceedingly complex due to degradation of the protonated species (Busch and Glish 1984). Although the FAB methods of oligonucleotide sequencing are rapid (Grotjahn et al. 1982; Panico et al. 1983), it is doubtful that these methods may achieve (at least in the near future) the level of sensitivity afforded by alternative DNA sequencing techniques (see e.g., Prober et al. 1987).

8 Structural Characterization
of Miscellaneous Secondary Plant Products

FAB-MS holds considerable promise for structural analysis of secondary plant metabolites, which were previously not amenable to analysis by conventional ionization methods. The potential utility of FAB-MS as a screening technique for mixtures of secondary metabolites is indeed illustrated by Paré et al. (1985) with respect to secondary products (tricothecene mycotoxins) derived from *Fusarium*. Recent applications of FAB-MS in plant secondary product biochemistry include the following range of uses; structural characterization of avenacins A-1 and B-1 in oats (Crombie et al. 1986); identification of enzymatic synthesis of nuatigenin 3-β-D-monoglucoside in oat leaves (Kalinowska and Wojciechowski 1986); characterization of two molluscicidal saponins from *Cussonia spicata* (Gunzinger et al. 1986); characterization of furostanol saponins from *Trigonella foenum-graecum* (Gupta et al. 1986); characterization of hydroxycinnamic esters in leaves of *Secale cereale* (Strack et al. 1986b); characterization of four new glycosides (3 cardenolides and a lignan) from *Asclepias subulata* (Jolad et al. 1986); identification of malonated anthocyanins in the Compositae (Takeda et al. 1986a), Liliaceae and Labiatae (Takeda et al. 1986b); and characterization of clerodane diterpenoids from *Portulaca* (Ohsaki et al. 1986), to name but a few representative examples. The above applications illustrate the use of negative ion FAB in polyethylene glycol 200 (Kalinowska and Wojciechowski 1986), negative ion FAB in glycerol (Strack et al. 1986b) and positive ion FAB in glycerol (Jolad et al. 1986; Takeda et al. 1986a) with quasimolecular ions ranging from $(M+H)^+ = 932$ (Crombie et al. 1986) to $(M-H)^- = 371$ (Strack et al. 1986b). Many of the above cited specific examples illustrate the use of FAB-MS in sugar linkage analysis of oligosaccharide-containing compounds (see e.g., sugar sequence analysis of saponins (Gunzinger et al. 1986; Gupta et al. 1986) and cardenolides (Jolad et al. 1986)). Because of the broad scope of the literature on FAB-MS applications in plant secondary metabolite analysis, I have attempted to summarize these diverse applications in Table 2, with reference to specific compounds or classes of analytes that have been successfully analyzed from specified plant sources by this methodology. The large number of references and diversity of products listed clearly precludes detailed discussion of each example, and the reader is referred to the original papers for more specific structural information on each compound or class of compounds. This list is likely incomplete; it is not intended to serve as a thorough bibliography of all FAB-MS applications in

Table 2. Applications of FAB-MS in structural analysis of secondary plant metabolites

Plant species	Compound or class of compounds	Reference
	Anthocyanins	
Viola tricolor		Saito et al. (1983)
Orchidaceae		Strack et al. (1986a)
Compositae		Takeda et al. (1986a)
Lilaceae and Labiatae		Takeda et al. (1986b)
Dianthus caryophyllus		Terahara and Yamaguchi (1986)
Camellia sp.		Saito et al. (1987)
Alstroemeria sp.		Saito et al. (1988)
Sinapsis alba		Takeda et al. (1988)
	Avenacins	
Avena sativa		Crombie et al. (1986)
	Cardenolides	
Asclepias subulata		Jolad et al. (1986)
	Chlorogenic acids	
Coffee bean		Sakushima et al. (1985)
	Clerodane diterpenoids	
Portulaca grandiflora		Ohsaki et al. (1986)
	Deoxycollybolidol	
Collybia peronata		Fogedal and Norberg (1986)
	1,2-Disinapolyl glucose	
Raphanus sativus		Strack et al. (1984)
	Flavonoids and flavonol glycosides	
Tamarix nilotica		Nawwar et al. (1984)
Securidaca diversifolia		Hamburger et al. (1985)
Ficus carica		Siewek et al. (1985)
Rubus sp.		Wald et al. (1986)
Triumfetta rhomboidea		Nair et al. (1986)

Table 2. *(Continued)*

Plant species	Compound or class of compounds	Reference
	Flavonoids and flavonol glycosides	
Iridaceae		Williams et al. (1986)
Mercurialis annua		Aquino et al. (1987)
miscellaneous		Stobiecki et al. (1988)
Dicranoloma robustum		Markham et al. (1988)
	Flavonoid sulphates	
Flaveria chloraefolia		Barron and Ibrahim (1987b)
Flaveria chloraefolia		Barron et al. (1986)
Flaveria chloraefolia		Barron and Ibrahim (1987a)
	Galloylsucroses	
Rheum sp.		Kashiwada et al. (1988)
	Gingerols	
Zingiber officinale		Chen et al. (1986)
	Humilixanthin	
Rivina humilis		Strack et al. (1987b)
	Hydroxycinnamic acid conjugates	
Raphanus sativus		Brandl et al. (1984)
Vigna radiata		Strack et al. (1985)
Secale cereale		Strack et al. (1986b)
miscellaneous		Schuster et al. (1986)
Spinacia oleraceae		Strack et al. (1987a)
	Iridoid glycosides	
Pentostemon richardsonii		Gering et al. (1987)
Randia dumetorum		Sati et al. (1986)
	Nuatigenin 3-β-D-monoglucoside	
Avena sativa		Kalinowska and Wojciechowski (1986)
	Quaternary indole alkaloids	
Strychnos usambarensis		Quetin-Leclercq and Angenot (1988)
	Saponins	
Paris polyphylla		Schulten et al. (1984)
Talinum tenuissimum		Gafner et al. (1985)
Cyamopsis tetragonoloba		Curl et al. (1986)
Phytolacca dodecandra		Dorsaz and Hostettmann (1986)
Cussonia spicata		Gunzinger et al. (1986)
Trigonella foenum-graecum		Gupta et al. (1986)
Polygala chamaebuxus		Hamburger and Hostettmann (1986)
Swartzia madagascariensis		Borel and Hostettmann (1987)
Dodonaea viscosa		Wagner et al. (1987)
Quillaja saponaria		Higuchi et al. (1988)
Dolichos kilimandscharicus		Marston et al. (1988)
	Sesquiterpene glycosides	
Lessingia glandulifera		Jolad et al. (1988)
Calendula arvensis		Pizza and De Tommasi (1988)
	Tannins	
Osbeckia chinensis		Su et al. (1988)
	Triterpenoid tetrasaccharides	
Androsace saxifragifolia		Waltho et al. (1986)

plant secondary product characterization to date, but rather as a representative sample of some of its major uses.

It seems pertinent to note that in many of these examples ^{1}HNMR, ^{13}CNMR, TLC, HPLC, GC-MS and/or alternative ionization, mass spectrometry methods have been employed along with FAB-MS to confirm structures. FAB-MS has been generally accepted as one of a battery of analytical methods now routinely available for structure elucidation. The FAB-MS method can augment, but is unlikely to completely replace other exceedingly useful methodologies. For example, Schulten et al. (1984) note that in the case of saponins, unambiguous sugar sequence analysis appears feasible using the combination of field desorption (FD) and fast atom bombardment MS.

9 Concluding Remarks

The use of FAB-MS in plant sciences has increased markedly over the last 8 years since the first announcement of this method (Barber et al. 1981), the vivid demonstrations of its capabilities for biomolecular analysis (Busch and Cooks 1982; Rinehart 1982), and the rapid acceptance of the methods. Within the last few years FAB-MS has played an important role in broadening the scope of methodologies available for structural elucidation of complex secondary plant products. The FAB method would seem ideal for applications designed to investigate stable isotope labeling behavior of the wide range of compounds found to be amenable to this desorption ionization procedure. A limited number of applications of FAB-MS in stable isotope studies of plant metabolism (drawn from the author's research interests) have been illustrated in this review article, with the view to encouraging further use of these approaches. First developed over 50 years ago (Schoenheimer and Rittenberg 1935), stable isotope tracer studies continue to provide powerful tools for elucidating fluxes and compartmentation of metabolism in vivo.

There seems to be little question that the FAB method is extremely sensitive and versatile, facilitating sequence and molecular weight analysis of diverse, fragile biopolymers at levels which were inconceivable a decade ago (Busch and Cooks 1982). FAB-MS has opened up new dimensions in precise structural analysis of complex biological molecules including polypeptides, oligosaccharides, and oligonucleotides (Rinehart 1982; Dell et al. 1988); methods which will likely continue to have a considerable impact on new developments in plant molecular biology, particularly when combined with MS/MS (Busch and Glish 1984; Biemann and Scoble 1987; Biemann 1988; Tecklenburg et al. 1989) and/or constant flow reaction monitoring (Caprioli 1988; Games et al. 1988). The scope of applications of this method appears almost unlimited.

References

Aquino R, Behar I, D'Agostino M, De Simone F, Schettino O, Pizza C (1987) Phytochemical investigation on *Mercurialis annua*. Biochem Syst Ecol 15:667–670

Barber M, Bordoli RS, Sedgwick RD, Tyler AN (1981) Fast atom bombardment of solids as an ion source in mass spectrometry. Nature 293:270–275

Barron D, Ibrahim RK (1987a) Quercetin and patuletin 3,3'-disulphates from *Flaveria chloraefolia*. Phytochemistry 26:1181–1184

Barron D, Ibrahim RK (1987b) 6-Methoxyflavonol 3-monosulphates from *Flaveria chloraefolia*. Phytochemistry 26:2085–2088

Barron D, Colebrook LD, Ibrahim RK (1986) An equimolar mixture of quercetin 3-sulphate and patuletin 3-sulphate from *Flaveria chloraefolia*. Phytochemistry 25:1719–1721

Beckner CF, Caprioli RM (1983) Protein N-terminal analysis using fast atom bombardment mass spectrometry. Anal Biochem 130:328–333

Benninghoven A, Sichtermann W (1977) Secondary ion mass spectrometry: a new analytical technique for biologically important compounds. Org Mass Spectrom 12:595–597

Biemann K (1988) Contributions of mass spectrometry to peptide and protein structure. Biomed Environ Mass Spectrom 16:99–111

Biemann K, Scoble HA (1987) Characterization by tandem mass spectrometry of structural modifications in proteins. Science 237:992–998

Borel C, Hostettmann K (1987) Molluscicidal saponins from *Swartzia madagascariensis* Desvaux. Helv Chim Acta 70:570–576

Bradley CV, Williams DH, Hanley MR (1982) Peptide sequencing using the combination of Edman degradation, carboxypeptidase digestion and fast atom bombardment mass spectrometry. Biochem Biophys Res Commun 104:1223–1230

Brandl W, Herrmann K, Grotjahn L (1984) Hydroxycinnamoyl esters of malic acid in small radish (*Raphanus sativus* L. var *sativus*). Z Naturforsch 39:515–520

Breaux EJ (1986) Identification of the initial metabolites of acetochlor in corn and soybean seedlings. J Agric Food Chem 34:884–888

Breaux EJ, Patanella JE, Sanders EF, Fujiwara H (1988) The identification of biologically important thiols. Biomed Environ Mass Spectrom 15:123–128

Brouquisse R, Weigel P, Rhodes D, Yocum CF, Hanson AD (1989) Evidence for a ferrodoxin-dependent choline monooxygenase from spinach chloroplast stroma. Plant Physiol 90:322–329

Brunk DG, Rich PJ, Rhodes D (1989) Genotypic variation for glycinebetaine among public inbreds of maize. Plant Physiol 91:1122–1125

Busch KL, Cooks RG (1982) Mass spectrometry of large, fragile and involatile molecules. Science 218:247–253

Busch KL, Glish GL (1984) New biological dimensions in mass spectrometry. BioTechniques 2:128–139

Busch KL, Unger SE, Vincze A, Cooks RG, Keough T (1982) Desorption ionization mass spectrometry: sample preparation for secondary ion mass spectrometry, laser desorption and field desorption. J Am Chem Soc 104:1507–1511

Caprioli RM (1988) On-line fast atom bombardment analysis of dynamic biological systems. Biomed Environ Mass Spectrom 16:35–39

Chen CC, Rosen RT, Ho CT (1986) Chromatographic analysis of gingerol compounds in ginger (*Zingiber officinale* Roscoe) extracted by liquid carbon dioxide. J Chromatogr 360:163–173

Connolly BA, Eckstein F, Grotjahn L (1984) Direct mass spectroscopic method for determination of oxygen isotope position in adenosine-5'-O-(1-thiotriphosphate). Determination of the stereochemical course of the yeast phenylanalyl-tRNA synthetase reaction. Biochemistry 23:2026–2031

Cotter RJ, Hansen G, Jones TR (1982) Mass spectral determination of long-chain quaternary amines in mixtures. Anal Chim Acta 136:135–142

Coughlan SJ, Wyn Jones RG (1982) Glycine betaine biosynthesis and its control in detached secondary leaves of spinach. Planta 154:6–17

Couso R, Van Halbeek H, Reinhold V, Kornfeld S (1987) The high mannose oligosaccharides of *Dictyostelium discoideum* glycoproteins contain a novel intersecting N-acetylglucasamine residue. J Biol Chem 262:4521–4527

Crombie WML, Crombie L, Green JB, Lucas JA (1986) Pathogenicity of 'take-all' fungus to oats: its relationship to the concentration and detoxification of the four avenacins. Phytochemistry 25:2075–2083

Curl CL, Price KR, Fenwick GR (1986) Isolation and structural elucidation of a triterpenoid saponin from guar, *Cyamopsis tetragonoloba*. Phytochemistry 25:2675–2676

Davis DV, Cooks RG, Meyer BN, McLaughlin JL (1983) Identification of naturally occurring quaternary compounds by combined laser desorption and tandem mass spectrometry. Anal Chem 55:1302–1305

Day RJ, Unger SE, Cooks RG (1979) Ionization of quaternary nitrogen compounds by secondary ion mass spectrometry. J Am Chem Soc 101:501–502

Dell A, Oates JE, Morris HR, Egge H (1983a) Structure determination of carbohydrates and glycosphingolipids by fast atom bombardment mass spectrometry. Int J Mass Spectrom Ion Phys 46:415–418

Dell A, York WS, McNeil M, Darvill AG, Albersheim P (1983b) The cyclic structure of β-D-(1→2)-linked D-glucans secreted by *Rhizobia* and *Agrobacteria*. Carbohydr Res 117:185–200

Dell A, Carman NH, Tiller PR, Thomas-Oates JE (1988) Fast atom bombardment mass spectrometric strategies for characterizing carbohydrate-containing biopolymers. Biomed Environ Mass Spectrom 16:19–24

Doner LW, Irwin PL, Kurantz MJ (1988) Preparative chromatography of oligogalacturonic acids. J Chromatogr 449:229–239

Dorsaz AC, Hostettmann K (1986) Further saponins from *Phytolacca dodecandra* L'Herit. Helv Chim Acta 69:2038–2047

Evans SV, Shing TKM, Aplin RT, Fellows LE, Fleet GWJ (1985) Sulphate ester of *trans*-4-hydroxypipecolic acid in seeds of *Peltophorum*. Phytochemistry 24:2593–2596

Field FH (1982) Fast atom bombardment study of glycerol: mass spectra and radiation chemistry. J Phys Chem 86:5115–5123

Fogedal M, Norberg T (1986) Deoxycollybolidol, a sesquiterpene from *Collybia peronata*. Phytochemistry 25:2661–2663

Forsberg LS, Dell A, Walton DJ, Ballou CE (1982) Revised structure for the 6-O-methylglucose polysaccharide of *Mycobacterium smegmatis*. J Biol Chem 257:3555–3563

Gafner F, Msonthi JD, Hostettmann K (1985) Molluscicidal saponins from *Talinum tenuissimum*. Helv Chim Acta 68:555–558

Games DE, Pleasance S, Ramsey ED, McDowall MA (1988) Continuous flow fast atom bombardment liquid chromatography/mass spectrometry: studies involving conventional bore liquid chromatography with simultaneous ultraviolet detection. Biomed Environ Mass Spectrom 15:179–182

Gering B, Junior P, Wichtl M (1987) Iridoid glycosides from *Penstemon richardsonii*. Phytochemistry 26:753–754

Gibson BW, Biemann K (1984) Strategy for the mass spectrometric verification and correction of the primary structures of proteins deduced from their DNA sequences. Proc Natl Acad Sci USA 81:1956–1960

Gorham J (1986) Separation and quantitative estimation of betaine esters by high-performance liquid chromatography. J Chromatogr 361:301–310

Grotjahn L, Frank R, Blöcker H (1982) Ultrafast sequencing of oligodeoxyribonucleotides by FAB-mass spectrometry. Nucl Acids Res 10:4671–4678

Gunzinger J, Msonthi JD, Hostettmann K (1986) Molluscicidal saponins from *Cussonia spicata*. Phytochemistry 25:2501–2503

Gupta RK, Jain DC, Thakur RS (1986) Two furostanol saponins from *Trigonella foenum-graecum*. Phytochemistry 25:2205–2207

Haavik S, Paulsen BS, Wold JK (1987) Glycoprotein allergens in pollen of Timothy. IV. Structural studies of a basic glycoprotein allergen. Int Arch Allergy Appl Immunol 83:225–230

Hamburger M, Hostettmann K (1986) New saponins and a prosapogenin from *Polygala chamaebuxus* L. Helv Chim Acta 69:221–227

Hamburger M, Gupta M, Hostettmann K (1985) Flavonol glycosides from *Securidaca diversifolia*. Phytochemistry 24:2689–2692

Hanson AD, Rhodes D (1983) [14]C Tracer evidence for synthesis of choline and betaine via phosphoryl base intermediates in salinized sugar beet leaves. Plant Physiol 71:692–700

Hanson AD, Ditz KM, Singletary GW, Leland TJ (1983) Gramine accumulation in leaves of barley grown under high temperature stress. Plant Physiol 71:896–904

Hanson AD, May AM, Grumet R, Bode J, Jamieson GC, Rhodes D (1985) Betaine synthesis in chenopods: localization in chloroplasts. Proc Natl Acad Sci USA 82:3678–3682

Haroldson PE, Reilly MH, Hughes H, Gaskell J, Porter CJ (1988) Characterization of glutathione conjugates by fast atom bombardment/tandem mass spectrometry. Biomed Environ Mass Spectrom 15:615–621

Higuchi R, Tokimitsu Y, Komori T (1988) An acylated triterpenoid saponin from *Quillaja saponaria*. Phytochemistry 27:1165–1168

Hitz WD, Hanson AD (1980) Determination of glycine betaine by pyrolysis-gas chromatography in cereals and grasses. Phytochemistry 19:2371–2374

Jin DF, West CA (1984) Characteristics of galacturonic acid oligomers as elicitors of casbene synthetase activity in castor bean seedlings. Plant Physiol 74:989–992

Jolad SD, Bates RB, Cole JR, Hoffmann JJ, Siahaan TJ, Timmermann BN (1986) Cardenolides and a lignan from *Asclepias subulata*. Phytochemistry 25:2581–2590

Jolad SD, Timmermann BN, Hoffmann JJ, Bates RB, Camou FA, Siahaan TJ (1988) Sesquiterpenoid glycosides and an acetogenin glucoside from *Lessingia glandulifera*. Phytochemistry 27:2199–2204

Kalinowska M, Wojciechowski ZA (1986) Enzymatic synthesis of nuatigenin 3β-D-glucoside in oat (Avena sativa) leaves. Phytochemistry 25:2525–2529

Kashiwada Y, Nonaka GI, Nishioka I (1988) Galloylsucroses from rhubarbs. Phytochemistry 27:1469–1472

Kingston EE, Beynon JH, Newton RP, Liehr JG (1985) The differentiation of isomeric biological compounds using collision-induced dissociation of ions generated by fast atom bombardment. Biomed Mass Spectrom 12:525–534

Klotz AV, Leary JA, Glazer AN (1986) Post-translational methylation of asparaginyl residues. Identification of β-71 γ-N-methylasparagine in allophycocyanin. J Biol Chem 261:15891–15894

Kulik W, Heerma W (1988) A study of the positive and negative ion fast atom bombardment mass spectra of α-amino acids. Biomed Environ Mass Spectrom 15:419–427

Lerma C, Hanson AD, Rhodes D (1988) Oxygen-18 and deuterium labeling studies of choline oxidation by spinach and sugar beet. Plant Physiol 88:695–702

Mabud MDA, Dekrey MJ, Cooks RG (1985) Surface-induced dissociation of molecular ions. Int J Mass Spectrom Ion Proc 67:285–294

Markham KR, Anderson ØM, Viotto ES (1988) Unique biflavonoid types from the moss *Dicranoloma robustum*. Phytochemistry 27:1745–1749

Marston A, Gafner F, Dossaji SF, Hostettmann K (1988) Fungicidal and molluscicidal saponins from *Dolichos kilimandscharicus*. Phytochemistry 27:1325–1326

Matthews DE, Hayes JM (1976) Systematic errors in gas chromatography-mass spectrometry isotope ratio measurements. Anal Chem 48:1375–1382

Meili J, Seibl J (1984) A new versatile matrix for fast atom bombardment analysis. Org Mass Spectrom 19:581–582

Miller KJ, Kennedy EP, Reinhold VN (1986) Osmotic adaptation by gram-negative bacteria: possible role for periplasmic oligosaccharides. Science 231:48–51

Morris HR, Panico M, Taylor GW (1983) FAB-mapping of recombinant-DNA protein products. Biochem Biophys Res Commun 117:299–305

Mulligan RM, Houtz RL, Tolbert NE (1988) Reaction-intermediate analogue binding by ribulose bisphosphate carboxylase/oxygenase causes specific changes in proteolytic sensitivity: the amino-terminal residue of the large subunit is acetylated proline. Proc Natl Acad Sci USA 85:1513–1517

Naidu BP, Jones GP, Paleg LG, Poljakoff-Mayber A (1987) Proline analogues in *Melaleuca* species. Responses of *Melaleuca lanceolata* and *M. uncinata* to water stress and salinity stress. Aust J Plant Physiol 14:669–677

Naik S, Oates JE, Dell A, Taylor GW, Dey PM, Pridham JB (1985) A novel mass spectrometric procedure for the rapid determination of the types of carbohydrate chains present in glycoproteins: application to α-galactosidase I from *Vicia faba* seeds. Biochem Biophys Res Commun 132:1–7

Nair AGR, Seetharaman TR, Voirin B, Favre-Bonvin J (1986) True structure of triumboidin, a flavone glycoside from *Triumfetta rhomboidea*. Phytochemistry 25:768–769

Nakano K, Tashiro Y, Kikumoto Y, Tagaya M, Fukui T (1986) Amino acid sequence of cyanogen bromide fragments of potato phosphorylase. J Biol Chem 261:8224–8229

Nawwar MAM, Souleman AMA, Buddrus J, Linscheid M (1984) Flavonoids of the flowers of *Tamarix nilotica*. Phytochemistry 23:2347–2350

Newton RP, Kingston EE, Evans DE, Younis LM, Brown EG (1984) Occurrence of guanosine 3′,5′-cyclic monophosphate (cyclic GMP) and associated enzyme systems in *Phaseolus vulgaris*. Phytochemistry 23:1367–1372

Nothnagel EA, McNeil M, Albersheim P, Dell A (1983) Host-pathogen interactions. XXII. A galacturonic acid oligosaccharide from plant cell walls elicits phytoalexins. Plant Physiol 71:916–926

Ohsaki A, Ohno N, Shibata K, Tokoroyama T, Kubota T (1986) Clerodane diterpenoids from *Portulaca* cv Jewel. Phytochemistry 25:2414–2416

Panico M, Sindona G, Uccella N (1983) Bioorganic applications of mass spectrometry. 3. Fast-atom-bombardment induced zwitterionic oligonucleotide quasimolecular ions sequenced by MS/MS. J Am Chem Soc 105:5607–5610

Paré JRJ, Greenhalgh R, Lafontaine P, ApSimon JW (1985) Fast atom bombardment mass spectrometry: a screening technique for mixtures of secondary metabolites from fungal extracts of *Fusarium* species. Anal Chem 57:1470–1472

Phillips LR, Nishimura O, Fraser BA (1983) The structure of the repeating oligosaccharide unit of the pneumococcal capsular polysaccharide type 18C. Carbohydr Res 121:243–255

Pizza C, De Tommasi N (1988) Sesquiterpene glycosides based on the alloaromadendrane skeleton from *Calendula arvensis*. Phytochemistry 27:2205–2208

Prober JM, Trainor GL, Dam RJ, Hobbs FW, Robertson CW, Zagursky RJ, Cocuzza AJ, Jensen MA, Baumeister K (1987) A system for rapid DNA sequencing with fluorescent chain-terminating dideoxynucleotides. Science 238:336–341

Quetin-Leclercq J, Angenot L (1988) 10-Hydroxy-N$_b$-methyl-corynantheol, a new quaternary alkaloid from the stem bark of *Strychnos usambarensis*. Phytochemistry 27:1923–1926

Reddy VA, Johnson RS, Biemann K, Williams RS, Ziegler FD, Trimble RB, Maley F (1988) Characterization of the glycosylation sites in yeast external invertase I. N-linked oligosaccharide content of the individual sequons. J Biol Chem 263:6978–6985

Rhodes D, Rich PJ (1988) Preliminary genetic studies of the phenotype of betaine deficiency in *Zea mays* L. Plant Physiol 88:102–108

Rhodes D, Handa S, Bressan RA (1986) Metabolic changes associated with adaptation of plant cells to water stress. Plant Physiol 82:890–903

Rhodes D, Rich PJ, Myers AC, Reuter CC, Jamieson GC (1987) Determination of betaines by fast atom bombardment mass spectrometry. Identification of glycine betaine deficient genotypes of *Zea mays*. Plant Physiol 84:781–788

Rhodes D, Rich PJ, Brunk DG (1989a) Amino acid metabolism of *Lemna minor* L. IV. [^{15}N] Labeling kinetics of the amide and amino groups of glutamine and asparagine. Plant Physiol 89:1161–1171

Rhodes D, Rich PJ, Brunk DG, Ju GC, Rhodes JC, Pauly MH, Hansen LA (1989b) Development of two isogenic sweet corn hybrids differing in glycinebetaine content. Plant Physiol 91:1112–1121

Rinehart KL Jr (1982) Fast atom bombardment mass spectrometry. Science 218:254–260

Rose ME, Veares MP, Lewis IAS, Goad J (1983) Analysis of steroid conjugates by fast atom bombardment mass spectrometry. Biochem Soc Trans 11:602–603

Rosenthal GA (1982) Plant nonprotein amino and imino acids. Biological, biochemical and toxicological properties. Academic Press, Lond NY

Ryan TM, Day RJ, Cooks RG (1980) Secondary ion mass spectra of diquaternary ammonium salts. Anal Chem 52:2054–2057

Saito N, Timberlake CF, Tucknott OG, Lewis IAS (1983) Fast atom bombardment mass spectrometry of the anthocyanins violanin and platyconin. Phytochemistry 22:1007–1009

Saito N, Yokoi M, Yamaji M, Honda T (1987) Cyanidin 3-*p*-coumaroylglucoside in *Camellia* species and cultivars. Phytochemistry 26:2761–2762

Saito N, Yokoi M, Ogawa M, Kamijo M, Honda T (1988) 6-Hydroxyanthocyanidin glycosides in the flowers of *Alstroemeria*. Phytochemistry 27:1399–1401

Sakushima A, Hisada S, Nishibe S, Brandenberger H (1985) Application of fast atom bombardment mass spectrometry to chlorogenic acids. Phytochemistry 24:325–328

Sano M, Ohya K, Kitaoka H, Ito R (1982) Field desorption mass spectrometry of betaine hydrohalides. Biomed Mass Spectrom 9:438–442

Sati OP, Chaukiyal DC, Nishi M, Miyahara K, Kawasaki T (1986) An iridoid from *Randia dumetorum*. Phytochemistry 25:2658–2660

Schoenheimer R, Rittenberg D (1935) Deuterium as an indicator in the study of intermediary metabolism. Science 82:156–157

Schulten HR, Singh SB, Thakur RS (1984) Field desorption and fast atom bombardment mass spectrometry of spirostanol and furostanol saponins from *Paris polyphylla*. Z Naturforsch 39e:201–211

Schuster B, Winter M, Herrmann K (1986) 4-O-β-D Glucosides of hydroxybenzoic and hydroxycinnamic acids — their synthesis and determination in berry fruit and vegetables. Z Naturforsch 41c:511–520

Siegel MM, Bauman N (1988) An efficient algorithm for sequencing peptides using fast atom bombardment mass spectral data. Biomed Environ Mass Spectrom 15:333–343

Siewek F, Herrmann K, Grotjahn L, Wray V (1985) Isomeric di-C-glycosylflavones in fig (*Ficus carica* L.). Z Naturforsch 40c:8–12

Smith LA, Caprioli RM (1983) Following enzyme catalysis in real-time inside a fast atom bombardment mass spectrometer. Biomed Mass Spectrom 10:98–102

Spellman MW, McNeil M, Darvill AG, Albersheim P, Dell A (1983) Characterization of a structurally complex heptasaccharide from the pectic polysaccharide rhamnogalacturonan II. Carbohydr Res 122:131–153

Steffens JC, Hunt DF, Williams BG (1986) Accumulation of non-protein metal-binding polypeptides (γ-glutamyl-cysteinyl)$_n$-glycine in selected cadmium-resistant tomato cells. J Biol Chem 261:13879–13882

Stobiecki M, Olechnowicz-Stepien W, Rzadkowska-Bodalska H, Cisowski W, Budko E (1988) Identification of flavonoid glycosides isolated from plants by fast atom bombardment mass spectrometry and gas chromatography-mass spectrometry. Biomed Environ Mass Spectrom 15:589–594

Stock BH, Schreiber J, Guenat C, Mason RP, Bend JR, Eling TE (1986) Evidence for a free radical mechanism of styrene-glutathione conjugate formation catalyzed by prostaglandin H synthase and horseradish peroxidase. J Biol Chem 261:15915–15922

Strack D, Dahlbender B, Grotjahn L, Wray V (1984) 1,2-Disinapolylglucose accumulated in cotyledons of dark grown *Raphanus sativus* seedlings. Phytochemistry 23:657–659

Strack D, Hartfeld F, Austenfeld FA, Grotjahn L, Wray V (1985) Coumaroyl-, caffeoyl- and feruloyltartronates and their accumulation in mung bean. Phytochemistry 24:147–150

Strack D, Busch E, Wray V, Grotjahn L, Klein E (1986a) Cyanidin 3-oxalylglucoside in orchids. Z Naturforsch 41c:707–711

Strack D, Engel U, Weissenböck G, Grotjahn L, Wray V (1986b) Ferulic acid esters of sugar carboxylic acids from primary leaves of rye (*Secale cereale*). Phytochemistry 25:2605–2608

Strack D, Heilemann J, Boehnert B, Grotjahn L, Wray V (1987a) Accumulation and enzymatic synthesis of 2-O-acetyl-3-O-(*p*-coumaroyl)-*meso*-tartaric acid in spinach cotyledons. Phytochemistry 26:107–111

Strack D, Schmitt D, Reznik H, Boland W, Grotjahn L, Wray V (1987b) Humilixanthin, a new betaxanthin from *Rivina humilis*. Phytochemistry 26:2285–2287

Su JD, Osawa T, Kawakishi S, Namiki M (1988) Tannin antioxidants from *Osbeckia chinensis*. Phytochemistry 27:1315–1319

Takeda K, Harborne JB, Self R (1986a) Identification and distribution of malonated anthocyanins in plants of the Compositae. Phytochemistry 25:1337–1342

Takeda K, Harborne JB, Self R (1986b) Identification of malonated anthocyanins in the Lilaceae and Labiatae. Phytochemistry 25:2191–2192

Takeda K, Fischer D, Grisebach H (1988) Anthocyanin composition of *Sinapsis alba*, light induction of enzymes and biosynthesis. Phytochemistry 27:1351–1353

Tecklenburg Jr RE, Castro ME, Russell DH (1989) Evaluation of pulsed fast-atom bombardment ionization for increased sensitivity of tandem mass spectrometry. Anal Chem 61:153–159

Terahara N, Yamaguchi M-A (1986) ^{1}H NMR spectral analysis of the malylated anthocyanins from *Dianthus*. Phytochemistry 25:2906–2907

Tomasz M, Lipman R, Chowdary D, Pawlak J, Verdine GL, Nakanishi K (1987) Isolation and structure of a covalent cross-link adduct between mitomycin C and DNA. Science 235:1204–1208

Tsai PK, Dell A, Ballou CE (1986) Characterization of acetylated and acetolyzed glycoprotein high-mannose core oligosaccharides by fast-atom-bombardment mass spectrometry. Proc Natl Acad Sci USA 83:4119–4123

Unger SE, Vincze A, Cooks RG, Chrisman R, Rothman LD (1981) Identification of quaternary alkaloids in mushroom by chromatography/secondary ion mass spectrometry. Anal Chem 53:976–981

Wagner H, Ludwig C, Grotjahn L, Khan MSY (1987) Biologically active saponins from *Dodonaea viscosa*. Phytochemistry 26:697–701

Wald B, Galensa R, Herrmann K, Grotjahn L, Wray V (1986) Quercetin 3-O-[6″-(3-hydroxy-3-methylglutaroyl)-β-galactoside] from blackberries. Phytochemistry 25:2904–2905

Waltho JP, Williams DH, Mahato SB, Pal BC, Barna JCJ (1986) Structure elucidation of two triterpenoid tetrasaccharides from *Androsace saxifragifolia*. J Chem Soc Perkin Trans I:1527–1532

Williams CA, Harborne JB, Goldblatt P (1986) Correlations between phenolic patterns and tribal classification in the family Iridaceae. Phytochemistry 25:2135–2154

Wyn Jones RG, Storey R (1981) Betaines. In: Paleg LG, Aspinall D (eds) Physiology and biochemistry of drought resistance in plants. Academic Press, Sydney, pp 171–204

Microdissection and Biochemical Analysis of Plant Tissues

R. Hampp, A. Rieger, and W.H. Outlaw, Jr.

1 Introduction

Biological tissues are not homogenous; instead they consist of cells having specific functions. A typical bifacial leaf, for example, contains not only photosynthetic mesophyll cells (palisade parenchyma plus spongy parenchyma) but also epidermal, guard, and bundle-sheath cells, as well as myriad minor cell types. Their specific functions indicate that profound biochemical differences exist among adjacent cells. These differences are obliterated by tissue homogenation, which precedes most analytical biochemistry. Biologists have thus been challenged to develop methods that allow for selective sampling of specific cell types and analysis of the resultant small amounts of material.

A general problem, irrespective of the method, is conservation of the metabolic state of the intact organ. Ideally, biochemical analysis would be noninvasive, but this goal is elusive. Unaltered cells are difficult to isolate, and even when they can be, intact membranes limit uptake of reaction components. This limitation can be overcome if razor-cut sections 10-μm-thick or less are taken (Gahan 1984). However, few tissues are amenable to this approach, and, inarguably, metabolite pools measured thus are artifactual (because of high turnover rates, reduced retention, and interference with cellular compartmentation).

Alternative histochemical approaches include various implementations of chemical fixation, embedding, or freeze stop. Because of its wider utility, we have often used the last method in our research. This method is the beginning point for our description. We refer the reader to *A Flexible System of Enzymatic Analysis* (Lowry and Passonneau 1972) for details of procedures we do not reference specifically.

2 Freeze Stop

Rapid freezing is the preferred method of quenching tissue. For subsequent measurement of metabolite pool sizes, it is imperative to stop endogenous reactions immediately; this requirement is sometimes less stringent for measurement of enzyme activities. Because of the high water content of plants, there is also a physical reason that freezing should be as rapid as possible: avoidance of disruptive formation of large ice crystals (Asahina 1956). Different methods of rapid freezing have been used (Jensen 1962; Pearse 1980); the choice depends on the purpose at hand and, equally important, the geometry of the tissue. Guard cells, on the leaf surface and lacking contact with the leaf interior, for instance, should be completely frozen through their 10-μm depth virtually instantaneously upon submersion in a coolant. On the other hand, parenchyma buried inside a tuber would be quenched only within seconds, obviously too slowly to permit the measurements of ATP or other metabolites that turn over rapidly.

For small samples, freezing in nitrogen-cooled freon-12 (CCl_2F_2) is recommended. Freon is kept in a metal container that is cooled with liquid nitrogen. The liquid is stirred during cooling to its freezing point ($-150°C$). The ratio of freon to tissue should be at least 25 ml/g. For larger samples (> 1 g), liquid nitrogen is recommended (Lowry and Passonneau 1972). For emphasis, we note that submersion in ordinary liquid nitrogen (which is at its boiling point) is inefficient because, as N_2 vaporizes, an insulating layer of gas encases the tissue (the "Leidenfrost effect"). To speed freezing, the liquid nitrogen is precooled to its melting point ($-210°C$) by evacuation. Briefly, a Dewar flask is sealed with a rubber stopper that is vented to a vacuum pump by insulated copper tubing. After evaporation of about 10% of the nitrogen, the liquid will be sufficiently cooled (as readily indicated by the loss of transparency that accompanies the phase transition). Without further evacuation, the temperature of the liquid nitrogen will stay well below the boiling point for at least 30 min.

With respect to the preservation of tissue integrity, less severe temperatures have been suggested for freezing of plant tissues (Chayen et al. 1960a,b; Gahan et al. 1967). These authors showed that the formation of ice crystals mainly occurred at temperatures between 0 and $-40°C$, which resulted in damaged cell cytoplasm, whereas freezing between -55 and $-65°C$ did not do any harm (for a discussion of this phenomenon see Gahan 1984).

The use of chilled propane or isopentane, suggested by some authors, is not recommended because of fire hazard. The latter is enhanced by the condensation of oxygen from the air.

3 Storage of Frozen Tissues

Storage temperatures must be low enough to prevent ice crystal growth, metabolic activities, and diffusion of solutes. Examples for the temperature dependence of metabolite turnover rates are given by Lowry and Passonneau (1972). In general, frozen tissue can be stored without significant losses or metabolic alterations for several months at $-50°C$ or below. However, warming to $-20°C$ results in significant metabolite losses (e.g., in mouse brain, ATP: 20% loss in 24 h; glucose 6-phosphate: 20% loss in 1h). Many enzymes, like metabolites, are unstable at $-20°C$ but retain their full activity after storage at $-80°C$ for several months. The prudent course is to sublime frozen material as soon as reasonably convenient. We have never documented loss of a metabolite or enzyme activity when dried tissue was stored under vacuum at $-80°C$.

4 Freeze-Drying

The use of frozen tissue sections is very old (Raspail 1825), and indeed freeze-drying has been used for a century (Altmann 1889). Wider application of this technique was stimulated by the work of Gersh (1932). The principles and equipment are so simple that the overview by Jensen (1962) remains pertinent. In outline, the samples are transferred to a –35° to –40°C compartment and dried by reduction of pressure to 0.01 mm of Hg or less. The vacuum pump and sample compartment are separated by a dry-ice trap, where moisture (or other volatiles, e.g., terpenes in conifer tissue) are condensed. The time required for complete tissue-water sublimation can vary between 6 h (20- to 100-μm sections) and 3 days (solid tissue samples). Sublimation progress is monitored by a vacuum meter. Ice crystals grown inside the evacuation tube evaporate only very slowly and can thus significantly prolong the drying process. On the other hand, the amount of water contained in a small ice crystal can completely ruin a sample. This danger can be minimized by a slow increase in temperature up to ambient conditions (ca. 20°C, where the vapor pressure of water is high) toward the end of the drying procedure.

The recommended sublimation temperature (–35° to –40°C) is a compromise between the demands of minimizing biochemical changes and being able to dry the tissue in a reasonable period.

Figure 1 shows a simple, effective, but inexpensive, freeze-drier. The system consists of (1) a commercial kitchen freezer that can maintain –35°C (e.g., Zanussi, Liebherr), (2) a vacuum pump that can maintain a pressure at or below 0.01 mm Hg, (3) a vacuum gauge, and (4) a dry ice trap. Important considerations are to prevent leaks, to minimize tubing lengths, and to maximize tubing diameter (molecular migration at these pressures is by "random walk").

Fig. 1. An effective freeze-drying apparatus. This custom-fabricated unit consists of a commercial deep freezer (–35°C under continuous operation), a vacuum pump and gauge, and a dry ice trap

5 Sample Containers

Large samples (e.g., pieces of leaf tissue) are dried in small glass vials, which are available in a wide range of sizes (e.g., from Whatman). To permit vapor exchange, the tops of the caps are machined off (Fig. 2a). The resulting hole is covered with nylon mesh (mesh size depends on the size of the individual samples, e.g., Züricher Beuteltuchfabrik, Zürich, Switzerland).

Tissue sections obtained by freeze-sectioning (e.g., Lowry and Passonneau 1972; Gahan 1984) are dried in specific sample holders; an example is shown in Figure 2b. The holder is kept at the cryostat temperature, and the tissue sections are transferred into the depressions. Deformation of the sections is routinely prevented by inverted thumb tacks used to cover and hold them down. As shown, the holder is covered by a glass slide (which prevents drying as well as tissue loss). Use of rubber-tipped forceps minimizes heat transfer during holder handling.

In freeze-drying per se, vials (or other sample holders) are transferred to precooled ($< -30°C$) drying assemblies (Fig. 3). Obviously, assemblies can be of any size and shape. We use a round-bottomed tube having a diameter of 4 cm and a length of 13 cm. The vessel fits its cap with a standard-taper ground-glass joint (40/50). Evacuation is through a stopcock on a 3- to 5-mm side arm. The stopcock surface is lightly coated with silicon stopcock grease.

Fig. 2a,b. Sample containers for freeze-drying. Larger pieces of tissue are contained in glass vials (**a**), whereas tissue sections are transferred to a cooled aluminium block (**b**); *arrow* sample covered by thumb tack

Fig. 3. Glass evacuation tube for drying and storage of samples

6 Storage of Freeze-Dried Material

Dried tissue, in contrast to frozen tissue, is stable at –20°C, if it is stored under vacuum. To avoid deleterious effects of ice crystals (see above), the orifice of the drying assembly is plugged, and the assembly is stored in a frost-free freezer (e.g. Liebherr "nofrost", F.R.G.). As a precaution, if one plans to assay for very sensitive enzymes, flushing the tubes with nitrogen gas is advised. Before samples are removed, the drying assembly is warmed to ambient temperature in an atmosphere of low humidity. Admission of air should be at a moderate rate. As the stability of the samples is primarily dependent on the time they are exposed to air and room temperature, sample handling under these conditions should be as short as possible.

7 Dissection of Tissue

7.1 Environmental Requirements

As we have mentioned, complete dryness of samples and only brief exposure to temperatures above –20°C are recommended. These requirements indicate the necessity for an especially controlled laboratory environment (relative humidity below 40%, temperature between 18° and 20°C). Air movement where ng-size samples are handled is controlled by lint-free curtains (Fig. 4).

7.2 Dissection Procedure

It is essential to ensure that the drying assembly has reached ambient temperature before it is opened. During acclimation, the stopcock orifice must remain plugged to avoid dew, which, if swept by the rush of air onto the samples, would render them worthless.

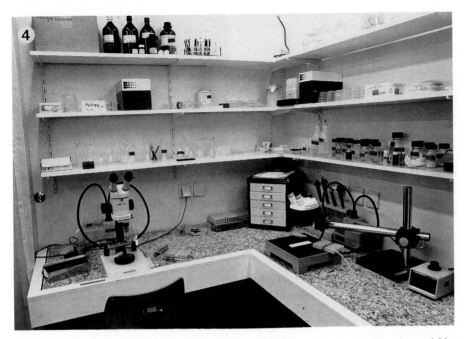

Fig. 4. Dissecting room. The overview shows different working areas for sample dissecting, weighing, and biochemical analysis. Curtains (pulled back for better visibility) prevent significant air movement. Conditioned air enters the room through pinholes in a false ceiling, resulting in a virtually undetectable laminar downdraft

Immediately after sample portions are removed, the drying assembly is again evacuated and returned to –20°C. For further handling, the sample is transferred to a piece of Plexiglas (translucent or black, depending on sample properties and the illumination system), which rests on the stage of a stereomicroscope (e.g., Leitz M6; Fig. 5).

Static electricity is a formidable force when acting on the small masses of single-cell samples. This problem is exacerbated by the required low humidities and the need to work on hydrophobic surfaces, which are insulators. There are two ways to minimize the aggravation: exposure of surfaces and tools to a radiation source and spraying equipment with ionized air. (The mechanism of dispelling the static charges is the same: both procedures surround the object with conducting ions).

Suitable radiation sources (^{241}Americium, ^{210}Polonium) are commercially available either as small-diameter discs (for microbalances see below) or as bars (e.g., Amersham-Buchler; Fig. 5). Ionizing compressed-air "guns" are sold as static eliminators (e.g., Simco, USA). Routinely, we keep the dissecting area low in static electricity by suspending a ^{210}Po bar over the Plexiglas (Fig. 5).

Sample transfer and microdissection require special tools. Large samples (> 1 μg) can be manipulated with commercial preparatory needles. Hair points can be used for intermediate-sized samples (> 500 ng; Fig. 6). To make such a tool, a Pasteur pipette is cut off. The cut end is fire polished and a length of hair is epoxyed

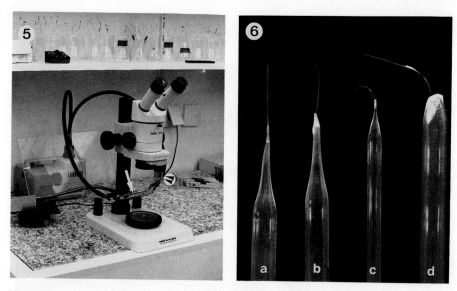

Fig. 5. Dissecting microscope. Tissue samples are transferred to a Plexiglas disc. Static charges are eliminated by a radiation source (Polonium bar, *arrow*). If samples require detailed evaluation, a switch from stereomicroscopic viewing to ordinary microscopy is helpful (Wild M3Z combistereo, microscope lens indicated by *2 arrowheads*)

Fig. 6a-d. Tools for sample handling. Depending on sample size, transfer of samples can be performed with a preparatory needle or different types of "hair points"; **a** simple quartz fiber, fixed to a modified Pasteur pipette; **b,c,d** hair points without (**b,c**) and with quartz fiber tips (**d**)

onto it. Gluing a fine quartz fiber (diameter 2 to 5 μm) to the hair point yields a small tip. [These quartz fibers are formed at the point of contact between two quartz rods heated to the melting point in a propane flame. Pulling the rods rapidly in opposite directions forms fine fibers (see Sect. 8.1).] A fiber about 2 cm in length is cut off with surgical scissors. The hair point is dipped in epoxy resin, and the tacky tip is aligned with one end of the quartz fiber. When the epoxy is sufficiently polymerized, the fiber is trimmed to a few millimeters (Fig. 6d). Such a quartz-tipped hair point is suitable for the transfer of samples weighing less than 20 ng (e.g., a pair of guard cells). Exposure of the tips to a radiation source facilitates manipulation of the specimen.

Different types of knives are desirable for different operations. Larger samples such as leaf pieces can be cut into smaller sections with a scalpel or an ordinary razor blade. Finer dissection, however, requires smaller knives, prepared thus: Slivers of the cutting edge of a razor blade (about 2 mm wide) are cut from the rest of the blade. (A paper cutter or a pair of scissors is suitable for this effort). Each sliver is then fragmented into pieces of 1 to 2 mm length (Fig. 7a). A fragment is epoxyed onto a nylon bristle (taken from a tooth brush), which in turn is glued to a copper wire that is fixed to a handle (Fig. 7a,b). The desired size of the blade fragment as well as the stiffness of the bristles depends on properties of the specimen. Finally, two points

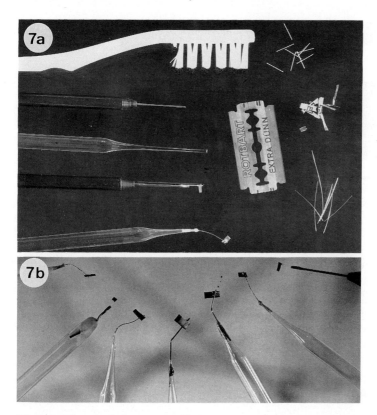

Fig. 7a,b. Dissecting knives. A razor blade is cut or broken into small splinters, which are affixed to a flexible nylon bristle. **a** Assembly of "raw materials"; (**b**) selection of microknives of different flexibility: blade splinters are glued to copper or steel wires of different diameter or to nylon bristles

about knife maintenance should be made: first, care should be taken that the blades are rust free and that spare fragments are stored under dry conditions; second, grease on the razorblade fragment can be removed by successive washes with ethanol and acetone.

Preference is the guiding principle for knife use. It can be held like a pencil, and the sharp tip, pressed obliquely onto the Plexiglas stage, is stabilized against erratic hand movements. Pressing the knife down further bends the bristle, and the tissue is cut as the knife blade closes down on the cutting surface. With some experience it is possible to dissect samples less than 20 μm in diameter.

7.3 Collection and Transfer of Dissected Samples

Dissected samples are collected on a transfer platform from the microscope stage. The platform (Fig. 8), as originally described by Lowry and Passonneau (1972), works well. The handles of the platform are milled from hardwood; the platform

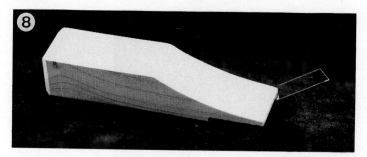

Fig. 8. Transfer platform for small samples. Handles are made of wood. A platform broken from a microscope coverslip is attached with epoxy resin. The angle between the wood and the glass strip facilitates sample transfer during weighing

itself is a 3- to 6-mm-wide strip of cover slip. Dissected samples are arranged in a single row (e.g., two guard-cell pairs/mm) parallel to the leading edge of the platform. From this position, samples are easily transferred by fiber tip into and out of a balance case or into an oil well for analysis, as will be described below. (To keep them clean and to prevent sample loss, the platforms are kept in Petri dishes).

8 Determination of Sample Mass

8.1 Microbalance

The entire microdissected sample is assayed. This fact implies that the experimental basis for expression of tissue amount must be mass, because its measurement is nondestructive. However, comparisons based on mass may confound a correct interpretation because mass (being mostly walls in plants) varies considerably and independently of protein, chlorophyll, or volume, the usual parameters of choice. Expressions based on these latter parameters are thus derived ones calculated by means of conversion factors (e.g., Outlaw et al. 1981).

Because of the small size of samples (*Vicia faba* mesophyll cells, 10–15 ng; Jones et al. 1977), weighing is usually not possible with commercially available balances, which have a sensitivity of about 100 ng (e.g., Sartorius balance MP 8-1). Fortunately, quartz fiber balances are quite suitable. Also descriptively called a "fishpole balance," a quartz fiber balance works as shown in Fig. 9 (i.e., the mass causes a deflection toward gravity). Small fibers (down to about 3 μm diameter) are commercially available. Very small fibers must be custom fabricated, and this process embodies also art. Fibers can be "jerked" from larger fibers (the preparation of which was described above). An alternative method for obtaining very small fibers is to "pull" them upward in the rush of a large, relatively cool flame. The balance case is made from a glass syringe with the closed end of the syringe barrel cut off. To the lower end of the plunger, a short piece of copper wire is attached with epoxy resin

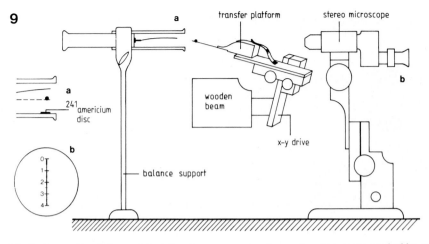

Fig. 9. Quartz fiber balance. Principles of arrangement and operation. Inserts are marked by *a* and *b*

(Fig. 10). The plunger is inserted all the way into the barrel so that the copper wire is exposed. The quartz fiber is then attached to the copper wire (again, with epoxy). The fiber is withdrawn into the body of the syringe for protection and to avoid disturbances from air currents.

The size of the balance case (volume of syringe barrel) and the size of the fiber depend on the required sensitivity. (Balance sensitivity and case volume are inversely correlated). Lowry and Passonneau (1972) recommend the following syringe volumes (ml) for the respective balance capacities (μg): 50, 5; 10, 0.5; 2, 0.05; 0.5, 0.005.

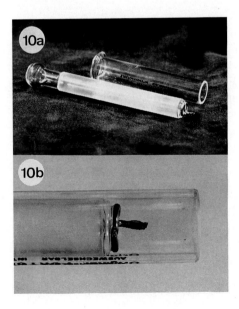

Fig. 10. Case of a quartz fiber balance disassembled (**a**) and with plunger and copper wire in place (**b**)

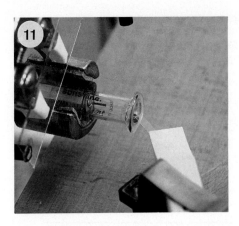

Fig. 11. Quartz fiber balance installed

8.2 Mounting the Balance

Dampening of vibration is essential, but the balance cannot be "isolated", because oscillations, once started, would persist. The horizontal balance case is fixed through a metal frame to a rubber-insulated ("floating") granite plate, about 5 cm thick (Fig. 11). In front of the balance case, a firm support (wooden beam, about 10 × 10 cm) is provided. This hand support is connected to the table [i.e., independent of the floating plate that supports the balance (cf. Fig. 9)]. The balance front is closed by flat glass (e.g., a portion of a microscope slide) that is held snug by springs. Finally, the plunger is rotated until the fiber is positioned advantageously. After this adjustment, the plunger is fixed to the barrel with a drop of sealing wax.

8.3 Illumination and Viewing

Illumination of the fiber tip is by a cold-light unit (halogen lamp with glass fiber optics; e.g., Schott "Kaltlichtleuchte"). Empirical orientation of the optical fiber quickly establishes the best position for viewing through a horizontal stereo-microscope (Fig. 12; e.g., American Optical Co.; Zeiss).

8.4 Handling

The viewing microscope is adjusted vertically so that the fiber tip matches the zero point of an ocular micrometer set in one of the eye pieces of the microscope. For sample loading, the balance case is opened and the transfer platform is advanced to a position near the fiber tip. (Note in Fig. 13 that the transfer platform is clipped to an X-Y drive that facilitates vertical and depth movements). With a quartz-tipped hair point, a small sample is transferred from the platform to the fiber tip. After withdrawal of the platform, the balance case is closed again. Bearing the weight of

Fig. 12. Illumination and viewing of the microbalance

Fig. 13. Double rock and pinion device (microscope stage micrometer). The sample holder is fixed by a spring clip, and under stereomicroscope control the samples are moved to a position close to but below the fiber tip

the sample, the tip is deflected downward. The tip position (with reference to the micrometer) is a measurement that can be related to sample mass. By reversal of the procedures described above, the sample is returned to its original position on the transfer platform. The remainder of the samples are weighed in the same way, but it is necessary to check the zero position only after several samples are weighed.

Electrostatic charges — an aid to sample transfer — can defeat any balance. In all cases, an alpha source (e.g., 20 μCi ^{241}Am disc; cf. Fig. 9 b) to dissipate static is needed.

8.5 Calibration

Balances with lower sensitivity ($>$ 500 ng) can be calibrated with p-nitrophenol crystals, as described by Lowry and Passonneau (1972). A crystal is placed on a fiber tip, and the deflection resulting from its mass is recorded. Then the crystal is removed and dissolved in carbonate buffer (100 mM, pH 10). From the absorbance (400 nm) of the solution, the mass of the crystal is inferred.

More sensitive balances are calibrated analogously, except that fluorescence, which is inherently more sensitive than absorbance, is used. We find that use of quinine hydrobromide (excitation 340; emission 450 nm) subsequently dissolved in 0.1 N H_2SO_4 at final concentration of 1 to 200 ng/ml gives satisfactory results.

8.6 Maintenance

The balance fiber must be cleaned occasionally, but otherwise no maintenance is required. Cleaning a balance fiber is tedious, and, unfortunately, fibers are easily broken during the process. Thus, cleaning should be conducted only on an "as-needed" basis, as indicated by adhering particles that cannot be removed or by excess fiber droop. For cleaning, a droplet of ethanol/H_2O (20/80) is applied to the base of the fiber (i.e., at its attachment to the copper wire) with a fine-tipped pipette and pulled along the fiber toward its tip. This procedure is repeated with 95% ethanol and, finally, with hexane.

Prudence requires that calibration be checked from time to time. This check is not intended to monitor fiber performance per se, which is stable. Instead, recalibration would reveal a shortened fiber (broken inadvertently during use). Occasional recalibration is desirable for an entirely different reason as well: it is a quantitative reminder to the user of the heedfulness required to attain acceptable precision. Elegantly simple, fiber balances have a narrow linear range, which requires that the initial and final readings be made as precisely as possible.

9 Biochemical Analysis of Samples

In principle, the microanalytical procedures we will describe are identical to conventional assays that are indicated by the absorbance of NAD(P)H. In the latter, the reaction product (perhaps through an enzymic couple) is measured directly, whereas in these microprocedures, the resulting NAD(P)(H) is measured only after additional amplification steps. There are particular pragmatic considerations, however. First, because of the very small sample size, the assay cocktail volume is diminished; the diminution is necessary to achieve a reasonable ratio of signal to "noise." Second, and notwithstanding the preceding, the tissue is more highly diluted by assay cocktail. Higher dilution implies less interference by endogenous substances, the requirement for stringent control of analytical contaminants, and special considerations to avoid denaturation of proteins. (The problem of protein denaturation is exacerbated by the high ratio of surface area to volume). This condition is ameliorated by inclusion of "excessive" exogenous protein (up to 10 mg/ml bovine serum albumin) and sulfhydryl reagent (up to 10 mM dithiothreitol).

9.1 Working with Small Assay Volumes

9.1.1 Assay Racks

As mentioned above, the assay volume is small (1 to 1000 nl) and, if exposed, the cocktail would evaporate rapidly. Thus, solutions are delivered into low-density oil that is held in the reaction wells of a Teflon rack (Fig. 14). (Teflon was chosen for its chemical resistance and hydrophobicity). The droplets sink in the oil, and they are

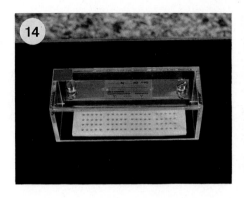

Fig. 14. Oil-well rack. The rack (120×30×3 mm) is made of Teflon. The reaction wells have a diameter of 3 mm. On top of a Plexiglas box, covering an oil-well rack, a radiation source (Polonium bar) is fixed. This is necessary to prevent or decrease electrostatic charges

illuminated from below and viewed through a stereomicroscope from above. To permit sufficient light transmission, the well bottoms must be thin. To make thin bottoms, the Teflon must be fiber free.

Commercial microtiter plates of polycarbonate are an alternative to custom-made Teflon racks. Plates as small as 10 μl/well are available and are resistant to the chemicals and temperatures employed. The oil must be less dense, remember, than the assay droplet. The problem of evaporation is aggravated at higher temperature (e.g., 80°C, used to destroy substrate pyridine nucleotide). Wide experience with mineral oil (e.g., Sigma "heavy white oil") thinned with n-hexadecane indicates its general utility.

Because of their high cost, Teflon racks are cleaned for reuse. The bulk of the oil can be removed by inversion and tapping off the racks on paper towels. The residual is washed away with petroleum ether. The racks are then submerged in ethanolic KOH or NaOH (700 ml 95% ethanol + 30 ml 5 N K(Na)OH) and heated at 95°C for 15 min. Finally, the racks are rinsed exhaustively with deionized water.

9.1.2 Pipetting

Modified Lang-Levy pipettes (Fig. 15) are used to deliver sub-μl-volumes into the oil wells. An oil-well pipette has an extremely fine tip, which facilitates transfer of aqueous solutions under oil. The bend, about 5 mm above the tip, improves handling. Larger pipettes (> 500 nl) function as other constriction pipettes do; small ones fill by capillarity, and their contents must be forcefully expelled, either by mouth or, in extreme cases, by a mechanical pneumatic device.

An oil-well pipette is used thus: The tip is inserted below the oil surface. Then the aqueous aliquot is expelled and remains hung as a bubble on the tip. Withdrawal of the pipette dislodges the bubble, which then sinks (see Fig. 16). A second reagent is added by direct pipetting into the previous droplet. Aliquots are removed in a similar fashion. It is advantageous to silanize (e.g., with Sigmacote, Sigma) the pipette tip exterior, onto which, otherwise, the expelled bubble creeps. It is, of course, necessary to keep the emptied pipette interior under pressure in order to exclude oil.

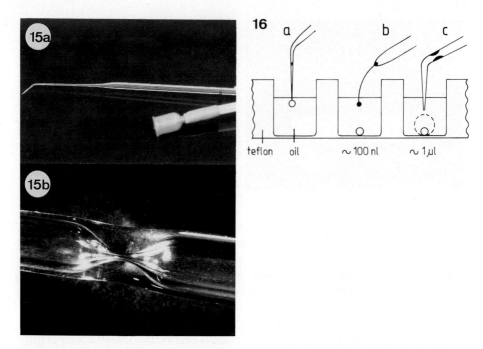

Fig. 15. Example of a constriction pipette as needed for loading and unloading of oil wells. The pipette shown is made from quartz and has a working volume of about 100 nl (**a**). The constriction area is shown at higher magnification in **b**. Pipetting is by suction and pressure application via a mouth piece, connected to the pipette with a silicone tubing

Fig. 16. Sequential scheme for different steps during an oil-well assay. **a** Release of the reagent droplet; **b** sample addition; **c** termination of the reaction by the addition of base or acid

9.1.3 Use of Oil Wells

Oil-well procedures are observed through a stereomicroscope (variable magnification, 7–30 x, is requisite). The overall set-up is shown in Fig. 17. The oil-well rack is positioned atop a dark Plexiglas plate, which in turn is mounted on a wooden box. A single 3 mm hole drilled through the Plexiglas plate provides a point source of light from below. The well under manipulation is positioned over the light and thus is illuminated brightly against a dark background. (In our implementation, a fiber-optic light is reflected by a substage mirror up into the hole.)

Addition of the dissected tissue samples to the oil-covered reagent droplets is the task most demanding of skill. To minimize the difficulty, an amount of oil just sufficient to cover the subsequently added reagent droplet is put into each well. After addition of the initial droplets to the oil well, the transfer platform is positioned above the oil-well rack. (The handle of the transfer platform rests on a thin block that is parallel to the oil-well rack.) The stereomicroscope — first focused on the droplet to inspect for gross contamination — is raised to focus on the samples on the transfer platform. Magnification is zoomed down to the lowest setting. The samples are

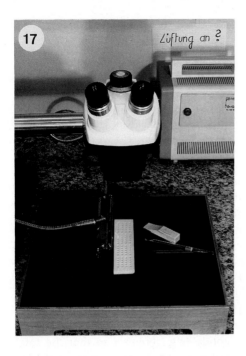

Fig. 17. Bench arrangement for working with oil-well racks

inspected to assure that no sample in the sequence has been lost. The quartz tip of a hair point is used to pick up the next sample. The microscope is zoomed to the maximum power and, while the sample is moved downward into the well, the microscope is kept focused on it. In one motion, the sample is pushed through the oil and into contact with the droplet. Contact with the droplet hydrates the sample, which can be clearly seen through the stereomicroscope. The fiber tip is removed from the well. The oil-well rack is moved to align the next well over the point source of light, and the process continues until each sample has been added to a well. The wells are topped off with oil, a precaution against droplet evaporation during the subsequent steps.

Static electricity is controlled by polonium bars mounted on a gooseneck fixture. Routinely, degassed oil is heated in the rack before use, and the rack covers also have polonium bars mounted on them (Fig. 14).

9.2 Enzymatic Cycling

The aliquot taken for a typical biochemical analysis contains the extract of 10^3 to 10^6 cells. Thus, if the small amounts of reaction products formed by single cells (attomole to picomole) are to be measured with conventional equipment, signal amplification is necessary. Enzymatic cycling is theoretically capable of limitless amplification of some chemical that is cycled between two "forms" (e.g., NAD^+ $\leftrightarrow$ NADH). Because they can be enzymically coupled to virtually any biochemical

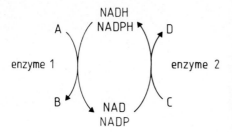

Fig. 18. Reaction scheme for the enzymatic cycling of pyridine nucleotides. Substrates (*A*,*C*) and enzymes (*1*,*2*) depend on the type of cycling (NAD or NADP). If, for example, the NAD system is used, *A* oxaloacetate; *C* ethanol; *1* malate dehydrogenase; *2* alcohol dehydrogenase (Kato and Lowry 1973)

reaction, and because the oxidized and reduced forms can be selectively destroyed, pyridine nucleotides are the most versatile.

In principle, a cycling reagent contains substrates (Fig. 18A,C) in excess and precisely controlled levels of enzymes (1 and 2). The respective pyridine nucleotide (NAD, NADH, NADP, or NADPH) is transferred to the cycling reagent, where — being alternatively oxidized and reduced — it causes the accumulation of products (Fig. 18B,D). It is particularly important to note that the cycling reagent does not distinguish between the oxidized and reduced forms. The rate of cycling depends on the total catalyst (here, $NAD^+ + NADH$). (This total amount does not change during the cycling step). Therefore, the step that precedes enzymic cycling is one to destroy unreacted reagent pyridine nucleotide (e.g., NAD^+) so that the total pyridine nucleotide added to the cycling reagent is just the reaction product (e.g., NADH).

Pyridine nucleotide concentrations are well below the Michaelis constants of the enzymes (Fig. 18, 1,2). Therefore the rate of accumulation of B and of D is proportional to the amount of the pyridine nucleotide added. The cycling rate can be adjusted by changing the enzyme concentration. We routinely run our assays at about 5000 cycles/h, which is less than 10% of demonstrated maximum rate. The precise number of cycles can be determined by pyridine nucleotide standards. Although knowing this number is not essential (sample readings are referred to those of standards), we check it routinely for control of activity of the cycle enzymes.

9.3 Indicator Reaction

The cycling step is terminated by heat. Then one of the products formed during cycling (malate, for the NAD-cycle; 6-P-gluconate, for the NADP-cycle) is measured by the addition of 1 ml of a reagent that couples the product enzymically to pyridine nucleotide. (It is important to recognize that this last step, an enzymic measurement of, e.g., malate, is conducted in 1-ml volumes at μmol concentrations. Thus there is no interference from substances, including the cycled pyridine nucleotide, carried over from the cycling reagent).

Conventionally, NADH (or NADPH) is measured spectrophotometrically; these reduced forms (but not NAD^+ or $NADP^+$) have an absorbance maximum at 340 nm. As discussed above, the product accumulated during cycling (e.g., malate) is coupled enzymically, in this case, to the reduction of NAD^+. In some implementations, the resulting NADH can be measured conventionally, as above. Usually, however, a more sensitive method, fluorescence, is required for precise determination of the (dilute) NADH found in the "indicator step". Over the analytical range — down to approximately 50 pmol/ml — fluorescence is proportional to NAD(P)H concentration. (The inherent sensitivity of fluorescence stems from the fact that this technique is a measure of light emitted by the sample, whereas spectrophotometry is a measure of attenuation of light as it passes through the sample cuvette.)

10 Example for the Complete Procedure: Determination of Fumarase Activity

Step 1. Specific Step. A microdissected sample is pushed into a 200-nl droplet (under oil; oil-well rack) that contains reagents that cause the following reactions to be limited by sample fumarase (Hampp et al. 1982):

$$\text{fumarate} + H_2O \xrightarrow{\text{fumarase}} \text{malate},$$ (1)

$$\text{malate} + \text{NAD} \xrightarrow{\text{malate dehydrogenase}} \text{oxaloacetate} + \text{NADH},$$ (2)

$$\text{oxaloacetate} + \text{glutamate} \xrightarrow{\text{GOT}} \text{aspartate} + 2\text{-oxoglutarate}.$$ (3)

Malate formed by the activity of sample fumarase (1) is oxidized with stoichiometric formation of NADH (2). In order to "pull" reaction (2), oxaloacetate is transaminated (3). Malate and NADH standards are carried through all steps of the assay. After a 30- to 60-min incubation at 25°C, the reaction is terminated by addition of 1 μl of 0.12 N NaOH to each sample well. Then the Teflon rack is heated to 80°C for 20 min. In addition to protein denaturation, this alkali treatment destroys (substrate) NAD^+ without affecting NADH formed during the incubation.

Step 2. Enzymatic Cycling. From each oil well, a 1-μl aliquot is withdrawn, transferred to a fluorometer tube containing 50 μl of reagent for the NAD cycle. After one hour at room temperature, this step is terminated (95°C, 3 min). Malate formed during cycling is 5000 times more abundant than that formed in reaction (1) in the specific step.

Step 3. Indicator Reaction. Malate accumulated in the previous step is oxidized with concomitant reduction of indicator-step NAD^+. This step is performed in a total volume of about 1 ml (1-μl sample aliquot + 50-μl cycling reagent + 1 ml indicator reagent). After about 20 min at room temperature the indicator reaction is terminated by heating of the fluorometry tubes (95°C, 3 min). With an appropriate fluorometer (e.g., Farrand, G.A.T.) the cold tubes can be used directly for NADH determination.

11 Examples for Application: Intercellular Compartmentation of Physiological and Biochemical Properties in Plant Tissues

11.1 Appearance of Enzymes of Major Pathways in Distinct Leaf Cells

A comparison of guard cells which regulate the gas exchange of a leaf with ordinary epidermal cells, palisade parenchyma cells, and spongy parenchyma cells of C3 dicotyledon leaves shows distinct morphological differences and functions (Pearson and Milthorpe 1974; Outlaw 1982, 1983; see also Fig. 19 for microdissected cells). This implies biochemical differentiation. Some of the biochemical differences are obvious, e.g., chlorophyll is abundant in palisade parenchyma, but is generally lacking in ordinary epidermal cells. Other differences are more subtle; e.g., phosphoenol pyruvate carboxylase is elevated in guard cells (Outlaw 1982). Investigators have attempted to determine differences by assaying enzyme distribution among various tissues, but few studies have limited the tissue sample to one particular cell (Outlaw 1982). In order to gain a better understanding of the overall biochemical

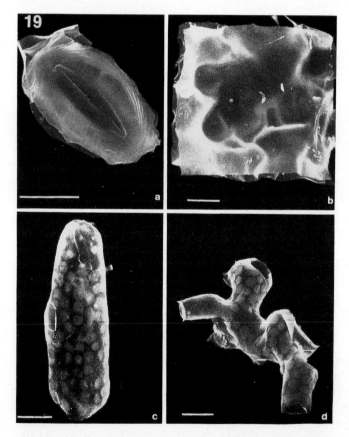

Fig. 19a-d. Microdissection of freeze-dried *Vicia faba* leaf tissue. Electronmicrographs of single cell preparations. **a** Guard cell pair; **b** epidermal cell; **c** palisade parenchyma cell; **d** spongy parenchyma cell

functioning of these different cells, we have assayed them for enzymes that catalyze steps in major carbon pathways. In Table 1 the levels of activity related to the protein content of the different cell types (Fig. 19; for conversion of dry weight into protein see Outlaw et al. 1981) are compared for ribulose-bisphosphate carboxylase (Rubisco; photosynthetic carbon reduction pathway), hydroxypyruvate reductase (photosynthetic carbon oxidation pathway), NAD-dependent glyceraldehyde-P dehydrogenase (TPDH; glycolysis), 6-P-gluconate dehydrogenase (oxidative pentose-P pathway, and fumarase (tricarboxylic acid pathway). Neither Rubisco nor hydroxypyruvate reductase could be detected in guard or epidermal cells. However, high activities of these enzymes were present in mesophyll cells. The specific activity of fumarase was about fourfold higher in guard cells than in epidermal, palisade or spongy parenchyma cells. In addition, TPDH and 6-P-gluconate dehydrogenase were also present at high protein specific activities in guard cells (two- to fourfold that in mesophyll cells). Thus these data indicate that, in accordance with stomatal function, the capacity for metabolic flux through the catabolic pathways is high in guard cells. On the other hand these results provide support for the notion that photoreduction of CO_2 by these guard cells is absent, or at least of extremely low rates.

Table 1. Protein-specific activities of enzymes involved in five major pathways in green plants[a]

	Rubisco	Hydroxypyr. reductase	TPDH	6-PGDH	Fumarase
Guard cell	0	0	30.8 ± 1.2	4.62 ± 0.2	16.5 ± 0.8
Epidermal cell	0	0	17.8 ± 0.8	2.60 ± 0.08	6.03 ± 0.3
Spongy par. cell	10.1 ± 1.1	3.47 ± 0.2	13.1 ± 1.9	1.17 ± 0.07	4.67 ± 0.1
Palisade par. cell	9.5 ± 0.5	3.18 ± 0.2	10.0 ± 0.5	0.73 ± 0.05	4.03 ± 0.2

[a] Photosynthetic carbon reduction pathway (ribulose-bisphosphate carboxylase, Rubisco), photosynthetic carbon oxidation pathway (hydroxypyruvate reductase), glycolysis (NAD-dependent glyceraldehyde-P dehydrogenase, TPDH), oxidative pentose-P pathway (6-P-gluconate dehydrogenase, 6-PGDH), and tricarboxylic acid pathway (fumarase). The activities [mol kg $^{-1}$ protein h^{-1} substrate consumed or product formed] are determined from single cells (cf. Fig. 19; Hampp et al. 1982). Values $\pm$ SE (n = 20).

11.2 Leaf Development: Quantitative Histochemical Analysis Along an Elongating Primary Leaf of *Hordeum vulgare*

Leaves of graminaceous plants develop from basal intercalary meristems which are surrounded by sheaths of fully developed leaves. Cell elongation occurs in a zone adjacent to the meristem that is also enclosed by sheaths of fully developed leaves. Thus, there is a continuum of cellular development ranging from achlorophyllous heterotrophic cells near the leaf base to fully expanded photosynthetically competent cells. In addition to studies on chloroplast development, graminaceous leaves have also been used to investigate relationships between growth and water potential (for ref. see Hampp et al. 1987). As growth requires water influx and turgor pressure (Morgan 1984), high growth rates must be coupled to rapid accumulation of solutes. In *Hordeum* primary leaves the basal intercalary meristem is very close to the grain and only extends to a fraction of 1 mm, followed by the elongation zone of a few millimeters in length. This very close spatial arrangement indicates that factors influencing growth should be studied on a scale where it actually occurs. We thus have employed quantitative histochemistry to examine the distribution of metabolites and enzymes possibly involved in leaf elongation (Hampp et al. 1987).

In Table 2 the content of starch, malate, potassium, and the activity of phosphoenolpyruvate carboxylase (PEPC) are compared in cells prepared from different zones of an elongating lyophylized primary leaf of barley. The results show that the amounts of potassium and malate were low in the region of the intercalary meristem (close to the point of grain attachment), but that concentrations of solutes increased abruptly and stoichiometrically (equivalent basis) in the elongation zone. As in parallel starch decreased, and with respect to the distribution of the activity of PEPC, the results indicate that cell expansion in barley could be augmented by a potassium malate osmoregulatory system that utilizes starch as a source of carbon skeletons.

Table 2. Levels of starch, malate and K^+, together with the activity of phosphoenolpyruvate carboxylase (PEPC) along an elongating primary leaf of barley[a]

Leaf zone	Starch	Malate (mmol kg^{-1} DW)	K^+	PEPC (mmol kg^{-1} DW h^{-1})
Point of leaf insertion	94.4	10.2	44.4	824
Elongation zone	54.0	98.7	211.5	1275
1 cm from leaf base	33.8	86.9	147.8	871
2 to 3 cm from leaf base	63.0	51.9	142.7	768
About 5 cm from leaf base	35.9	28.8	163.1	750
About 10 cm from leaf base	44.2	3.7	155.1	536

[a] Cells were microdissected from leaf areas at different distances from the grain, representing distinct stages of leaf differentiation. Starch is given as hexose units.

11.3 Leaf Movement: Examples for Biochemical Properties of Cells Along a Cross Section Through a *Phaseolus coccineus* Pulvinus

Certain plants, such as *Phaseolus coccineus*, change the position of their leaves in a day-night cycle (e.g., Satter and Galston 1981; Ruge and Hampp 1987). The nearly horizontal orientation of the lamina during the day, followed by a close to vertical position at night, is known to be a consequence of antagonistic volume changes in

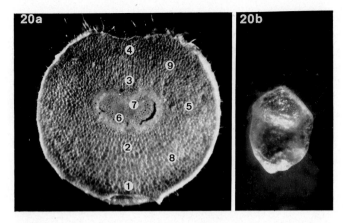

Fig. 20a. Cross section through a laminar pulvinus of *Phaseolus coccineus*. The extensor is located in the *lower*, the flexor in the *upper part*. The central vascular core (bundle) is surrounded by collenchyma at the inner sides of the cortex. The tissue areas to the *left* and to the *right* of the bundle are named flanks. *Numbers* refer to the data given in Table 3. A single cell microdissected from that area of the flexor containing "motor cells" is shown in **b**

the extensor and flexor cells of the laminar pulvini (for a pulvinus cross section see Fig. 20). As in guard cells, these volume changes are due to changes in the osmotic potential, mainly owing to K^+ influx and efflux. In order to identify those pulvinar cells which are involved in leaf movement and to characterize them biochemically we use the techniques described in this chapter for the determination of metabolite levels and enzyme activities. To give an example, the amounts of adenine nucleotides contained in cell clusters (about five to ten cells) microdissected from different areas of a pulvinus cross section (see Fig. 20) are compared in Table 3.

Table 3. Levels of adenine nucleotides (mmol/kg DW) and ratio of ATP/ADP in cells microdissected from different areas of a cross section of a freeze-dried leaf pulvinus (*Phaseolus coccineus*)[a]

Origin of cell		ATP	ADP	AMP	Sum	ATP/ADP
Extensor	(1)	0.93	0.20	0.74	1.87	4.59
	(2)	0.73	0.17	0.63	1.53	4.32
Flexor	(3)	1.48	0.26	0.85	2.59	5.73
	(4)	2.02	0.23	1.00	3.25	8.89
Flank	(5)	1.71	0.22	0.26	2.19	7.64
Bundle	(6)	1.11	0.29	0.67	2.07	3.85
	(7)	1.03	0.19	0.55	1.77	5.39
Flank/extensor interspace	(8)	1.44	0.22	0.67	2.33	6.45
Flank/flexor interspace	(9)	2.07	0.40	0.84	3.31	5.13

[a] The pulvinus was cut from a leaf in day position (3.5 h after onset of illumination) and adenylates were determined by luminometry (Verhoek-Köhler et al. 1983; Hampp 1985). Cell identification numbers refer to the areas designated in Fig. 20.

Interestingly, the different cell types show distinct differences in both their total adenine nucleotide pool and the ratio of ATP/ADP. As the data refer to dry weight, some of the differences in pool size might be due to differences in cell wall thickness. The ATP/ADP ratios, however, a measure of the energy status of a cell, should indicate functional peculiarities. Here especially, some of the peripheral cell layers (Fig. 20, *4,5,8*) show increased values. As K^+ plays an important role in leaf movement and because of its active transport between apoplast and symplast (Starrach et al. 1985), these specific differences in local ATP/ADP ratios (plasma membrane ATPase) could reflect a distinct participation in potassium transport between flexor, extensor, and flanks.

Final Remarks

Although it is beyond the scope of this chapter, the power of enzymatic cycling has stimulated many modifications. Our description here is of an indicator step that couples an accumulated product enzymically. For examples of diversity, the accumulated product itself may be "amplified" chemically (e.g., Outlaw and Kennedy 1978), or the disappearing substrate (microfluorometry, Outlaw et al. 1985) or accumulated product (enzyme-amplified immunoassay, Harris et al. 1988) may be measured directly. We also note, for completeness, that other methods for measuring sub-picomole quantities of pyridine nucleotides are available (e.g., luminometry; Guder et al. 1984) and have been used by our laboratories.

Acknowledgements. As far as own results are concerned we gratefully acknowledge financial support from the Deutsche Forschungsgemeinschaft, the Alexander von Humboldt foundation, NSF, and DOE.

References

Altmann R (1889) Die Elementarorganismen und ihre Beziehungen zu Zellen. Verlag W Engelmann, Leipzig

Asahina E (1956) The freezing process in plant cells. Contrib Inst Low Temp Sci, Hokkaido Univ Ser A, No 10:83–126

Chayen J, Cunningham GJ, Gahan PB, Silcox AA (1960a) Life-like preservation of cytoplasmic detail in plant cells. Nature 186:1068–1069

Chayen J, Cunningham GJ, Gahan PB, Silcox AA (1960b) Newer methods in cytology. Bull Res Counc Isr 8D:273–279

Gahan PB (1967) Freeze sectioning of plant tissues: the technique and its use in histochemistry. J Exp Bot 18:151–159

Gahan PB (1984) Plant histochemistry and cytochemistry. An introduction. Academic Press, Lond NY

Gahan PB, McLean J, Kalina M, Sharma W (1967) Freeze sectioning of plant tissues: The technique and its use in histochemistry. J Exp Bot 18:151–159

Gersh I (1932) The Altmann technique for fixation by drying while freezing. Anat Rec 53:309

Guder WG, Pürschel S, Vandewalle A, Wirthensohn G (1984) Bioluminescence procedures for the measurement of NAD(P) dependent enzyme catalytic activities in submicrogram quantities of rabbit and human nephron structures. J Clin Chem Clin Biochem 22:129–140

Hampp R (1985) ADP, AMP; Luminometric method. In: Bergmeyer J, Graßl M (eds) Methods in enzymatic analysis, vol 7. Verlag Chemie, Weinheim, pp 370–379

Hampp R, Outlaw WH Jr, Tarczynski MC (1982) Profile of basic carbon pathways in guard cells and other leaf cells of *Vicia faba* L. Plant Physiol 70:1582–1585

Hampp R, Outlaw WH Jr, Ziegler H (1987) Quantitative histochemical analysis of starch, malate, and K^+, together with the activity of phosphoenolpyruvate carboxylase along an elongating primary leaf of *Hordeum vulgare*. Z Naturforsch 42c:1092–1096

Harris MJ, Outlaw WH Jr, Mertens R, Weiler EW (1988) Water-stress-induced changes in the abscisic acid content of guard cells and other cells of *Vicia faba* L. leaves as determined by enzyme-amplified immunoassay. Proc Natl Acad Sci USA 85:2584–2588

Jensen WA (1962) Botanical histochemistry. Freeman, San Francisco

Jones MGK, Outlaw WH Jr, Lowry OH (1977) Enzymic assay of 10^{-7} to 10^{-14} moles of sucrose in plant tissues. Plant Physiol 60:379–383

Kato T, Lowry OH (1973) Distribution of enzymes between nucleus and cytoplasm of single nerve cell bodies. J Biol Chem 248:2044–2048

Lowry OH, Passonneau JV (1972) A flexible system of enzymatic analysis. Academic Press, Lond NY

Morgan JM (1984) Osmoregulation and water stress in higher plants. Ann Rev Plant Physiol 35:299–319

Outlaw WH Jr (1982) Carbon metabolism in guard cells. In: Creasy LL, Hrazdina G (eds) Cellular and subcellular localization in plant metabolism. Plenum Press, NY, pp 185–222

Outlaw WH Jr (1983) Current concepts on the role of potassium in stomatal movements. Physiol Plant 59:302–311

Outlaw WH Jr, Kennedy J (1978) Enzyme and substrate basis for the anaplerotic step in guard cells. Plant Physiol 62:648–652

Outlaw WH Jr, Manchester J, Zenger VE (1981) The relationship between protein content and dry weight of guard cells and other single cell samples of *Vicia faba* L. leaflet. Histochem J 13:329–336

Outlaw WH Jr, Springer SA, Tarczynski MC (1985) Histochemical technique. A general method for quantitative enzyme assays of single cell 'extracts' with a time resolution of seconds and a reading precision of femtomoles. Plant Physiol 77:659–666

Pearse AGE (1980) Histochemistry: theoretical and applied. Vol 1. Churchill, Lond

Pearson CJ, Milthorpe FL (1974) Structure, carbon dioxide fixation and metabolism of stomata. Aust J Plant Physiol 1:221–236

Raspail FV (1825) Developpement de la fécule dans les organes de la fructification des céréales et analyse microscopique de la fécule, suivie d'expériences propres à enexpliquer la conversion engomme. Ann Sci Nat 6:224–239

Ruge WA, Hampp R (1987) Leaf movements of nyctinastic plants — a review. Plant Physiol (Ind Acad Sci, Life Sci Adv) 6:149–158

Satter RL, Galston AW (1981) Mechanisms of control of leaf movements. Annu Rev Plant Physiol 32:83–110

Starrach N, Flach D, Mayer WE (1985) Activity of fixed negative charges of isolated extensor cell walls of the inner laminar pulvinus of primary leaves of *Phaseolus*. J Plant Physiol 120:441–455

Verhoek-Köhler B, Hampp R, Ziegler H, Zimmermann U (1983) Electro-fusion of mesophyll protoplasts of *Avena sativa*. Determination of the cellular adenylate-level of hybrids and its influence on the fusion process. Planta 158:199–204

Photoacoustic Spectroscopy — Photoacoustic and Photothermal Effects

C. BUSCHMANN and H. PREHN

1 Introduction

Definition. Photoacoustic (PA) or optoacoustic measurements in the original, strict sense comprise the detection of acoustic waves which arise from non-radiative de-excitation (radiationless transition) of the sample as a consequence of absorption of continuously amplitude-modulated light. Today this term is extended to measurements of acoustic waves induced by other factors, e.g. gas exchange, surface deformation, evaporation. Furthermore photothermal (PT) techniques have been developed which do not determine acoustic waves but measure the temperature rise induced by non-radiative de-excitation or thermal effects on the sample itself or on the medium surrounding the sample. The excitation of the sample may be carried out not only by light but also by any other type of energetic radiation. A general scheme of PA and PT measurements is given in Fig. 1. In this contribution the main principles, detecting techniques and different applications are summarized with special emphasis on plant analysis.

New, Unique Features. The measurement of PA (or PT) effects represents a non-destructive technique which offers a variety of new, unique features:

1. With increasing light-scattering and absorption of the sample PA measurements become more advantageous, since PA effects are not sensitive to light scattering and even totally opaque materials can be examined.
2. Non-radiative de-excitation processes can be studied in PA spectra of pigments as well as in action spectra of different photobiological processes.
3. Heat production can be compared to other competing ways of de-excitation (e.g. luminescence or photochemistry) with the possibility of monitoring changes of energy balance.
4. Photochemical reactions, e.g. photo-induced gas evolution (or consumption) may be measured. Direct detection of brutto photosynthesis can thus be obtained.
5. Photo-induced mass transfer (particle, liquid, gas) and/or heat transfer of physicochemically active samples may be studied. Thus diffusive and convective transport parameters may be evaluated.
6. Depth profiling allows non-invasive subsurface studies for gathering spectral and thermal information from layered samples which cannot be achieved by any other non-destructive method.
7. Thermal properties can be sensed with the possibility of calorimetric calibra-

tion. This may be of importance for studies of thermochemistry, thermo-regulation and thermotolerance of plants.

8. PA imaging techniques and PA microscopy can provide in vivo information about the thermal, optical or elastic properties of surface and subsurface layers.

Historical Development. The first description of the PA effect by Alexander Graham Bell in 1880 stimulated several scientists to study this phenomenon. Within only 1 year, a dozen publications appeared, among them studies by Lord Rayleigh, W.C. Röntgen and J. Tyndall. Later on, experimentation with the PA effect practically ceased. In the 1930s some observations of the PA gas detection were carried out by A.H. Pfund and M.L. Viengerov. In 1971 the PA effect was "rediscovered" by L.B. Kreuzer and became a modern spectroscopic technique, strongly propagated further by the work of A. Rosencwaig. Since then, the total number of publications in a wide range of natural and technical sciences has enormously increased.

The first PA spectra with biological material were described by Harshbarger and Robin in 1973. The studies of photosynthetic parameters were started by D. Cahen, S. Malkin and E.I. Lerner in 1978. In 1979 Y. Inoue, A. Watanabe and K. Shibata first published photosynthetic induction kinetics of the PA effect, determined simultaneously with those of chlorophyll fluorescence. In the same year, heat production directly detected by means of PA measurements was compared to photochemical energy conversion and fluorescence emission (Vacek et al. 1979). A first theoretical treatment of the PA effect with special emphasis on photosynthetic energy conversion was given by S. Malkin and D. Cahen (1979). PA in vivo spectra of intact leaves were systematically carried out by C. Buschmann and H. Prehn (1981, 1983). In 1982 the hypothesis was raised that at low modulation frequencies of the excitation light the PA signal of intact leaves detected in a closed PA cell is determined by pulsed O_2 evolution (Bults et al. 1982a). Many studies on photosynthetic O_2 evolution dependent on a variety of endogenous and exogenous factors have been carried out in the last few years (e.g. by O. Canaani, M. Havaux and D. Yakir). Heat production detected simultaneously with fluorescence emission has been studied as a thermal PA signal in order to evaluate changes in energy balance, e.g. during photoinhibition (Buschmann 1987) and energy quenching (Buschmann and Kocsányi 1989).

2 Physics of the Photoacoustic Effect

The theory of PA measurements has been summarized earlier (e.g. Rosencwaig 1978; Rosencwaig and Gersho 1976; Pao 1977; Prehn 1979; Tam 1986).

2.1 Excitation

The common element in all excitation processes is that heat is produced by interaction of an incident beam with the sample. The absorption of the beam transfers the sample into an excited state. The following thermal de-excitation (relaxation) induces a temperature rise. Subsequently heat may be distributed throughout a large region of the sample or confined to a small region such as the surface of an opaque solid. The heat source gives rise to corresponding thermal or acoustic fields, to which the different detection schemes are sensitive (McDonald 1985; Fig. 1).

In most cases the excitation of the sample is achieved by illuminating the sample with light. In general, a high PA signal is achieved with samples characterized by a

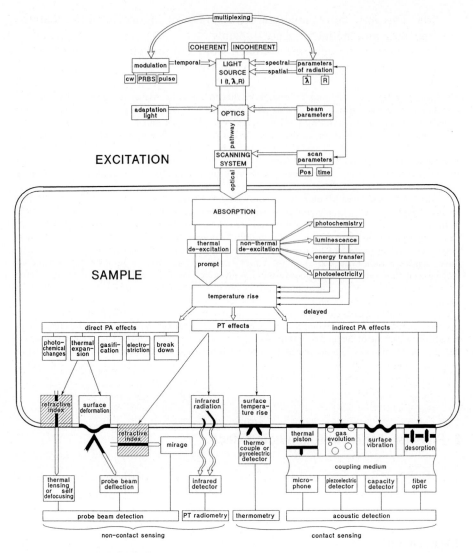

Fig. 1. General scheme of photoacoustic and photothermal measurements from the excitation source via the temperature rise, and the generation of photoacoustic and photothermal effects in the sample towards the different types of detection

high light absorption, which guarantees the uptake of high amounts of energy. The penetration of light into different depths of the sample is determined by the absorption coefficient β of the sample (Fig. 2a).

Part of the absorbed energy produces heat in a prompt thermal de-excitation process. The energy may, however, also be transferred into four different channels of non-thermal de-excitation: photochemistry, luminescence (i.e. in leaves mainly

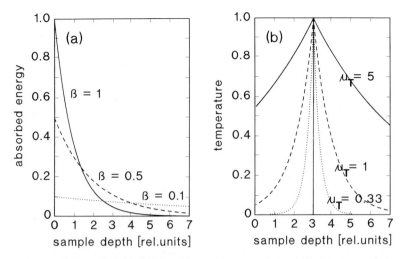

Fig. 2a. Energy of absorbed light at different depths of homogenous samples with different absorption coefficients β. It reflects the penetration of light into the sample from the sample surface (sample depth = 0). **b** Amplitude of the temperature modulation in samples with different thermal diffusion length μ_T for an absorption process at a relative sample depth of 3. It shows how heat is propagated inside the sample (Prehn 1979)

chlorophyll fluorescence), energy transfer or photoelectricity, which may again induce delayed heat production (Fig. 1). Radiation-induced heat production within the solid or semisolid sample will lead to a temperature rise which then induces PA and PT effects (Fig. 1). PA effects may be directly evoked within the sample at the site where absorption has taken place or indirectly in the medium surrounding the sample (e.g. air or water). In the following pages, only those effects which are already applied to plant research or exhibit a potential to future applications are discussed.

2.2 Generation and Propagation of the Photoacoustic Signal

Various processes may be detected as direct PA effects. In order of descending efficiency of PA generation these are: break-down or plasma formation, gasification or ablation, photochemical changes, thermal expansion and electrostriction (Tam 1986). The indirect PA effects in the medium surrounding the sample comprise the "thermal piston" (see Sect. 2.2.1), gas evolution, surface vibration, desorption (Tam 1986). For liquid or wet samples evaporation may also be taken into account (Korpiun et al. 1986). In the following pages the generation and propagation of a PA signal measured via the indirect PA effect of the thermal piston is described. Since this effect is used most frequently, theoretical treatments are most elaborated.

2.2.1 Signal Generation

Modulated photo-induced heat produced in the solid or semisolid sample causes pressure changes in a thin layer of the surrounding "coupling" medium. This active layer induces a "thermal piston", which transfers acoustic waves to the rest of the coupling medium where they are finally monitored by an acoustic transducer, e.g. microphone, piezoelectric or capacity detector or a fiber optic (Fig. 1). A semiquantitative theory of a one-dimensional three-layer model of PA signal generation from a solid sample mounted on a backing material with the sample surface in contact with the coupling gas was first presented by Rosencwaig and Gersho (1976). This so-called RG-theory has been further developed by various theoretical elaborations with respect to specific applications. The one-dimensional analytical solution to the generation of the PA signal has been simplified by Poulet et al. (1980), extended to three-dimensional heat flow by McDonald (1980, 1981) and later applied to layered structures by Opsal et al. (1983). A theoretical treatment for impulse response experiments has been given by Mandelis and Royce (1979). A one-dimensional finite difference model employing a direct digital simulation as a method of modelling signal generation and propagation was described by Miller (1985).

Regarding the numerous applications in plant physiology with rather complex sample matrix, a more intuitive approach based on the RG-theory seems to be appropriate to predict the various cases of PA signal generation. Of main interest for the user is the fundamental dependence of the PA signal on the optical and thermal properties of the sample and of the coupling gas as well as on the parameters of excitation and on different designs of detector, sample compartment and instrument. From this the user can conclude whether this method can be applied to meet his analytical purpose.

When the sample enclosed in a PA cell is periodically heated by the absorption of the modulated light, it is assumed that the primary source of a PA signal results from a periodic heat flow from the sample to the surrounding gas. A simple, cylindrical cell with three linear and homogeneous layers (backing material, sample and boundary layer of coupling gas) is modelled with further intuital assumptions of the appropriate boundary conditions (Pao 1977). A brief summary for practical applications has been given by Prehn (1979). The basic formulation consists of Lambert's law of optical absorption and the thermal diffusion equation in the solid, taking into account that monochromatic light sinusoidally modulated at the angular frequency ω will be absorbed as given in equation (1). The dependence of light absorption in different depths of a homogeneous sample on the absorption coefficient β is shown in Fig. 2a.

$$I(x) = \frac{1}{2} I_0 (1 + \cos \omega t) e^{-\beta x}, \tag{1}$$

$I(x)$ = intensity of light inside the sample at depth x,   [W m^{-2}]
I_0 = intensity of the incident light,   [W m^{-2}]
ω = angular modulation frequency,   [rad s^{-1}]
    = $2 \pi \nu$,

ν = modulation frequency,   [Hz = s^{-1}]
t = time of one cycle of the modulated light,   [s]
β = absorption coefficient,   [m^{-1}]
x = distance from the sample surface.   [m]

2.2.2 Signal Propagation

The photo-induced heat will be propagated following the laws of thermal conductivity (Eq. 2). Acoustic waves generated inside the sample are propagated according to the laws of sound. Since these acoustic waves make only a small contribution to the PA signal they will not be discussed here.

$$\frac{\partial^2 T}{\partial x_i^2} = \frac{1}{\alpha_i} \frac{\partial T}{\partial t} - \frac{\eta}{\kappa_i} \frac{\partial I(x)}{\partial x_i} \; , \tag{2}$$

T = absolute temperature,   [K]
α_i = thermal diffusivity,   [m^2 s^{-1}]
η = yield for non-radiative de-excitation $(0 < \eta < 1)$,
κ_i = thermal conductivity.   [J s^{-1} m^{-1} K^{-1}]

The subscript i can take s, g or b for sample, gas or backing material. The propagation of photo-induced heat is mainly determined by the thermal properties and the density summarized in the term of the thermal diffusivity α (Eq. 2a).

$$\alpha_i = \frac{\kappa_i}{\rho_i c_i} \; , \tag{2a}$$

ρ_i = density,   [kg m^{-3}]
c_i = specific heat,   [J kg^{-1} K^{-1}]

After absorption of light only a fraction of the energy taken up will be converted into heat by non-radiative de-excitation. Equations (1) and (2) can thus be transformed to:

$$\frac{\partial^2 T}{\partial x_i^2} = \frac{1}{\alpha_i} \frac{\partial T}{\partial t} - \frac{\eta}{\kappa_i} \beta \frac{1}{2} I_0 (1 + \cos \omega t) e^{-\beta x}. \tag{3}$$

The solutions of this differential equation represent thermal waves, which are damped by the factor $e^{-a_T x}$ after travelling the length x away from their local generation. The term a_T is defined as thermal diffusion coefficient. Its reciprocal value represents the thermal diffusion length $\mu_T (= 1/a_T)$ analogous to the definition of the optical absorption length $(1/\beta = \mu_\beta)$. The thermal diffusion length μ_T reflects the thickness of the sample in which heat is decreased to $1/e$ of its original value. Fig. 2b shows the temperature profile in samples with different μ_T. With increasing values for μ_T, the heat transfer towards the sample surface becomes more efficient. The term μ_T is dependent on the modulation frequency:

$$\mu_T(\omega) = \frac{1}{a_T(\omega)} = \sqrt{\frac{2\alpha}{\omega}} = \sqrt{\frac{\alpha}{\pi \nu}} \; , \tag{4}$$

Table 1. Thermal diffusion length μ_T and thermally active layer L_s of water calculated for different modulation frequencies of the excitation light[a]

Modulation frequency ν (Hz)	Modulation frequency ω (rad s^{-1})	Thermal diffusion length μ_T (μm)	Thermally active layer L_s (μm)
1	6	214	1343
2	13	151	950
4	25	107	672
8	50	76	475
16	101	53	336
32	201	38	237
64	402	27	168
128	804	19	119
256	1608	13	84
512	3217	9	59
1024	6434	7	42

[a] The calculations are based on the equations given in Section 2.2.2. The values of the thermal properties of water were taken from Berber et al. (1977): thermal conductivity $\kappa = 0.6$ J s^{-1} m^{-1} K^{-1}, density $\rho = 1000$ kg m^{-3}, specific heat $c = 4180$ J kg^{-1} K^{-1}. The thermal diffusivity of water then amounts to $\alpha = 1.435 \; 10^{-7}$ m^2 s^{-1}.

$\mu_T(\omega)$ = thermal diffusion length,   [m]
$a_T(\omega)$ = thermal diffusion coefficient.   [m^{-1}]

This fact is of great practical importance because of the ability to vary the thermal diffusion length within a certain range by choosing an appropriate modulation frequency ω. The reason for this is that during a period of the modulated light, only a definite thermally active layer of the thickness

$$L_s(\omega) = 2\pi\,\mu_T(\omega) \tag{5}$$

will be heated, and thermal oscillations in a depth $x > L_s(\omega)$ will not reach the surface within one period of light and are therefore not able to contribute to the PA signal. PA measurements offer the unique possibility of non-destructive determination of a depth profile of optical absorption within a simple layered, opaque sample. Higher modulation frequency is tantamount to probing the sample closer to its surface. By decreasing ω one increases the thermal diffusion length μ_T and the thermally active layer L_s. Thus optical absorption data from further within the material are obtained. Since the thermally active layer decreases with increasing modulation frequency the PA signal itself also decreases (Fig. 3). Because of the complexity of the different solutions of the differential equation (3) with its specific boundary conditions, a semiquantitative consideration of the main effects and parameters affecting the PA signal will be given here. The generated PA signal which corresponds to an acoustic wave with the intensity S_{PA} is correlated with the parameters of the excitation (exc), the instrument (inst) and sample (sam) which are discussed in the following:

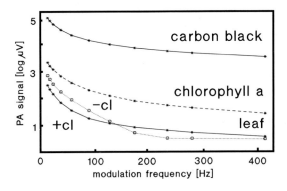

Fig. 3. Photoacoustic signals (logarithmic scale) of different samples measured in a closed photoacoustic cell dependent on the modulation frequency of the excitation light (632.8 nm, He/Ne laser). Isolated chlorophyll a was measured from a filter paper after evaporation of the diethylether solvent. As leaf a *Raphanus cotyledo* was taken without (−cl) or with addition of non-modulated light saturating photosynthesis (+cl, 2000 $\mu E\ m^{-2}\ s^{-1}$) (Buschmann 1987)

$$S_{PA} \approx [I_0(\lambda)]_{exc} \left[\frac{p_g}{T}\ \mu_{T,g}(\omega)\ \frac{1}{x_g} \right]_{inst} \left[\eta\ \frac{K_s}{\kappa_s}\ x_s\ \frac{\mu_{T,s}(\omega)}{\mu_{\beta,s}(\lambda)} \right]_{sam}, \tag{6}$$

p_g =   pressure of the gas above the sample,
$\mu_{\beta,s}(\lambda)$ =   optical absorption length of the sample,
K_s =   heat transmission coefficient.

2.2.2.1 Influence of the Sample Parameters. Assuming that all instrumentation terms remain constant, the physical properties of the sample essentially determine the PA signal. High PA signals are created when the sample possesses a high yield (η) for non-radiative de-excitation. A high absorption coefficient β, i.e. a small absorption length $\mu_{\beta,s}(\lambda)$, leads to a good energy uptake. Furthermore, the thermal properties and the density of the sample, which are summarized in Eq. (2a) and contained in the term $\mu_{T,s}(\omega)$, play an important role. A low thermal diffusivity α, density ρ and specific heat c of the sample will create optimal PA signals. A high coefficient of heat transmission K_s, which correlates with a large sample surface (spongy or fine grained material), may further improve the transition of heat from the sample to the surrounding medium (Adams et al. 1976) and thus enhance the PA signal. This effect of the porous sample relative to that of the solid form was explained by Monchalin et al. (1984) and McGovern et al. (1985). The important feature in porous samples is that the interstitial gas may expand to the main gas region, contributing to the signal as would the surface expansion. This should always be considered, especially when measuring plant tissues with intercellulars.

According to Rosencwaig and Gersho (1976) six cases can be grouped with different optical and thermal transparency (or opaqueness) as determined by the three lengths μ_T, μ_β and x_s characterizing a sample (Fig. 4). Transparent samples with high optical absorption length μ_β where light is absorbed throughout the length of the sample are contrasted to optically opaque samples where most of the light is absorbed within a distance small compared to x_s (i.e. small optical absorption length $\mu_\beta(\lambda)$). For both optical transparency and optical opaqueness, three examples are distinguished: thermally thin (high μ_T), thermally medium and thermally thick samples (low μ_T). A good PA signal can only be obtained from a thermally thick

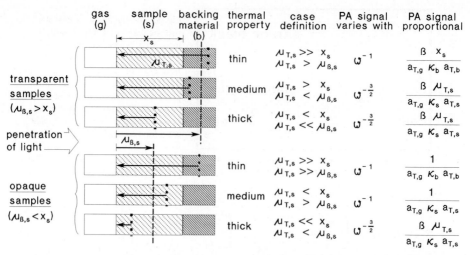

Fig. 4. Six theoretical cases given by Rosencwaig and Gersho (1976) to demonstrate the dependence of the photoacoustic signal on the optical and thermal characteristics of the sample, the material of the sample holder (backing material) and on the gas above the sample. Optically transparent samples (*upper* three examples) are contrasted to optically opaque samples (*lower* three examples) each with high, medium and low thermal diffusion length μ_T (Coufal and Lüscher 1978)

sample on condition that μ_T is smaller than x_s and smaller than μ_β. In these cases the signal depends only on the absorption characteristic μ_β of the sample. If light penetrates to the backing material (transparent samples), the PA signal is affected by the backing material, provided that the thermal diffusion length μ_T is longer than the thickness of the sample x_s. For a sample where the thermal diffusion length μ_T is longer than the optical absorption length μ_β the PA signal becomes "saturated" (McClelland and Kniseley 1976). The dependence of the PA signal on the modulation frequency indicates whether the PA signal is saturated (S_{PA} proportional to ω^{-1}) or not (S_{PA} proportional to $\omega^{-3/2}$). A typical sample with a saturated signal is the case of a very black absorber, e.g. carbon black. In such a sample there is no increase of the PA signal when light absorption is enhanced. Saturation in one part of a spectrum is indicated by the flattening of the peaks. The PA spectrum of carbon black may be used as a reference to compensate for the spectral emission of the light source and the optical characteristics of the spectrometer.

If the PA magnitude is proportional to $1/\mu_\beta(\lambda)$ or $\beta(\lambda)$ the normalized PA signal measured for a range of excitation wavelengths λ can provide the absorption spectrum $\beta(\lambda)$. This is the case with all types of optically transparent samples and of the optically opaque, thermally thick sample. In the transparent sample with thin or medium thermal property the spectrum of the sample is partially superimposed on the spectrum of the backing material. The dependence of μ_T on the modulation frequency of the excitation light (Eqs. (4) and (5)) offers the unique application of non-destructive determination of a depth profile of optical absorption within a simple layered opaque sample. A higher modulation frequency ω is tantamount to probing the sample closer to the surface. By decreasing ω one increases the thermal diffusion length μ_T, and optical absorption data from further inside the material are obtained.

Due to Eq. (6) other specific parameters of the sample can be studied by PA detection for various non-spectroscopic applications. As long as all sample parameters except one are known and/or varied in a controlled fashion, information about the sample parameters can be achieved. For instance, monitoring the time dependence of the quantum yield of non-radiative de-excitation processes is made possible because under certain experimental conditions the time course of the PA signal directly follows $\eta(t)$. Furthermore, measurements of the thermal conductivity are made possible, and this is often very difficult to obtain from biological tissue. Other thermal properties of the sample may be evaluated with the aid of Eq. (2a). Specific features of the sample surface may be analyzed by determining the heat transmission coefficient K_s or changes of its time course. Finally, measurements of the thickness of layered material or substances can be achieved by sensing x_s.

2.2.2.2 Influence of Instrument Parameters. In contrast to the sample parameters, which are invariable for a given sample, instrument parameters in most cases may be varied by the experimenter. The modulation frequency ω affects the value of the diffusion length of both the sample and the coupling gas (Eq. (6)). Signal saturation (see Sect. 2.2.2.1) can be omitted by increasing the modulation frequency and thus decreasing the thermal diffusion length. However, the decrease of the thermal diffusion length is smaller than the decrease of the signal height. In cases close to saturation a PA spectrum is preferably to be measured at higher modulation frequencies. However, one has to keep in mind that the PA signal decreases with increasing modulation frequency (Fig. 3). Table 1 summarizes the values of the thermal diffusion length μ_T and the thermally active layer L_s depending on the modulation frequency of the excitation light calculated for water, which represents in most cases the main component of plant material. According to Table 1 and Eq. (6) one can estimate the appropriate frequency for each type of measurement.

The first term of Eq. (6) with the subscript $_{exc}$ is due to the instrument parameters of the light source. High intensities of irradiation will generate strong PA signals. Therefore, if no other limitations (e.g. denaturation) are expected, high-intensity light sources are required. From the second term of Eq. (6) with the subscript $_{inst}$, it can be concluded that the PA signal amplitude will increase with higher pressure p_g in the PA cell, lower temperature T and shorter length of the gas columns x_g. The PA signal of carbon black was shown to increase with the decrease of the temperature from 320 K to 90 K (Bechthold and Campagna 1981). However, the increase was somewhat stronger than the $1/T$ dependence written in Eq. (6). The thermal diffusion length of the gas $\mu_{T,g}(\omega)$ can be controlled by ω according to Eq. (4). If the length x_g of the gas column equals about 2 $\mu_{T,g}$ the largest signals are expected. The essential prescription for the optimization would then be to have the cell radius and gas length as small as physically practical (as reviewed by McDonald (1985)), but with $x_g > 2 \mu_{T,g}$ at the lowest modulation frequency used, and a radius of the PA cell $r_c > 2 x_g$.

2.2.2.3 Influence of the Parameters of the Medium Surrounding the Sample. When measuring indirect PA effects in the medium surrounding the sample (e.g. gas or water) the physical characteristics of these media influence the height of the signal. In a closed PA cell the thermal properties of the "coupling" gas included in $\mu_{T,g}(\omega)$, pressure p_g and thickness x_g of the gas layer adjacent to the sample determine the PA signal (Eq. (6) term with the subscript $_{inst}$). The medium should be fully transparent for the excitation light. In a closed PA cell the pressure changes determined by the microphone become more pronounced when the gas is optimal in taking up the heat and transferring it to the microphone. By choosing a gas with appropriate characteristics the PA signal may be enhanced (Adams et al. 1976; Thomas et al. 1978). Helium has been shown to be approximately twice as efficient as nitrogen, the main component of air (Eaton and Stuart 1977). When measuring changes of the refractive index outside the sample (see Sect. 3.2.3) a variation of the medium may equally improve the PA signal.

2.3 Physicochemical and Biological Photoacoustic Effects

In PA measurements of (passive) physical samples it is assumed that during spectral measurements all other factors of the sample remain constant. This cannot, however, be assumed in an (interactive) sample with physicochemical and biological activity. Various time-related processes may also contribute to the PA signal, which may be related to different physicochemical processes due to compound phase transitions. Chemical reactions may also dominate the PA signal in various manners. Furthermore, metabolic or thermoregulative (e.g. stomatal regulation) processes which originate from photobiological activity of the sample may contribute to the PA signal. These contributions can be investigated by making use of new PA or PT monitoring possibilities.

Diffusive and convective transport of mass was extensively studied by Korpiun et al. (1988). The Rosencwaig-Gersho model without mass transport was compared with their model including mass transfer. It could be shown that convective as well as diffusive transport significantly affects the PA amplitude, which was explained by oscillating mass transfer from the sample into the gas. Busse et al. (1988) have demonstrated that the formalism used for thermal wave transmission can also be applied to monitor diffusion processes by concentration wave analysis. Mass transport experiments with modulated input have been postulated, and the determination of the diffusive coefficient have been made possible. This may be of great advantage in membrane research.

A laser-induced acoustic source to measure the flowing fluid of a pure, particle-free gas and liquid by displaced excitation and probe beam was demonstrated by Zapka and Tam (as reviewed by Tam 1986). This represents a unique method evaluating both flow velocity and fluid temperature simultaneously with the same experimental set-up. Enhancement of PA signals from condensed phases due to vapour of volatile liquids in the PA cell has been studied extensively by Ganguly and Somasundaram (1988), who indicated that the PA technique may be extended to the study of liquid films on surfaces. Further mass transfer processes have been reviewed by McDonald (1985).

Photochemical effects may also cause acoustic emission, as reviewed by Tam (1986). Photochemical or photoelectrochemical processes influencing the magnitude of the PA or PT signal may be: the complementarity of a chemical reaction, reactive gas evolution or consumption, photochemical chain PA amplification and kinetics of reactant concentrations (for application to photosynthesis, see Sect. 4.3.2).

3 Measuring Systems

In general, a PA measuring system consists of a specific excitation source (e.g. a light source), a mechanism which modulates the amplitude of the excitation and the detecting system with its instruments for the signal processing. Figure 5 shows a universal PA instrumentation which is in use in our institute at Karlsruhe and has been described in detail for detection with a microphone (Nagel et al. 1987). A general description of different PA measuring systems has been given earlier (Rosengren 1975; Adams et al. 1976; Tam 1986). As in other spectroscopic instruments, a single beam set-up with exclusive detection of the sample or a double beam set-up (Blank and Wakefield 1979), which simultaneously measures the signal of a reference, can be used. In most cases, carbon black serves as a reference sample. It possesses an optimal light absorption in the whole visible spectrum and a high, wavelength-independent yield of non-radiative de-excitation. Since its PA signal is saturated (see Sect. 2.2.2.1), the PA spectrum of carbon black reflects the spectrum of the light source including the characteristics of the optical components (e.g. monochromator) needed for normalization of the sample spectrum.

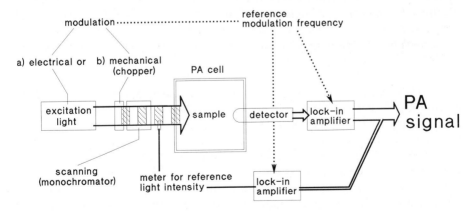

Fig. 5. General scheme of a single-beam photoacoustic spectrometer

3.1 Excitation Sources

Any type of energetic radiation including particles (e.g. electrons, protons, ions) or electromagnetic radiation from x-ray to radio frequency may serve as a source of excitation. The energy of the excitation source should be chosen as high as possible; it is, however, limited by damage effects to the sample. The range of the radiation depends on the absorption characteristic of the sample. As light source for the excitation a high-intensity, polychromatic incandescent lamp or monochromatic lasers are currently used. The light source can be specified by its parameters of irradiation (e.g. spectral and spatial emission: λ, R) its coherence of emission and its type of modulation [e.g. cw, pulse, pseudorandomized binary sequence (PRBS)] or by the different characteristics of spatial or spectral multiplexing (Fig. 1). Furthermore, parameters of the optical pathway or image will be determined by the optical system, whereas the beam position on the sample and the time sequence of irradiation will be controlled by a scanning system.

For the measurement of PA spectra a monochromator with a scanning motor has to be placed behind the polychromatic light source. It is advisable to measure the light intensity of the excitation light in front of the sample as reference, in order to eliminate changes in light intensity eventually occurring during the measuring procedure. The amplitude-modulation of the excitation light is achieved by means of a mechanical chopper or by electric power regulation of the light source. In most cases a continuous sinusoidal, amplitude-modulation (cw) between 1 Hz and 1 kHz is used, but a PRBS modulation may also be applied (Kirkbright et al. 1984). Furthermore, short single (laser) light pulses are used as excitation light (Patel and Tam 1981). A non-modulated light (adaptation light) inducing photochemical reactions (e.g. Bults et al. 1982a) or light modulated at a modulation frequency different from that of the modulated measuring light (e.g. Dau and Hansen 1989) may be added to the modulated measuring light used for exciting the PA signal. Heat produced by the absorption of these additional lights cannot contribute to the PA signal, since the direct response is eliminated by the signal processing (see Sect. 3.3). Additional light may be an intense white light (e.g. for saturation of photosynthesis) or light with only one wavelength (laser) or a narrow wavelength band which leads to specific photoreactions (e.g. exciting exclusively PS I or II; Canaani and Malkin 1984).

Excitation Sources for Fourier-Transform Photoacoustic Spectroscopy. Fourier-transform (FT) spectroscopy is already a well established technique in infrared spectroscopy. FTPAS (Fourier-transform photoacoustic spectroscopy) has permitted the combination of the capabilities of PA measurements with the multiplex advantages of FT spectroscopy (Mandelis and Power 1988; Royce and Alexander 1988). Central component of FTPAS is a Michelson interferometer as a spectral modulator of a broad-band light source. Radiation entering the interferometer is sent along two different optical paths and then recombined by a beam splitter to generate the exit beam. This beam has an intensity which depends on the spectral characteristic of the source and the optical path difference between the two arms of the interferometer. The great difference of FTPAS compared to other PAS techniques is that it represents a dynamic method compared to serially applied spectral excitation, offering the additional, multiplexing advantage of a much shorter measuring time.

Broad-band, modulated excitation overcomes the main disadvantage of pulsed laser excitation with its large peak power possibly causing damage to the sample. Further advantages result from the FT-specific signal processing (see Sect. 3.3). However, there are difficulties in obtaining a quantitative spectrum with this technique. Although the positions of the absorption peaks are correct, the relative amplitude of the bands within the spectrum is biased, since the PA signal is not a direct measurement of the absorption coefficient β (see Eq. (6)). In fact, the PA signal is a function of the modulation frequency ω, which is in this case related to the velocity of the interferometer and thus to the wavelength of the excitation light, resulting in the production of some bias to the spectral bands. Theoretical considerations about this effect and a method to correct errors have been presented by Choquet et al. (1985). Royce and Alexander (1988) have discussed some parameters which are important for the interpretation of FTPAS-data. Palmer et al. (1988) have solved the inherent disadvantage of FTPAS mentioned above by replacing the rapid scanning mode with step and integrate scanning.

3.2 Detectors

Among the various detector systems developed for sensing the thermal or acoustic waves generated by different effects (Fig. 1), four different methods are generally applied: (1) sensing in direct contact to the sample without a coupling medium, (2) contact monitoring of acoustic waves via a coupling medium, (3) non-contact sensing by probe beam detection and (4) remote sensing by photothermal radiometry. Each measuring system may be well adapted to the specific application. The most frequently used different types of PA detector systems are shown in Fig. 1.

3.2.1 Contact Detectors

Pyroelectric detectors directly sense a temperature rise causing PT effects (Baumann et al. 1983; Coufal 1984; Mandelis and Power 1988; Mandelis et al. 1988; Kocsányi and Giber 1989). This thermometry detector is best suited for wide-band (i.e. Fourier transform) excitation and response. The main advantage of the pyroelectric detector is its possibility of calorimetric calibration by carbon black. Its main disadvantage for in vivo studies is the direct coupling which has to be optimized to each tissue and which may cause distortion of the physiological environment. A numerical simulation of the signal generation (Fromm and Coufal 1988) provided insight into the theoretical understanding of this transducer. Kocsányi et al. (1988) found good agreement between PA spectra of plant materials taken with pyroelectric and acoustic detection.

Piezoelectric detectors are special crystals or polyvinylidene difluoride (PVF_2) films which detect vibration of the sample induced by the thermal expansion caused by the indirect PA effects (Farrow et al. 1978; Jackson and Amer 1980). They have to be fixed in close contact with the sample or need a contact medium, e.g. water (Kitamori et al. 1983; Jabben and Schaffner 1985). Piezoelectric detectors are preferred for PA detection after pulse excitation (with plant material: Jabben et al. 1984; Jabben and Schaffner 1985; Nitsch et al. 1988), because of their fast risetime and good impedance matching. PVF_2 films can easily be cut to different sizes; they show, however, a lower sensitivity. Piezoelectric crystals may also possess pyro-electric properties, i.e. may also detect PT heating.

3.2.2 Acoustic Detectors

Acoustic waves caused by indirect PA effects (Fig. 1) can be sensed by a microphone, a piezoelectric detector (see Sect. 3.2.1.), a capacity detector or a fiber optic. In many cases the most sensitive and most frequently used detector system is the microphone enclosed in a tightly closed sample compartment (the PA cell, Fig. 5), which makes it possible to measure pressure changes of the gas adjacent to the sample induced by an indirect PA effect (thermal piston, see Sect. 2.2.1). The microphone as well as the material and shape of the sample compartment must be adapted to the needs of the PA measurements. The sample should fill nearly the entire PA cell and leave only little space for a small layer of "coupling" gas above the sample (see Sect. 2.2.2.2). A fully transparent glass window on top of the PA cell allows the entrance of the modulated excitation light. Apart from the high sensitivity of the microphone, it is of importance to prevent illumination of the cell walls and of the microphone diaphragm and to isolate the PA cell against acoustic and seismic disturbances from outside. An advantage of the PA microphone might be its high sensitivity and the expanded range of detection, since sound is propagated far away (much more than several μ_T) from the excitation region. The slow propagation speed of the acoustic wave makes a microphone disadvantageous, especially for pulse excitation. Furthermore, the measurements are thus restricted to a limited range of modulation frequencies, which results in a limited range of thermal wavelengths.

   Different types of cell design have been proposed for the measurement of solids, semisolids and liquids (e.g. Cahen 1981). A set-up for measuring at very low temperatures has been described (Bechthold et al. 1982). For specific applications the PA signal may be amplified by constructing the PA cell as a Helmholtz resonator (Fernelius 1979). For PA measurements of pigments separated on a thin layer chromatography plate (Fishman and Bard 1981) and for in vivo measurements of leaves (in medicine also for the in vivo measurements of skin and blood), a PA cell was constructed with one side open, which is closed by contact with the sample (Poulet and Chambron 1983).

3.2.3 Probe Beam Detection

The deviation of a non-modulated light beam during illumination with modulated or pulsed excitation light is used in probe beam detection techniques. They are based on monitoring direct PA effects when directly measuring the sample. However, when measured in the medium surrounding the sample, a PT effect is sensed (Fig. 1). Photo-induced surface deformation may be measured by detecting direction changes of the probe beam reflected on the sample surface (probe beam deflection). Photo-induced heat production may cause a change of the refractive index gradient in the sample, which affects the propagation of the excitation beam itself ("self-defocusing") or of the additional probe beam (probe beam refraction = PBR). The deviation of the probe beam induced by the modulated excitation light can be determined as change of light intensity by means of a photodetector or directly by a beam position sensor. A probe beam penetrating the sample within the excitation beam is used for "thermal lensing" (TL). Time-resolved TL techniques were shown to deliver information regarding quantum yields and lifetimes of energy storing species (Redmond and Braslavsky 1988).

The probe beam may also be parallel to the sample surface. Especially for opaque samples and for spatial resolution of the signal, the "mirage effect" has been developed as a PBR method with the excitation beam falling perpendicularly on the sample or passing orthogenally to the probe beam parallel to the sample surface. This orthogenal or perpendicular PT PBR is based on photo-induced changes of the refraction index of the medium surrounding the sample. The thermal and refractive properties of the deflecting medium (liquid or gas) above the sample will influence the signal and must be chosen suitably. A theoretical estimation of the essential parameters dominating the amplitude of the deflection, and thus the intrinsic sensitivity of this method, has been given by Palmer and Smith (1985). Aamodt et al. (1988) have studied the validity of a simple theory of mirage detection under actual working conditions. According to their theory the PT PBR signal depends on the thermal and optical properties of the sample and can be used to measure absorption spectra and/or other optical and non-optical parameters. The propagation speed depends on the velocity of the elastic waves within the medium surrounding the sample (frequency range up to 100 kHz). Probing by means of the mirage effect is only possible with the probe beam positioned a very short distance from the sample surface, and theoretical analysis is hampered by the complexity of the problems. In an experiment a large number of parameters affecting the signal have to be controlled. The disadvantage of PT PBR detection is that the sample surface must be reasonably flat so that the probe beam can skim the surface at a low height above it. Its sensitivity to external sources of vibration is much more crucial than for measurements with a PA cell. It also needs sophisticated instrumentation to control the positions of two laser beams and the distance to the sample surface.

3.2.4 Photothermal Radiometry

Heat production can be detected as IR radiation (induced as one type of PT effect) by specific IR detectors. This photothermal radiometry (PTR) can be carried out remotely from distances of up to about 1 km (Tam 1986), and calorimetric calibration can be obtained. PTR has been carried out with plant material (Bults et al. 1982b; Kanstad et al. 1983; Nordal and Kanstad 1985). In contrast to the PA signal obtained by acoustic detectors where heat is sensed only from the thermally active layer L_s, the PTR signal may be detected also from deeper inside the sample, provided the sample is transparent for IR radiation. One of the advantages of PTR is its fast response, ultimately limited only by the thermal relaxation time (10^{-11}–10^{-10} s), whereas acoustic signals rely on a diffusive or elastic transfer mechanism. This should make PTR most valuable in experiments with single-pulse excitation. A theoretical approach to thermal radiation has also been given (Kanstad and Nordal 1979; Tam 1986). Four common experimental variants of excitation mode have been proposed: excitation with continuous wave (cw) or pulses and measuring in the transmission or backscatter mode. For exactly reproducible PTR-measurements, a stable temperature environment has to be established. The spectral region for IR detection must fall outside of strong resonance in emissivity, since coherent noise is the dominant noise source which must be avoided or compensated.

3.2.5 Photoacoustic and Photothermal Imaging

PA and PT imaging is an emerging technology for the non-destructive evaluation (NDE) of surface and subsurface features of opaque samples (i.e. irregularities, inhomogeneities of material, defects in substrates, compositional variations, inclusions, doping or staining concentrations, flaws, water content, non-uniform thermal properties in a sample with uniform optical properties). Three types of contrasts are possible: a) changes of the thermal properties, b) changes of the optical properties and c) ultrasonic scattering. The image obtained is due either to optical absorption variations with a resolution of the volume of μ_β or thermoelastic property variations with a resolution of the volume μ_T (approximately a hemispere of radius μ_β respectively μ_T). Thus a reasonable rule of thumb is that an irregularity may be detected at depths up to its size.

Reviews on PA/PT imaging have been presented by Thomas et al. (1985), Tam (1986) and McDonald (1985). For site-specific resolution of a sample PA/PT, imaging methods and PA microscopes have been developed (Busse 1980; Luukkala 1980; Rosencwaig 1980; Möller et al. 1983; Wada et al. 1986; Masujima et al. 1988). In general, all imaging techniques involve the interaction of a highly damped thermal wave with surface and subsurface features. Different excitation and detection schemes can be employed. Either the excitation source or the detection must be localized to form an image. Commonly highly localized excitation laser sources are used at a wavelength which is strongly absorbed by the sample. Local or non-local detection is applied. Spatial resolution depends on the ability to localize the illumination at the site of absorption. It is possible to localize the source much better than by the length of one wavelength. By bringing it very close to the sample the resolution may become many times better than one wavelength.

Much effort has been undertaken in order to avoid damage by high-intensity sources. Coufal et al. (1982) were the first to propose the use of 1D-Hadamard or Fourier coding of the excitation light in order to reduce the light intensity without decreasing the signal-to-noise ratio. More recently, 2D-Hadamard transform masks have successfully been used by Wei et al. (1985, reviewed by Fournier and Bandoz 1988). Fournier and Bandoz (1988) applied spatially multiplexed excitation by the linear geometry of the mirage detection. Spatially multiplexed PT detection using a linear photodiode array has been reported by Charbonnier et al. (1988a, 1988b). According to the sensing technique various detection schemes have been developed, e.g. contact or non-contact, local or non-local. Each scheme is characterized by inherent advantages which fit the specific application.

A gas *microphone* can be used for area-detection. It requires bonding between sample and transducer via the gaseous coupling medium. Because of the planar symmetry, a theoretical analysis can be carried out by plane wave scattering theory. As a consequence of this symmetry, detection of vertical irregularities is precluded. A typical remote local point detection can be achieved by a focussed *IR detector*. A disadvantage is that variations in the emissivity of the surface can obscure the image. Remote line sensing can be carried out by PT PBR using the mirage effect (Murphy and Aamodt 1980; Dovichi et al. 1984). This technique permits study of vertical defects because of its intrinsic detection symmetry. Furthermore, other detection schemes can be used. Williams and Wickramasinghe (1988) have demonstrated a thermal microscope with a resolution below 100 nm based on a non-contact nearfield thermal probe (nearfield imaging) which consists of a thermocouple sensor tip with 100 nm dimension.

3.3 Signal Processing

The signal measured by the detector has to be processed by different instruments to yield the final PA signal. In most cases the pre-amplified signal is transformed in a lock-in amplifier. By comparing the signal with a reference of the modulation frequency, the lock-in amplifier eliminates the contribution of processes occurring with modulation frequencies other than that of the excitation light. If the light intensity falling upon the sample is detected as a reference, a second lock-in amplifier has to be used for data processing of this reference signal. The final PA signal is then given by the division of the sample signal with the reference signal (Fig. 5). By means of a lock-in amplifier one can also determine the time delay between the excitation and the appearance of the signal, i.e. the phase. The phase expresses the time delay as part of the repetition cycle in angular degrees. The time of a full excitation light cycle is represented by 360°. A phase angle of 90° means that the signal is detected within the first quarter of the repetition cycle. With the conventional one-phase lock-in amplifiers the phase angle in which the signal is detected remains constant and has to be optimized prior to the measurement.

If the phase is adjusted to an optimum signal the detection is "in phase", whereas with the phase shifted to 90° from the "in phase" value the detection is "out of phase" or in "quadrature". By means of a two-phase lock-in amplifier the phase angle is automatically chosen for "in phase" detection to reach an optimal signal. For PA/PT measurements with single laser pulses no lock-in amplifiers are needed. The signal detected in the μs range must be subtracted from the signal of a reference with about the same light-scattering properties as that of the sample (e.g. Jabben and Schaffner 1985). A computer analysis with signal averaging and data correlation may finally complete detection of the signal.

In PA/PT measurements with Fourier transformation (see Sect. 3.1), during the signal processing the signal can be filtered to achieve a better signal-to-noise ratio and/or averaged in the time or frequency domain within a wide speed range of the measurement. The FT PA signal is decomposed by Fast Fourier Transform processes (FFT). Signal processing of wide-band methods in general represents correlation analysis and spectral analysis. Dodgson et al. (1985) utilized monochromatic light of pseudorandom pulse duration (PRBS) and correlate the signal with the random light input at different time delays. The experiment is repeated using different wavelengths, and a cross-correlation spectrum is finally produced.

4 Applications in Plant Analysis

The application of PA measurements in plant sciences has been reviewed before (Malkin and Cahen 1979; Balasubramanian and Rao 1981; Moore 1983; Buschmann et al. 1984; Braslavsky 1986, Buschmann and Prehn 1986). A first theoretical treatment of PA effects related to photosynthetic energy conversion has been given by Malkin and Cahen (1979).

4.1 Detection of Spectral Properties

4.1.1 Detection of Pigment Composition

PA spectra show about the same characteristics as the absorption spectra. In general, PA spectra have been applied to identify or characterize certain pigments. However, PA spectra were also taken to examine the quantity of phytobiomass in a river (Merle et al. 1983) or of active microbial DNA (Prehn 1981). In some cases physiological intactness (Thomasset et al. 1982; Nagel et al. 1987) or food quality (Kocsányi et al. 1988) was evaluated. The composition of pigments in vivo has been studied by taking PA spectra from leaves (e.g. Buschmann and Prehn 1981, 1983; Moore 1983; Kirkbright et al. 1984; Jabben and Schaffner 1985; Nagel et al. 1987; Nagel 1988; Nagel and Lichtenthaler 1988; Szigeti et al. 1989), fruits (Kocsányi et al. 1988), flower petals (Li et al. 1983), tissue cultures (Farringer et al. 1985), phytoplankton, fungi, bacteria (Prehn 1981; Boucher et al. 1983), lichens (Moore 1983; O'Hara et al. 1983), cyanobacteria (Schubert et al. 1980; Carpentier et al. 1983, 1984; Moore 1983), isolated cells (Popovic et al. 1988), membranes (Cahen et al. 1978; Bechthold et al. 1982), chloroplasts, immobilized organelles, membranes and particles (Frackowiak et al. 1985b; Thomasset et al. 1982) and pigment complexes (Vacek et al. 1979; Dienstbier et al. 1984; Fragata et al. 1987; Camm et al. 1988). Figure 6 gives an example of a PA spectrum of an intact radish cotyledo.

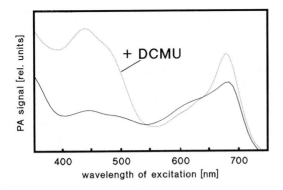

Fig. 6. In vivo photoacoustic spectra of the upper side of a *Raphanus* cotyledo before and after incubation with 10^{-4} M DCMU. The spectra were taken with a closed photoacoustic cell at a modulation frequency of 20 Hz. They are normalized to the same height at 540 nm (Buschmann and Prehn 1981)

4.1.2 Detection of Particular Components

Pigments were measured in vitro from thin layer chromatography plates (Rosencwaig and Hall 1975; Castleden et al. 1979; Fishman and Bard 1981), extracts, stained electropherograms (Möller et al. 1983), polymer films (Frackowiak et al. 1985a, 1985b; Drabent et al. 1985; Inagaki et al. 1985), liquid crystal matrices and powders (Cesar et al. 1984; Kocsányi et al. 1988; Nagel 1988). Gases have been detected both quantitatively and qualitatively, and PA sensors have been proposed for gas chromatography detection (Kreuzer 1978). Control of air pollution is currently being carried out (Sigrist 1988). Recently, the detection of ethylene exhalation of flowers was demonstrated with a low detection limit (Woltering et al. 1988).

4.1.3 Depth Profile Analysis

Depth profile analysis was carried out with leaves of a variety of different plants (Buschmann and Prehn 1983; Moore 1983; Nagel 1988) as well as with lichens (Moore 1983; O'Hara et al. 1983), stromatolites (Schubert et al. 1980) and tissue cultures (Farringer et al. 1985). Depth profile analysis may be used for in vivo detection of pigments and for testing intactness under changing environmental conditions. Measurements of the thickness of ultrathin layers covering a material have been achieved.

One principle of this technique is clearly demonstrated in the PA spectra of a leaf of *Tradescantia* (Fig. 7). As mentioned above (see Sect. 2.2.2), with decreasing modulation frequency of the excitation light the thermally active layer L_s is increased (Table 1). This means for a leaf of *Tradescantia* that at higher modulation frequencies only the peaks of the anthocyanins contained in the epidermis are detected. By lowering the modulation frequency the chlorophylls of the mesophyll below the epidermis contribute to the PA signal. Consequently, a chlorophyll absorption peak in the red-light region of the spectrum appears. The measurement of a restricted depth of the sample makes it also possible to monitor the pigment characteristics of the two sample sides separately (Buschmann and Prehn 1981).

When increasing the modulation frequencies into ranges of several hundred Hz, reflection characteristics of the deeper (mesophyll) layers may become

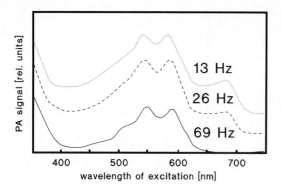

Fig. 7. In vivo photoacoustic spectra of the lower side of a *Tradescantia* leaf taken with a closed photoacoustic cell at different modulation frequencies of the excitation light. For a better comparison the spectra are plotted with the zero-line slightly raised with decreasing modulation frequency (Buschmann and Prehn 1983)

predominant (Nagel et al. 1987; Nagel 1988). Under these conditions absorption of light reflected from the mesophyll obviously induces a PA signal in the outer epidermal layer. The PA spectrum of a green leaf then shows peaks (as in a reflectance spectrum) in the green and in the far red region.

A depth profile analysis may also be carried out by measuring spectra at different, constant phase angles. Signals from the interior of the sample are detected later than signals close to the sample surface. Thus a spectrum of the cuticula of a leaf may be separated from that of the chlorophyll containing mesophyll (Moore 1983; Farringer et al. 1985). Impulse response PA measurements are a further method for depth profiling (Kirkbright et al. 1984). This technique makes use of the fact that PA signals from close to the sample surface disappear earlier than PA signals from inside the sample. The time-course of the PA signal induced by several measurements with excitation at different wavelengths reflects the absorption characteristics in different depths of the sample.

4.1.4 Photoacoustic Imaging of Plant Material

PA imaging and PA microscopy have, up to now, rarely been used in studying biological material. The distribution of biological material was examined both in vitro from gels (Möller et al. 1983) and in vivo from tissue sections (Wada et al. 1986) as well as immunoassay gels (Masujima et al. 1988). The image of a leaf both in PA amplitude and phase angle has been demonstrated (Busse 1980).

4.2 Detection of Thermal Properties

Thermal properties of a sample can be detected based on the equations given in Sect. 2.2.2. Specific features of the surface of the sample may be analyzed by the determination of the coefficient of the heat transmission K_s, or changes of its time course can be monitored. When decreasing the modulation frequency of the excitation light the first appearance of the absorption characteristics of a pigment indicates that the thermally active layer L_s equals the distance between the pigment and the sample surface. However, only relative changes of thermal properties will thus become detectable. By measuring the dependence of the phase angle (see Sect. 3.3) on the modulation frequency, thermal diffusivity can be determined, provided the thickness (Adams and Kirkbright 1977) or the optical absorption coefficient (Roark et al. 1978) of the sample is known. In plant material these properties are certainly not homogeneous for the whole sample. This makes it even difficult to calculate the thermally active layer inversely. Since water represents the main component of plant material, the thermal properties of water should lead to a rather close approximation of the thermal diffusion length μ_T or of the thermally active layer L_s (Table 1) of plants. The thickness of leaves varies within a wide range depending on the species but also on growth conditions, e.g. sun or shade (e.g. Nagel 1988).

4.3 Detection of Physiological Parameters

In plant science most PA measurements are carried out in photosynthesis research. Since photosynthetically active pigments are characterized by an optimized light absorption, they are most suited to be studied by the light-induced PA processes.

4.3.1 Action Spectra

PA spectra of intact leaves represent spectra of excitation, i.e. a kind of action spectrum (with the exception that usually the intensity of the excitation light is not constant). Because of the high signal and of the thick thermally active layer, PA spectra are normally measured at a low modulation frequency. The low-frequency PA spectrum of photosynthetically active leaves determined in a closed PA cell exhibits a relatively lower peak in the blue light region than in the red light region (Fig. 6). This effect is in contrast to the absorption spectrum and may be explained by the contribution of O_2 evolution to the PA signal (see Sect. 4.3.2), which should be higher in the red-light than in the blue-light part of the PA spectrum (Nagel 1988; Szigeti et al. 1989). If O_2 does not contribute to the PA signal, the PA spectrum more resembles the absorption spectrum. This is the case when measurements are not taken in a PA cell, when measuring in vivo after inhibition of photosynthesis with DCMU (Fig. 6), during illumination with additional, non-modulated light (Nagel 1988; Szigeti et al. 1989) or when measuring in vitro. The ratio between the PA signal at 675 and that at 475 nm has been used to characterize the photosynthetic intactness of leaves (Nagel et al. 1987). Treatment with DCMU leads to higher changes for the upper leaf side than for the lower leaf side (Buschmann and Prehn 1981), confirming an earlier hypothesis that the palisade parenchyma of the upper leaf half has a higher photosynthetic activity than the spongy parenchyma of the lower leaf half.

Spectra of relative quantum yield for photosynthetic O_2 evolution and for Emerson enhancement have been demonstrated (Canaani and Malkin 1984). From comparison of the action spectrum of the PA O_2 component (low-frequency acoustic wave determined in a PA cell) and the thermal component (determined by PTR) of the PA signal it was concluded that PS I has a higher contribution to heat production than to the O_2 evolution (Nordal and Kanstad 1985). Action spectra of "photochemical loss" (see Sect. 4.3.4) and of photosynthetic activity both for PS I and PS II have been determined with cyanobacteria by specifically inducing PS I or PS II activity (Carpentier et al. 1984). Changes in the PA spectrum induced by suppressing or activating photosynthetic activity demonstrated energy conversion of PS II particles and broken chloroplasts (Camm et al. 1988) and the involvement of pheophytin in an energy storage process of isolated PS II particles (Fragata et al. 1987).

4.3.2 Photosynthetic Parameters

PA measurements in plant science have been used most frequently in photosynthesis research. Heat production, which can always be measured by the PA method, occurs in the antenna systems parallel to energy transfer and fluorescence emission as well as in the reaction centres parallel to the photochemistry, and to energy transfer back to the antenna chlorophyll a (Butler 1978). When measuring with a PA cell at low modulation frequencies (i.e. in radish cotyledons below 125 Hz), pulses of modulated O_2 evolution are obviously superimposed on the thermal component and essentially determine the PA signal (Bults et al. 1982a). This hypothesis is based on the fact that pressure changes induced by pulses of photoinduced gas evolution or gas consumption may be sensed by the microphone of the PA cell (Gray and Bard 1978). It has been shown earlier (Joliot and Joliot 1968) that O_2 evolution of chloroplasts is pulsed at modulation frequencies below 200 Hz. Above 200 Hz it becomes more and more homogeneous.

The PA signal seems not to be affected by photosynthetic CO_2 fixation (Bults et al. 1982a), since this process is not so strictly related to the light absorption as the production of O_2. Obviously, there is a delay of the CO_2 fixation due to the numerous enzymatic steps between light absorption and the Calvin cycle. Respiratory O_2 consumption is not light-induced and thus cannot be sensed by PA measurements. Under certain conditions, e.g. during an early state of the induction kinetic (Malkin 1987) or after heat shock treatment (Havaux et al. 1987b), light-induced O_2 uptake has been reported. The PA signal measured in a PA cell at a low modulation frequency can be used to detect brutto photosynthesis directly. The portion of non-radiative de-excitation included in the low-frequency PA signal may be excluded by the "phase separation" technique introduced by Poulet et al. (1983) (see Sect. 4.3.3). At high modulation frequencies, the effect of O_2 on the PA signal measured with a PA cell decreases with increasing modulation frequency, and the thermal component becomes more and more predominant.

4.3.2.1 Induction Kinetics. PA induction kinetics represent a transient of light-induced heat production which follows in tendency that of fluorescence emission (Kautsky effect) (Buschmann 1987). When measuring with a closed PA cell this kinetic can only be measured at a high modulation frequency. The kinetic of light-induced heat production shows a fast rise and a subsequent slow decline (Fig. 8). The fluorescence signal measured simultaneously somewhat precedes the thermal component of the PA signal (Buschmann and Kocsányi 1989). This shows that during the induction kinetic, the yield of fluorescence and heat does not change

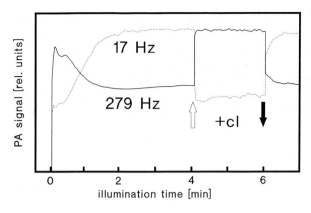

Fig. 8. Induction kinetics of the photoacoustic signal of a dark-adapted *Raphanus* cotyledo measured with a closed photo-acoustic cell at a high and a low modulation frequency. Four min after the onset of the kinetic by illumination with modulated light, non-modulated light saturating photosynthesis (2000 μE m^{-2} s^{-1}) was added for 2 min (Buschmann 1987)

in parallel (Buschmann 1987). The induction kinetic of the PA signal is also different from that of the energy quenching of fluorescence (Buschmann and Kocsányi 1989), which is usually taken as a measure of non-radiative de-excitation during energy conversion related to fluorescence.

At low modulation frequencies the induction kinetic measured in the PA cell shows a rise to a high steady state (Fig. 8) parallel to the fluorescence decrease (e.g. Buschmann 1987). This behaviour has been interpreted in terms of an induction transient of photosynthetic O_2 evolution (Bults et al. 1982a).

In certain cases the PA signal of an intact organism is not determined by O_2 even when measured at low modulation frequencies in a PA cell. In the green alga *Bryopsis maxima* this seems to be due to the special tissue structure (Katoh and Yamagishi 1984). Only light-induced heat can also be detected from isolated bundle-sheath cells of C_4 plants (Popovic et al. 1988) and heterocysts of cyano-bacteria (Carpentier et al. 1984), since these cells lack photosynthetic water splitting.

4.3.2.2 Measurements Under Steady State Conditions. In many cases changes of the PA signals are studied when the induction kinetic has reached its steady state. Addition of substances affecting photosynthetic electron transport (Lasser-Ross et al. 1980; Katoh and Yamagishi 1984; Popovic et al. 1988) or photophosphorylation (Katoh and Yamagishi 1984) have been carried out. Strong, non-modulated light ("background light", Bults et al. 1982a) added to the modulated measuring light in order to saturate photosynthesis leads to a decrease of the PA signal when measuring with a PA cell at a low modulation frequency but to a rise when measuring at a high modulation frequency (Fig. 8). No changes can be detected when the sample is photosynthetically inactive. The absorption of non-modulated light itself does not contribute to the PA signal, since the signal processing (see Sect. 3.3) only amplifies processes occurring with the modulation frequency of the measuring light. If photosynthesis is saturated with the non-modulated light the modulated light induces an increase of non-radiative de-excitation and the PA signal rises. This rise is always achieved when measuring by means of PTR using an IR detector (Bults et al. 1982b; Kanstad et al. 1983; Nordal and Kanstad 1985). With a closed PA cell the rise only shows up at high modulation frequencies (Bults et al. 1982b). At low modulation frequencies, the PA signal decreases when non-modulated light is added, since O_2 evolution is no longer modulated and cannot contribute to the PA signal. Under these conditions the PA signal is only determined by the heat production. Such a negative "background light"-effect may be taken as an indicator of O_2 mainly determining the PA signal, whereas with a positive "background light" effect the thermal component is predominant (Fig. 3).

4.3.2.3 Energy Transfer. Changes of the energy transfer induced by addition of magnesium have been deduced from the changes of the PA spectrum of isolated chlorophyll-protein complexes (Dienstbier et al. 1984). Energy transfer between carotenoids and bacteriochlorophyll was studied by comparing PA spectra with absorption spectra (Boucher et al. 1983). The function and orientation of antenna pigments of thylakoids have been studied in isolated pigments fixed in polyvinyl alcohol films using excitation light with different polarizations (Frackowiak et al. 1985b). State 1/state 2 transition and Emerson enhancement have been demonstrated by measuring the O_2 component of the PA signal (Canaani and Malkin 1984). The measurement was carried out with modulated PS II light (640 nm), which leads to the establishment of state 2. During several minutes of illumination with additional non-modulated PS I light (710 nm) the leaf is transferred into state 1. By means of a short additional illumination with non-modulated PS I light, Emerson enhancement during state 2 can be detected. Emerson enhancement during state 1 can be shown after turning off the non-modulated PS I light.

4.3.2.4 Phenomenological Studies. During greening in the light the development of photosynthetic functions has been studied in a leaf by simultaneously measuring induction kinetics of fluorescence and of the PA signal taken at low or high modulation frequency (Buschmann and Prehn 1986). Changes of photosynthetic activity of the algal component of a lichen have been examined during hydration and dehydration (O'Hara et al. 1983). PA signals in dependence of substances affecting the photosynthetic electron transport have been reported for bundle-sheath cells of maize (Popovic et al. 1988). After photoinhibition the decrease of photosynthetic activity is reflected in the lower rise of the induction kinetic of the O_2 component of the PA signal and in a lower decrease of the fluorescence and of the high-frequency PA signal (Buschmann 1987). During the photoinhibitory treatment the decrease of the fluorescence is paralleled by an increase of the thermal PA signal (Buschmann and Prehn 1988). The inhibition of photosynthetic O_2 evolution after chilling (Yakir et al. 1985), water stress (Canaani et al. 1986; Havaux et al. 1986), heat (Havaux et al. 1987a, 1987b) or SO_2 treatment (Ronen et al. 1984) have been determined by the PA O_2 measurements. Screening of heat tolerance by PA detection has been proposed (Havaux et al. 1987a). Changes of photosynthetic O_2 evolution during Emerson enhancement and state 1/state 2 transitions (see Sect. 4.3.2.3) have been taken to monitor imbalance of excitation between the two photosystems (Canaani and Malkin 1984).

4.3.3 Kinetic Analysis

Time constants of photochemical processes have been detected with isolated particles of PS I and PS II (Nitsch et al. 1988), phytochrome (Jabben et al. 1984) and isolated purple membranes of *Halobacterium* (Cahen et al. 1978; Bechthold et al. 1982). The fast adaptation to changes of light intensity in times ranging from 1 s to 2 h has been determined in leaves by measuring the low-frequency PA signal (Dau and Hansen 1989). From these experiments 6 different time constants of photosynthetic electron transport have been calculated which lie in the range of milliseconds to minutes.

The time delay between the excitation and the generation of the PA signal (phase) can be measured with two-phase lock-in amplifiers (see Sect. 3.3). In a photochemically inactive sample heat is produced nanoseconds after light absorption. Photochemistry increases this time delay. First, the heat production of a photosynthetically active leaf should emanate from the two photosystems separately, later from the two photosystems in series (Malkin and Cahen 1979). The phase of the PA signal is used to better characterize the origin of the PA signal (e.g. in depth profile analysis, see Sect. 4.1.3) and to exclude the partition of non-radiative de-excitation involved in the low-frequency PA signal detected in a PA cell together with photosynthetic O_2 evolution. The O_2 component of the PA signal shows a phase which is different from the thermal component.

The "phase separation" technique (Poulet et al. 1983) is applied to detect only O_2 with the low-frequency PA signal. The phase of non-radiative de-excitation is first determined from the low-frequency PA signal during the addition of non-modulated light saturating photosynthesis ("background light", see Sect. 4.3.2.2), when only the thermal component of the PA signal is present. The phase is adjusted to yield a maximal PA signal and then turned by $90°$, changing from the detection of the "in phase" component to that of the "quadrature" component. The measurement carried out without "background light" at that particular (quadrature) phase angle is assumed to result in a PA signal of "pure oxygen".

Photochemical events occurring with time constants below 1 ms can only be detected by PA measurements with single nanosecond laser pulses. They provide information on very early primary photoproducts. This technique was used for studies of photochemistry PS I and PS II (Nitsch et al. 1988), and of phytochrome and chlorophyll (Jabben et al. 1984).

4.3.4 Quantum Yields

PA measurements have been carried out to determine the absolute quantum efficiency of fluorescence for different samples (Adams et al. 1980). PA measurements made it possible, for the first time, to really detect radiationless transition. With P700-enriched particles it could be shown that 80% of the absorbed energy is used for photosynthesis, 3% in fluorescence and 17% in non-radiative de-excitation (Vacek et al. 1979). The kinetic behaviour of non-radiative de-excitation processes needs further examination, since during the photosynthetic induction kinetic non-radiative de-excitation is only in tendency parallel to the fluorescence (Buschmann 1987). Energy storage in PS II particles and broken chloroplasts (Camm et al. 1988) as well as in isolated bundle sheath cells of *Zea mays* (Popovic et al. 1988) has been evaluated by measuring the difference between PA spectra taken with and without electron acceptors and electron donors.

The "photochemical loss" was proposed as a direct indicator of photosynthetic activity representing the amount of energy inducing photosynthesis before the addition of non-modulated "background light" (Malkin and Cahen 1979). It is calculated from the high-frequency PA signals as the ratio between the increase caused by addition of non-modulated light saturating photosynthesis and the PA signal measured without "background light". A method to determine the relative quantum yield for photosynthetic O_2 evolution was introduced by Poulet et al. (1983). It is expressed as the ratio between the O_2 component (A_{ox}) and the thermal component (A_{pt}) of the PA signal, where A_{pt} is assumed to be directly proportional to the total light energy absorbed by the leaf. A_{pt} is detected by correcting the low-frequency PA signal (PA_+) taken with "background light" by the "photochemical loss" PL: $A_{pt} = (1-PL) PA_+$. A_{ox} is given by the A_{pt} value subtracted from the low-frequency PA signal taken without "background light" (PA_-): $A_{ox} = PA_- - A_{pt}$.

4.3.5 Other Parameters

The permeability of cuticles for water has been determined by PA measurements (Büchner et al. 1983). Water penetrating the sample from a reservoir below the cuticle to the gas phase above can be sensed, since the PA signal seems to be dependent on the vapour pressure of the water in the gas. Moisture of single-cell protein of defined particle size has been quantified by utilising a ratio of PA signals at 1900 and 1550 nm (Castleden et al. 1980). Detection of adulterated or altered coffee powder to be used in food quality control has been described by measuring PA signals excited at four different wavelengths (Cesar et al. 1984). Light-induced electron donation and acceptance of flavins have been compared to PA spectra in order to better understand the activity of flavins as photoreceptors (Drabent et al. 1985). PA measurement of agglutination in an immunoassay has been proposed as a rapid and highly sensitive detection method (Kitamori et al. 1988). Immunoassays (Masujima et al. 1988) and enzyme localisation on tissue cross section (Wada et al. 1986) have been carried out on a microscopic scale.

5 Further Development

There is still a fast development in the wide field of PA and PT research. Excitation and detection are specifically designed to meet different sample purposes. Thus the sample handling will be further facilitated. New information is gathered by simultaneously examining processes competing with heat production. Action spectra derived from various parameters (e.g. photosynthetic activity, metabolic rates, physicochemical or chemical changes) will give more clear insights into complex processes. PA and PT techniques will be further applied for calorimetric studies (ΔG, ΔH etc.). Progress in monitoring of transport processes (velocity, diffusivity), chemical reactions and trace gas variations as well as in dosimetry of radiation over a wide range of the electromagnetic spectrum can be expected. Especially in plant science, PA and PT methods will become a routine technique for screening in quality control, production, toxicity and evaluation of photosynthetic activities and other types of bioactivity. An interesting potential lies in the application of high-intensity shock PA effects inside the sample inducing mechanical or thermal stress, controlled photoablation or destructive depth profiling. In the near future further PA or PT instruments and/or options to spectrometers will become commercially available, and compact and mobile instrumentation will be developed. The transfer of the technique into other fields (e.g. broad-band and tissue dosimeters) will further stimulate interdisciplinary research activities.

Abbreviations. cw = continuous wave, DCMU = 3-(3,4-dichlorophenyl)-1,1-dimethyl-urea, FFT = Fast Fourier transform, FT = Fourier transform, FTPAS = Fourier transform photoacoustic spectroscopy, IR = infrared, PA = photoacoustic, PAS = photoacoustic spectroscopy, PBR = probe beam refraction, PRBS = pseudorandomized binary sequence, PS I = photosystem I, PS II = photosystem II, PT = photothermal, PTR = photothermal radiometry, PVF_2 = polyvinylidene difluoride, RG-theory = Rosencwaig-Gersho theory, TL = thermal lensing

Acknowledgements. Thanks are due to László Kocśanyi (Budapest) and Eckehard Nagel (Karlsruhe) for valuable discussion. One of the authors (C.B.) gratefully acknowledges the possibility of carrying out PA measurements with the apparatus of H.K. Lichtenthaler (Karlsruhe), financed by the "Projekt Europäisches Forschungszentrum für Maßnahmen zur Luftreinhaltung" (PEF) and assembled by Eckehard Nagel.

References

Aamodt LC, Murphy JC, Maclachlan JW (1988) Image distortion in optical-beam-deflection imaging. In: Hess P, Pelzl J (eds) Photoacoustic and photothermal phenomena. Springer, Berlin Heidelberg New York Tokyo, pp 385–388

Adams MJ, Kirkbright GF (1977) Thermal diffusivity and thickness measurements for solid samples utilising the optoacoustic effect. Analyst 102:678–682

Adams MJ, King AA, Kirkbright GF (1976) Analytical optoacoustic spectrometry – Part I. Instrument assembly and performance characteristics. Analyst 101:73–85

Adams MJ, Highfield JG, Kirkbright GF (1980) Determination of the absolute quantum efficiency of luminescence of solid materials employing photoacoustic spectroscopy. Anal Chem 52:1260–1264

Balasubramanian D, Rao CM (1981) Photoacoustic spectroscopy of biological systems. Photochem Photobiol 34:749–752

Baumann T, Dacol F, Melcher RL (1983) Transmission thermal-wave microscopy with pyroelectric detection. Appl Phys Lett 43:71–73

Bechthold PS, Campagna M (1981) Variable temperature photoacoustic spectroscopy. II. Temperature characteristic and application. Optics Commun 36:373–377

Bechthold PS, Kohl KD, Sperling W (1982) Low temperature photoacoustic spectroscopy of the purple membrane of *Halobacterium halobium*. Appl Opt 21:127–132

Berber J, Karcher H, Meyer H (1977) Formeln und Tabellen zur Physik. Voigt, Hamburg

Blank RE, Wakefield T (1979) Double-beam photoacoustic spectrometer for use in the ultraviolet, visible, and near-infrared spectral regions. Anal Chem 51:50–54

Boucher F, Lavoie L, Antippa AF, Leblanc RM (1983) Energy transfer in photosynthetic bacteria as examined by photoacoustic spectroscopy. Can J Biochem Cell Biol 61:1117–1122

Braslavsky SE (1986) Photoacoustic and photothermal methods applied to the study of radiationless deactivation processes in biological systems and in substances of biological interest. Photochem Photobiol 43:667–675

Büchner B, Korpiun P, Lüscher E, Schönherr J (1983) Influence of water in plant cuticles on the PA signal. J Phys 44:C6 125–C6 129

Bults G, Horwitz BA, Malkin S, Cahen D (1982a) Photoacoustic measurements of photosynthetic activities in whole leaves – Photochemistry and gas exchange. Biochim Biophys Acta 679:452–465

Bults G, Nordal PE, Kanstad SO (1982b) In vivo studies of gross photosynthesis in attached leaves by means of photothermal radiometry. Biochim Biophys Acta 682:234–237

Buschmann C (1987) Induction kinetics of heat emission before and after photoinhibition in cotyledons of *Raphanus sativus*. Photosynth Res 14:229–240

Buschmann C, Kocsányi L (1989) Light-induced heat production correlated with fluorescence and its quenching mechanisms. Photosynth Res 21:129–136

Buschmann C, Prehn H (1981) In vivo studies of radiative and non-radiative de-excitation processes of pigments in *Raphanus* seedlings by photoacoustic spectroscopy. Photobiochem Photobiophys 2:209–215

Buschmann C, Prehn H (1983) In vivo photoacoustic spectra of *Raphanus* and *Tradescantia* leaves taken at different chopping frequencies of the excitation light. Photobiochem Photobiophys 5:63–69

Buschmann C, Prehn H (1986) Photosynthetic parameters as measured via non-radiative de-excitation. In: Marcelle R, Clijsters H, Van Poucke M (eds) Biological control of photosynthesis. Nijhoff, Dordrecht, pp 83–91

Buschmann C, Prehn H (1988) Inverse yield changes of heat and fluorescence during photoinhibition of photosynthesis. In: Hess P, Pelzl J (eds) Photoacoustic and photothermal phenomena. Springer, Berlin Heidelberg New York Tokyo, pp 523–526

Buschmann C, Prehn H, Lichtenthaler HK (1984) Photoacoustic spectroscopy (PAS) and its application in photosynthesis research. Photosynth Res 5:29–46

Busse G (1980) The optoacoustic and photothermal microscope: the instrument and its application. In: Ash EA (ed) Scanned image microscopy. Academic Press, Lond NY, pp 341–345

Busse G, Twardon F, Müller R (1988) Diffusion waves: analogy to thermal waves. In: Hess P, Pelzl J (eds) Photoacoustic and photothermal phenomena. Springer, Berlin Heidelberg New York Tokyo, pp 329–332

Butler WL (1978) Energy distribution in the photochemical apparatus of photosynthesis. Annu Rev Plant Physiol 29:345–378

Cahen D (1981) Photoacoustic cell for reflection and transition measurements. Rev Sci Instrum 52:1306–1310

Cahen D, Garty H, Caplan SR (1978) Spectroscopy and energetics of the purple membrane of *Halobacterium halobium*. FEBS Lett 91:131–134

Camm EL, Popovic R, Lorrain L, Leblanc RM, Fragata M (1988) Photoacoustic characterization of energy storage of photosystem 2 core-enriched particles from barley isolated with octyl-β-D-glucopyranoside detergent. Photosynthetica 22:27–32

Canaani O, Malkin S (1984) Distribution of light excitation in an intact leaf between the two photosystems of photosynthesis – changes in absorption cross-sections following state 1-state 2 transitions. Biochim Biophys Acta 766:513–524

Canaani O, Havaux M, Malkin S (1986) Hydroxylamine, hydrazine and methylamine donate electrons to the photooxidizing side of photosystem II in leaves inhibited in oxygen evolution due to water stress. Biochim Biophys Acta 851:151–155

Carpentier R, LaRue B, Leblanc RM (1983) Photoacoustic spectroscopy of *Anacystis nidulans*. II. Characterization of pigment holochroms and thermal deactivation spectrum. Arch Biochem Biophys 222:411–415

Carpentier R, LaRue B, Leblanc RM (1984) Photoacoustic spectroscopy of *Anacystis nidulans*. III. Detection of photosynthetic activities. Arch Biochem Biophys 228:534–543

Castleden SL, Elliott CM, Kirkbright GF, Spillane DEM (1979) Quantitative examination of thin-layer chromatography plates by photoacoustic spectroscopy. Anal Chem 51:2152–2153

Castleden SL, Kirkbright GF, Menon KR (1980) Determination of moisture in single-cell protein utilising photoacoustic spectroscopy in the near-infrared region. Analyst 105:1076–1081

Cesar CL, Vargas H, Lima CAS, Filho JM, Miranda LCM (1984) On the use of photoacoustic spectroscopy for investigating adulterated or altered powdered coffee samples. J Agric Food Chem 32:1355–1358

Charbonnier F, Fournier D, Boccara AC (1988a) Spatially multiplexed photothermal detection using a charge coupled linear photodiode array. In: Hess P, Pelzl J (eds) Photoacoustic and photothermal phenomena. Springer, Berlin Heidelberg New York Tokyo, pp 478–480

Charbonnier F, Nerozzi M, Le Liboux M, Fournier D, Boccara AC (1988b) Laser diodes and optical fibers: Two new approaches for mirage detection sensors. In: Hess P, Pelzl J (eds) Photoacoustic and photothermal phenomena. Springer, Berlin Heidelberg New York Tokyo, pp 481–483

Choquet M, Rousset G, Bertrand L (1985) Fourier-transform photoacoustic spectroscopy: a more complete method for quantitative analysis. Can J Phys 64:1081–1085

Coufal H (1984) Photothermal spectroscopy using a pyroelectric thin-film detector. Appl Phys Lett 44:59–61

Coufal H, Lüscher E (1978) Photoakustische Spektroskopie — Eine vielversprechende neue Technik? Physik in unserer Zeit 9:46–52

Coufal H, Möller U, Schneider S (1982) Photoacoustic imaging using the Hadamard transform technique. Appl Opt 21:116–120

Dau H, Hansen UP (1989) Studies on the adaptation of intact leaves to changing light intensities by a kinetic analysis of chlorophyll fluorescence and of oxygen evolution as measured by the photoacoustic signal. Photosynth Res 20:59–83

Dienstbier M, Il'ina MD, Borisov AY, Ambroz M, Vacek K (1984) Radiative and non-radiative losses of radiant energy in chloroplasts and pigment-protein complexes studied by fluorescence and photoacoustic techniques. Photosynthetica 18:512–521

Dodgson JT, Mandelis A, Andreetta C (1985) Optical-absorption coefficient measurements in solids and liquids using correlation photoacoustic spectroscopy. Can J Phys 64:1074–1080

Dovichi NJ, Nolan TG, Weimer WA (1984) Theory for laser-induced photothermal refraction. Anal Chem 56:1700–1704

Drabent R, Frackowiak D, Jadzyn C, Hotchandani S, Leblanc RM (1985) Photoacoustic spectra and photovoltaic effects of flavins. Biochim Biophys Acta 843:25–28

Eaton HE, Stuart JD (1977) Optoacoustic spectrometry of solid materials: Effect of the filler gas on the observed signal. Analyst 102:531–534

Farringer EL, O'Hara EP, Moore TA (1985) A Photoacoustic study of morphological changes occuring in plant tissue cultures accompanying differentiation. Photochem Photobiol 41:417–419

Farrow MM, Burnham RK, Auzanneau M, Olsen SL, Purdie N, Eyring EM (1978) Piezoelectric detection of photoacoustic signals. Appl Opt 17:1093–1098

Fernelius NC (1979) Helmholtz resonance effect in photoacoustic cells. App Opt 18:1784–1787

Fishman VA, Bard AJ (1981) Open-ened photoacoustic spectroscopy cell for thin-layer chromatography and other applications. Anal Chem 53:102–105

Fournier D, Bandoz J (1988) Photothermal imaging with a spatially multiplexed excitation. In: Hess P, Pelzl J (eds) Photoacoustic and photothermal phenomena. Springer, Berlin Heidelberg New York Tokyo, pp 475–477

Frackowiak D, Hotchandani S, Leblanc RM (1985a) Photoacoustic spectra of phycobiliproteins and chlorophyll in isotropic and anisotropic polyvinyl alcohol films. Photochem Photobiol 42:559–565

Frackowiak D, Lorrain L, Wrobel D, Leblanc RM (1985b) Polarized photoacoustic, absorption and fluorescence spectra of chloroplasts and thylakoids oriented in polyvinyl alcohol films. Biochem Biophys Res Commun 126:254–261

Fragata M, Popovic R, Camm EL, Leblanc RM (1987) Pheophytin-mediated energy storage of photosystem II particles detected by photoacoustic spectroscopy. Photosynth Res 14:71–80

Fromm J, Coufal H (1988) Numerical simulation of the signal generation and detection process in a pyroelectric calorimeter. In: Hess P, Pelzl J (eds) Photoacoustic and photothermal phenomena. Springer, Berlin Heidelberg New York Tokyo, pp 464–465

Ganguly P, Somasundaram T (1988) The importance of an adsorbed liquid layer for the enhancement of photoacoustic signals. In: Hess P, Pelzl J (eds) Photoacoustic and photothermal phenomena. Springer, Berlin Heidelberg New York Tokyo, pp 316–320

Gray RC, Bard AJ (1978) Photoacoustic spectroscopy applied to systems involving photo-induced gas evolution or consumption. Anal Chem 50:1262–1265

Havaux M, Canaani O, Malkin S (1986) Photosynthetic responses of leaves to water stress, expressed by photoacoustics and related methods – II. The effect of rapid drought on the electron transport and the relative activities of the two photosystems. Plant Physiol 82:834–839

Havaux M, Canaani O, Malkin S (1987a) Rapid screening for heat tolerance in *Phaseolus* species using the photoacoustic technique. Plant Sci 48:143–149

Havaux M, Canaani O, Malkin S (1987b) Oxygen uptake by tobacco leaves after heat shock. Plant Cell Environ 10:677–683

Inagaki T, Ito A, Motosuga M, Hieda K, Kobayashi K, Maezawa H, Ito T (1985) Vacuum-ultraviolet photoacoustic spectroscopy of biological materials using synchrotron radiation as a light source. Photochem Photobiol 41:527–533

Jabben M, Schaffner K (1985) Pulsed-laser induced optoacoustic spectroscopy of intact leaves. Biochim Biophys Acta 809:445–451

Jabben M, Heihoff K, Braslavsky SE, Schaffner K (1984) Studies of phytochrome photoconversions in vitro with laser-induced optoacoustic spectroscopy. Photochem Photobiol 40:361–367

Jackson WB, Amer NM (1980) Piezoelectric photoacoustic detection: Theory and experiment. J Appl Phys 51:3343–3353

Jackson WB, Amer NM, Boccara AC, Fournier D (1981) Photothermal deflection spectroscopy and detection. Appl Opt 20:1333–1344

Joliot P, Joliot A (1968) A polarographic method for detection of oxygen production and reduction of Hill reagent by isolated chloroplasts. Biochim Biophys Acta 153:625–634

Kanstad SO, Nordal PE (1979) Infrared photoacoustic spectroscopy of solids and liquids. Infrared Phys 19:413–422

Kanstad SO, Cahen D, Malkin S (1983) Simultaneous detection of photosynthetic energy storage and oxygen evolution in leaves by photothermal radiometry and photoacoustics. Biochim Biophys Acta 722:182–189

Katoh S, Yamagishi A (1984) Parallel inductive kinetics of fluorescence and photoacoustic signal in dark-adapted thalli of *Bryopsis maxima*. Biochim Biophys Acta 767:185–191

Kirkbright GF, Miller RM, Spillane DEM, Vickery IP (1984) Impulse response photoacoustic spectroscopy of biological samples. Analyst 109:1443–1447

Kitamori M, Fujii M, Sawada T, Goshi Y (1983) Theoretical aspects of photoacoustic signal detection with a direct coupling cell for liquid. J Phys 44:C6 209–C6 210

Kitamori T, Suzuki K, Sawada T, Gohshi Y (1990) Basic study of photoacoustic immunoassay and determination of trace rheumatoid factor. In: Hess P, Pelzl J (eds) Photoacoustic and photothermal phenomena. Springer, Berlin Heidelberg New York Tokyo, pp 561–562

Kocsányi L, Giber J (1990) Photothermal spectroscopy by a closed photoacoustic cell and an open contact detector. Sci Instrum (in press)

Kocsányi L, Richter P, Nagel E, Buschmann C, Lichtenthaler HK (1988) The application of photoacoustic spectroscopy in food investigation. Acta Aliment 17:277–237

Korpiun P, Herrmann W, Kindermann A, Rothmeyer M, Büchner B (1986) Sorption of water investigated with the photoacoustic effect. Can J Phys 64:1042–1048

Korpiun P, Herrmann W, Osiander R (1988) Effect of diffusive and convective transport of mass on the photoacoustic effect of binary mixtures of liquids. In: Hess P, Pelzl J (eds) Photoacoustic and photothermal phenomena. Springer, Berlin Heidelberg New York Tokyo, pp 325–328

Kreuzer LB (1978) Laser optoacoustic spectroscopy for GC detection. Anal Chem 50:597A–606A

Lasser-Ross N, Malkin S, Cahen D (1980) Photoacoustic detection of photosynthetic activities in isolated broken chloroplasts. Biochim Biophys Acta 593:330–341

Li X, Brücher KH, Görtz W, Perkampus HH (1983) PA-spectroscopic investigation on flower petals. J Phys 44:C6 137–C6 143

Luukkala M (1980) Photoacoustic microscopy at low modulation frequencies. In: Ash EA (ed) Scanned image microscopy. Academic Press, Lond NY, pp 273–289

Malkin S (1987) Fast photoacoustic transients from dark-adapted intact leaves: Oxygen evolution and uptake pulses during photosynthetic induction – A phenomenology record. Planta 171:65–72

Malkin S, Cahen D (1979) Photoacoustic spectroscopy and radiant energy conversion: Theory of the effect with special emphasis on photosynthesis. Photochem Photobiol 29:803–813

Mandelis A, Power JF (1988) Frequency modulation time delay photopyroelectric spectrometry (FM-TD P^2ES). In: Hess P, Pelzl J (eds) Photoacoustic and photothermal phenomena. Springer, Berlin Heidelberg New York Tokyo, pp 456–463

Mandelis A, Royce BSH (1979) Time-domain photoacoustic spectroscopy of solids. J Appl Phys 50:4330–4338

Mandelis A, Lo W, Wagner RE (1988) Photopyroelectric spectroscopy (P^2ES) of electronic defect centers in crystalline n-CdS. In: Hess P, Pelzl J (eds) Photoacoustic and photothermal phenomena. Springer, Berlin Heidelberg New York Tokyo, 35–40

Masujima T, Munekane Y, Kawai C, Yoshida H, Imai H, Juing-Yi L, Sato Y (1988) Photoacoustic imaging immunoassay for biological component microanalysis. In: Hess P, Pelzl J (eds) Photoacoustic and photothermal phenomena. Springer, Berlin Heidelberg New York Tokyo, pp 558–560

McClelland JF, Kniseley RN (1976) Signal saturation effects in photoacoustic spectroscopy with applicability to solid and liquid samples. Apply Phys Lett 28:467–469

McDonald FA (1980) Three-dimensional heat flow in the photoacoustic effect. Apply Phys Lett 36:123–125

McDonald FA (1981) Three-dimensional heat flow in the photoacoustic effect – II. Cell wall conduction. J Appl Phys 52:381–385

McDonald FA (1985) Photoacoustic, photothermal, and related techniques: a review. Can J Phys 64:1023–1029

McGovern S, Royce BSH, Benziger JB (1985) The importance of instital gas expansion in infrared photoacoustic spectroscopy of powders. J Appl Phys 57:1710–1718

Merle AM, Cherqaoui A, Brzezinski L (1983) Evolution of particulate chlorophyll in the Gironde estuary. J Phys 44:C6 361–C6 366

Miller RM (1985) Digital simulation of photoacoustic impulse responses. In: Technical digest of the 4th International topical meeting on photoacoustic, thermal and related sciences. Esterel, Canada, pp MA9.1–MA9.4

Möller U, Köst HP, Schneider S, Coufal HJ (1983) Evaluation of stained electropherograms by photoacoustic spectroscopy. J Phys 44:C6 121–C6 124

Monchalin JP, Bertrand L, Rousset G, Lepoutre F (1984) Photoacoustic spectroscopy of thick powdered or porous samples at low frequency. J Appl Phys 56:190–210

Moore TA (1983) Photoacoustic spectroscopy and related techniques applied to biological materials. Photochem Photobiol Rev 7:187–221

Murphy JC, Aamodt LC (1980) Photothermal spectroscopy using optical beam probing: Mirage effect. J Appl Phys 51:4580–4588

Nagel E (1988) Photoakustische Untersuchungen an Pflanzen. PhD thesis, Univ Karlsruhe, Karlsruhe Contrib Plant Physiol, vol 16

Nagel E, Lichtenthaler HK (1988) Photoacoustic spectra of green leaves and of white leaves treated with the bleaching herbizide Norflurazon. In: Hess P, Pelzl J (eds) Photoacoustic and photothermal phenomena. Springer, Berlin Heidelberg New York Tokyo, pp 568–569

Nagel E, Buschmann C, Lichtenthaler HK (1987) Photoacoustic spectra of needles as an indicator for photosynthetic capacity of healthy and damaged conifers. Physiol Plant 7:427–437

Nitsch C, Braslavsky SE, Schatz GH (1988) Laser-induced optoacoustic calorimetry of primary processes in isolated photosystem I and photosystem II particles. Biochim Biophys Acta 934:201–212

Nordal PE, Kanstad SO (1985) New developments in photothermal radiometry. Infrared Phys 25:295–304

O'Hara EP, Roderick DT, Moore TA (1983) Determination of the in vivo absorption and photosynthetic properties of the lichen *Acarospora schleicheri* using photoacoustic spectroscopy. Photochem Photobiol 38:709–715

Opsal J, Rosencwaig A, Willenborg DL (1983) Thermal-wave detection and thin-film thickness measurements with laser beam deflection. Appl Opt 22:3169–3176

Palmer RA, Smith MJ (1985) Rapid-scanning Fourier-transform infrared spectroscopy with photothermal beam-deflection (mirage effect) detection at the solid-liquid interface. Can J Phys 64:1086–1092

Palmer RA, Smith MJ, Manning CJ, Chao JL, Boccara AC, Fournier D (1988) Step-and-integrate interferometry in the mid-infrared with photothermal beam deflection and sample-gas-microphone detection. In: Hess P, Pelzl J (eds) Photoacoustic and photothermal phenomena. Springer, Berlin Heidelberg New York Tokyo, pp 50–52

Pao YH (1977) Optoacoustic spectroscopy and detection. Academic Press, Lond NY

Patel CKN, Tam AC (1981) Pulsed optoacoustic spectroscopy of condensed matter. Rev Mod Phys 53:517–550

Popovic R, Leblanc RM, Beauregard M (1988) Photoacoustic studies of bundle sheath cell photosynthesis in *Zea mays*. J Plant Physiol 132:94–97

Poulet P, Chambron J (1983) Conception and realisation of a photoacoustic detector for in situ spectroscopy. J Photoacoust 1:329–346

Poulet P, Chambron J, Unterreiner R (1980) Quantitative photoacoustic spectroscopy applied to thermally thick samples. J Appl Phys 51:1738–1742

Poulet P, Cahen D, Malkin S (1983) Photoacoustic detection of photosynthetic oxygen evolution from leaves − Quantitative analysis by phase and amplitude measurements. Biochim Biophys Acta 724:433–446

Prehn H (1979) Photoakustische Spektroskopie. GIT-Fachz Lab 23:281–289

Prehn H (1981) Medical and biological applications of the photoacoustic effect. Abstr DTG Conf Photoacoust, Bad Honnef (23.–26.2. 1981), vol 1, p 8

Redmond RW, Braslavsky (1988) A time-resolved thermal-lensing investigation of photosensitization of singlet oxygen. In: Hess P, Pelzl J (eds) Photoacoustic and photothermal phenomena. Springer, Berlin Heidelberg New York Tokyo, pp 95–98

Roark JC, Palmer RA, Hutchison JS (1978) Quantitative absorption spectra via photoacoustic phase angle spectroscopy. Chem Phys Lett 60:112–116

Ronen R, Canaani O, Garty J, Cahen D, Malkin S, Galun M (1984) The effect of air-pollution and bisulfite treatment in the lichen *Ramalina duriaei* studied by photoacoustics. In: Sybesma C (ed) Advances in photosynthesis research, vol. 4 Nijhoff/Dr W Junk, The Hague, pp 251–254

Rosencwaig A (1978) Theoretical aspects of photoacoustic spectroscopy. J Appl Phys 49:2905–2910

Rosencwaig A (1980) Thermal-wave imaging and microscopy. In: Ash EA (ed) Scanned image microscopy. Academic Press, Lond NY, pp 291–317

Rosencwaig A, Gersho A (1976) Theory of the photoacoustic effect with solids. J Appl Phys 47:64–69

Rosencwaig A, Hall SS (1975) Thin-layer chromatography and photoacoustic spectrometry. Anal Chem 47:548–549

Rosengren LG (1975) Optimal optoacoustic detector design. Appl Opt 14:1960–1976

Royce BSH, Alexander J (1988) Fourier transform photoacoustic spectroscopy of solids. In: Hess P, Pelzl J (eds) Photoacoustic and photothermal phenomena. Springer, Berlin Heidelberg New York Tokyo, pp 114–121

Schubert W, Giani D, Rongen P, Krumbein WE, Schmidt W (1980) Photoacoustic in vivo spectra of recent stromatolites. Naturwissenschaften 67:129–132

Sigrist MW (1988) Atmospheric trace gas monitoring by laser photoacoustic spectroscopy. In: Hess P, Pelzl J (eds) Photoacoustic and photothermal phenomena. Springer, Berlin Heidelberg New York Tokyo, pp 114–121

Szigeti Z, Nagel E, Buschmann C, Lichtenthaler HK (1989) In vivo photoacoustic spectra of herbicide-treated bean leaves. J Plant Physiol 134:104–109

Tam AC (1986) Application of photoacoustic sensing techniques. Rev Mod Phys 58:81–431

Thomas LJ, Kelly MJ, Amer NM (1978) The role of buffer gases in optoacoustic spectroscopy. Appl Phys Lett 32:736–738

Thomas RL, Favro LD, Kuo PK (1985) Thermal-wave imaging for nondestructive evaluation. Can J Phys 64:1234–1237

Thomasset B, Thomasset T, Barbotin JN, Véjux A (1982) Photoacoustic spectroscopy of active immobilized chloroplast membranes, Appl Opt 21:124–126

Vacek K, Lokaj P, Urbanová M, Sladky P (1979) Radiative and non-radiative transitions in subchloroplast particles highly enriched in P 700. Biochim Biophys Acta 548:341–347

Wada K, Masujima T, Yoshida H, Murakami T, Yata N, Imai H (1986) Application of photoacoustic microscopy to analysis of biological components in tissue sections. Chem Pharm Bull 34:1688–1693

Williams CC, Wickramasinghe HK (1988) Photothermal imaging with sub-100-nm spatial resolution. In: Hess P, Pelzl (eds) Photoacoustic and photothermal phenomena. Springer, Berlin Heidelberg New York Tokyo, pp 114–121

Woltering EJ, Harren F, Boerrigter HAM (1988) Use of laser-driven photoacoustic detection system for measurement of ethylen production in *Cymbidium* flowers. Plant Physiol 88:506–510

Yakir D, Rudich J, Bravdo BA (1985) Photoacoustic and fluorescence measurements of the chilling response and their relationship to carbon dioxide uptake in tomato plants. Planta 164:345–353

Membrane Operational Impedance Spectra of Plant Cells

F. HOMBLÉ and A. JENARD

1 Introduction

The use of electricity in plant physiology is natural from the biophysical point of view, since many substances transported in the plant are charged, and since electrical signals have properties suitable for both transfer and transduction of informations. Moreover, it is now well established that most cells use the flow of electric current to perform some of their natural functions.

It is a common practice to analyze the properties of a membrane in terms of an electrical equivalent circuit consisting of a resistor in parallel with a capacitor (Fig. 1). It is now generally accepted that the unit membrane can be described by the "Lipid-Protein Mosaic" model proposed by Singer and Nicolson (1972). According to this model the structure of the unit membrane consists of a fluid lipid bilayer, to the surface of which may be attached functional proteins (the extrinsic proteins) and which may contain intrinsic proteins embedded in the bilayer. Some of the intrinsic proteins can span the membrane; others are exposed only on one of its surfaces. A protein will penetrate the lipid bilayer to different depths, depending upon the relative number and location of its hydrophobic and hydrophilic groups.

The presence of transmembrane proteins helps account for some specific biophysical and biochemical properties of the membrane. For instance, active and passive transports of specific ions are some of the electrical functions attributable to these proteins. These transport functions are responsible for the conductance properties of the membrane.

An active transport is the transfer of a particle against its electrochemical gradient, and it requires energy which is usually provided by the cellular metabolism, whereas a passive transport occurs when a particle moves down its electrochemical gradient. More recently, these two kinds of transport have been distinguished on the basis of their rate of ion transport. For instance, the rate of passive transport of potassium ions through a single K^+–channel of *Chara corallina* is $5 \ 10^7$ ions/s at -5 mV (Homblé et al. 1987), whereas the rate of active transport through the sodium-potassium pump is $5 \ 10^2 \ Na^+$/s (Jorgensen 1975).

The fact that biological membranes behave as capacitors implies that a steady state membrane potential difference is associated with a charge separation. The lipid bilayer of membranes separates both internal and external electrolyte solutions by a thin insulating layer, which impedes the movement of ions from one side of the membrane to the other. Such interface between two conducting solutions forms a significant electrical capacitor.

It is interesting to point out that some authors put an electromotive force in series with the resistor of the electrical equivalent circuit of the membrane to

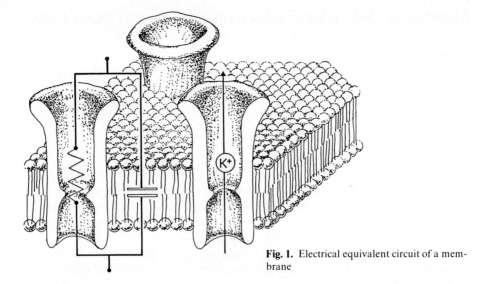

Fig. 1. Electrical equivalent circuit of a membrane

emphasize that the net driving force on an ion, i, is the difference between the membrane potential (E^M) and the equilibrium potential of that ion (E^i) and not simply the membrane potential, as suggested by Ohm's law. For instance, for the case of potassium ions the steady state current-voltage law is given by:

$$I_K = g_K(E^M - E^K), \tag{1}$$

where I_K is the specific potassium current and g_K is the specific potassium conductance.

The general approach used for impedance spectroscopy experiments on plant cells is to apply an electrical stimulus (a known voltage or current) to the membrane and to observe the response (the resulting current or voltage respectively). Three different types of electrical method are used in impedance spectroscopy (Fig. 2):

a) the transient method: a step function of current may be applied at $t = O$ to the membrane, and the resulting time-varying voltage measured. The electrical signals are usually mathematically transformed into the frequency domain using the Laplace transform in order to calculate the frequency dependent impedance. For this reason we have called this method the Laplace transform analysis (Homblé and Jenard 1986, 1987);
b) the white noise analysis: in this case the stimulus is an electrical current composed of random (white) noise, and the resulting voltage is measured. One generally Fourier-transforms the electrical signals to pass into the frequency domain (Marmarelis and Marmarelis 1978; Ross et al. 1985);
c) the sinusoidal analysis: this is the most common method used for impedance spectroscopy measurements. The impedance is measured directly in the frequency domain by applying a single-frequency sinusoidal current to the membrane and measuring the phase shift and amplitude of the resulting voltage signal at that

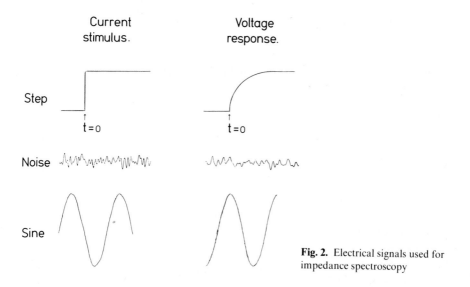

Fig. 2. Electrical signals used for impedance spectroscopy

frequency (Kishimoto 1974; Coster and Smith 1977; Vorobiev and Musaev 1979).

This chapter is not intended to serve as a reference on the topic of impedance spectroscopy but rather to introduce one of the methods used for impedance spectroscopy measurements: the Laplace transform analysis.

2 The Laplace Transform

The transform theory is developed in many textbooks on mathematics (Boas 1966), and we will here only mention definitions and transform properties which are related to our purpose.

The bilateral Laplace transform $(X(s))$ of a function $x(t)$ is defined by the following equations:

$$X(s) = L\{x(t)\} = \int_{-\infty}^{+\infty} x(t)\exp(-st)dt, \tag{2}$$

where the symbol L is a linear operator which satisfies the important property of superposition:

$$L(uV + vW) = L(uV) + L(vW) = uL(V) + vL(W), \tag{3}$$

where u is a constant and V and W are functions of t. When the linear operator L applies upon variables and functions which have a physical meaning, t and s have inverse dimensions. In the most frequent case, and especially when we are dealing with the study of an electrical signal, t has the dimension of a time and s has the dimension of a frequency.

The variable s is a complex number:

$$s = \sigma + j\omega \quad \text{where } j = \sqrt{-1}. \tag{4}$$

When s is purely imaginary ($s = j\omega$), the Laplace transform of $x(t)$ reduces to the Fourier transform of $x(t)$; that is:

$$X(j\omega) = \int_{-\infty}^{+\infty} x(t)\exp(-j\omega)dt. \tag{5}$$

Transforms have their reciprocal counterpart (called inverse transforms) which permit the recovery of $x(t)$ when $X(s)$ is known. In general cases the inverse Laplace transform is given by:

$$x(t) = L^{-1}\{X(s)\} = \frac{1}{j2\pi} \int_{\sigma - j\omega}^{\sigma + j\omega} X(s)\exp(st)ds. \tag{6}$$

The inverse Laplace transform is much more difficult to perform than the direct one, for reasons related essentially to the condition of convergence of the integral. Thus, if you are not a trained expert, you should refer to a published table of transform pairs (e.g., Abramowitz and Stegum 1970) if necessary.

Practically, the lowest time limit of the integral in Eq. (2) is kept at $-\infty$ in the Fourier transform and is set to zero in other cases. Therefore, a somewhat different form of the Laplace transform, often referred to as the unilateral Laplace transform or simply the Laplace transform, can be defined, which plays an important role in analyzing linear systems. The (unilateral) Laplace transform $L(s)$ of a signal $x(t)$ is written:

$$L(s) = \int_0^\infty x(t)\exp(-st)dt. \tag{7}$$

The main difference in the definition of the bilateral and unilateral Laplace transforms [Eqs. (2) and (7) respectively] lies in the lower limit on the integral. The unilateral Laplace transform depends on the signal from $t = 0$ to $t = \infty$, whereas the bilateral Laplace transform depends on the whole signal from $t = -\infty$ to $t = +\infty$. The unilateral Laplace transform must not be seen as a new kind of transform, but simply as a bilateral Laplace transform of a function whose value is set to zero for $t \leq 0$. Therefore, two signals which are identical for $t > 0$ but differ for $t \leq 0$ will have the same unilateral Laplace transform but a different bilateral Laplace transform. From the experimental point of view the Laplace transform (unilateral) is particularly relevant when a linear system is stimulated by a step function (Fig. 2), because in this case both input and output signals and their derivatives will be equal to zero for $t \leq 0$ but not necessarily for $t > 0$.

3 Measuring Techniques

Impedance spectroscopy measurements require electronic devices for recording bioelectric signals and for the stimulation of plant cells. Any arbitrary time domain excitation can be used to measure the membrane impedance, provided that both applied excitation and recorded response are over a sufficiently long time to complete the Laplace transform over the desired frequency band. In order to study the electrical properties of plant cells, different kind of electrical stimulus may be used. When the stimulus is a voltage signal, the membrane current is the response. Conversely, one gets a membrane voltage response when a current stimulus is applied to the membrane.

3.1 Current Clamp and Voltage Clamp Techniques

Both time- and voltage-dependent ionic conductances and voltage-independent ionic conductances are found in biological membranes. Two experimental methods have been developed in order to determine the relationship between the voltage applied across the membrane and the current flowing through it: these are the constant current or current clamp and the constant voltage or voltage clamp techniques. In the case of the current clamp, one applies a constant current step to the membrane and records the resulting membrane potential changes. In the case of the voltage clamp, the magnitude of the membrane potential is imposed, and one monitors the resulting current which flows through the membrane. The current clamp method is of limited use when ionic conductances change at a threshold potential, because in this case unstable potentials arise which are not attainable by a constant current.

A current clamp system is easier to set up than a voltage clamp system. A constant current source is obtained with a power supply V_s in series with high value resistor R_0 (Homblé et al. 1988). The series resistor must be much more larger than the membrane resistance in order to avoid any current change in response to a change in membrane conductance. This system is depicted in Figure 3.

A voltage clamp system involves the use of a negative feedback circuit. Different electronic circuits have been described in detail in the literature (Kishimoto 1961; Beilby and Coster 1979; Homblé 1988). In Figure 4 a simple and general scheme of a voltage clamp set up is displayed. In this diagram, the properties of the electrometer which feels the membrane potential and the current to voltage converter (I/V) are both assumed to be ideal (no phase shift and potential perturbation). The

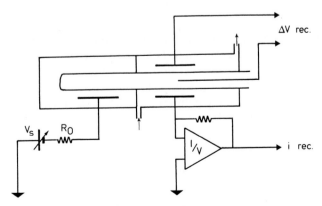

Fig. 3 Schematic diagram of a simple current clamp device

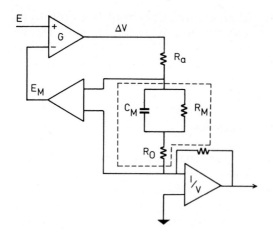

Fig. 4. Schematic diagram of a voltage clamp device

membrane potential difference (E^M) is compared to a command potential (E) at the input of a large gain amplifier (G). The polarities are so arranged that the output voltage of G (ΔV) force the membrane potential to be equal to the command potential. The access resistance R_A arises from the electrodes, and R_0 is the serial resistance which accounts for the solution conductivity. Assuming that the values of the circuit elements are at steady state, the operation of the voltage clamp circuit can be readily understood. The output voltage of the feedback amplifier G is equal to:

$$\Delta V = (E - E^M)G, \tag{8}$$

where G is the amplifier gain. Moreover, this output voltage is distributed between the voltage drop across the access resistance (R_A) and the voltage drop across the membrane, so that:

$$\Delta V = IR_A + E^M. \tag{9}$$

Combining these two equations, we find:

$$E^M = \frac{EG}{(1+G)} - \frac{IR_A}{(1+G)}. \tag{10}$$

Hence, we observed that a large gain for the feedback amplifier minimizes the effect of the access resistance and forces the membrane potential to be equal to the command potential. It is worth stressing the fact that E^M should be close to E. Because of the presence of R_s between the membrane and the voltage recording electrodes, E^M will differ from E by the voltage drop produced by the membrane current flowing across the series resistance R_0.

3.2 Instrumentation

Most biophysicists build at least part of their measuring equipment by themselves. Modern equipment is built with integrated circuits. Each integrated circuit may contain many transistors, along with capacitors and resistors, all fabricated on a single slab of highly purified silicon. We will briefly describe later the main electronic components which are now in common use in electrophysiological equipment. The brief account in this section can only be considered a survey. No attempt is made to derive the fundamental mathematical relations; a more complete treatment can be found in numerous textbooks (Graeme et al. 1971).

3.2.1 Operational Amplifier

The term "operational amplifier" was originally introduced by workers in the analogue computer field to denote an amplifier circuit which performed various mathematical operations such as integration, differentiation, summation, subtraction and multiplication.

It is not essential that the user of operational amplifiers be familiar with the intricacies of the internal circuit details, but he must understand the function of the external terminals provided by the manufacturer. In a first consideration of operational amplifiers it is convenient to assume that the amplifier has ideal characteristics: infinite gain, infinite input impedance, zero output impedance and infinite frequency bandpass.

Most operational amplifiers have two input terminals, only one of which produces an inversion of sign. The terminals are conventionally marked + and −, as in Fig. 5. These designations do not mean that terminals are to be connected only to potentials of the indicated sign, but rather that the one marked − gives sign inversion and the other (marked +) does not. In those operational amplifiers which have only a single input, it is always the non-inverting one which is omitted. In case the non-inverting input is not required in a particular application, it should be grounded to avoid instability.

The basic connections of an operational amplifier are shown in Fig. 5.

Most circuits using operational amplifiers depend on negative feedback: a connection is made through a suitable impedance Z_f from the output to the inverting input. If the signal to be sensed by the amplifier is a voltage, then it must be applied through an impedance Z_1.

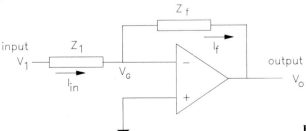

Fig. 5. The operational amplifier

Since the input to the ideal amplifier draws negligible current (because of the infinite input impedance), the current flowing in Z_1, namely, $(V_1-V_G)/Z_1$, must be equal to that in the feedback loop, given by $(V_G-V_0)/Z_f$. But the current in the feedback loop can come only from the output of the amplifier. Therefore, when an input signal is applied, the amplifier must adjust itself so that the feedback and input currents are precisely equal or:

$$I_{in} = I_f,\tag{11}$$

which leads to:

$$V_0 = \frac{V_G (Z_f + Z_1) - V_1 Z_f}{Z_1}.\tag{12}$$

This relation can be greatly simplified by taking into consideration the infinite gain of the ideal amplifier (often called its open-loop gain). This means that the potential V_G at the summing junction must be very small compared to V_0. Practically, V_G is so small ($< 10\ \mu V$ compared to ground) that the summing junction is commonly said to be at virtual ground. Hence, in Eq. (12) the term involving V_G can be neglected, giving:

$$V_0 = -V_1\ \frac{Z_f}{Z_1},\tag{13}$$

which is the basic working equation of an operational amplifier.

If Z_f and Z_1 are purely resistive, Eq. (13) shows that the output voltage will be the negative of the input multiplied by a constant. In case the value of Z_f is higher than Z_1, the ratio Z_f/Z_1 will be higher than 1, and the amplifier operates like a multiplicator; if Z_f is lower than Z_1 then the ratio Z_f/Z_1 will be lower than 1, and the amplifier operates as a divider.

Addition or subtraction of several potentials may be accomplished by connecting different voltage sources to the summing junction through an appropriate resistor as shown in Fig. 6.

The relation between the inputs currents and the feedback current is:

$$-I_f = I_1 + I_2 + I_3,\tag{14}$$

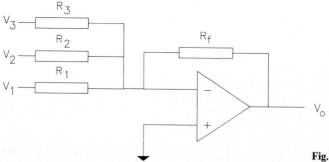

Fig. 6. The summing amplifier

or, using Ohm's law:

$$-V_0 = \frac{R_f}{R_1}V_1 + \frac{R_f}{R_2}V_2 + \frac{R_f}{R_3}V_3. \tag{15}$$

The differential amplifier (Fig. 7) is widely used in instrumentation.

In this configuration it is the difference between the two input signals which is amplified. Following the same procedure as for Eq. (13) it can be shown that the equation governing this circuit is given by:

$$V_0 = -\frac{R_f}{R_1}(V_1 - V_2), \tag{16}$$

where the suffix 1 and 2 refer to the negative and positive inputs respectively. The differential amplifier is advantageous because it discriminates against direct (zero frequency) current variations, drifts and noise.

Operational amplifiers are not only used to operate upon voltages. A current source may be connected to the inverting input of the amplifier (Fig. 8).

Since the input impedance is infinite, no current flows into the operational amplifier, and the input current must be equal to the feedback current. Therefore, the output voltage will be equal to the product of the input current and the feedback resistor:

$$V_0 = -R_f I_{in}. \tag{17}$$

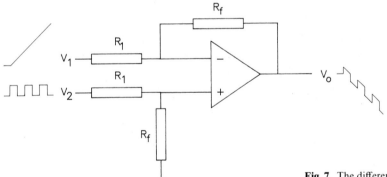

Fig. 7. The differential amplifier

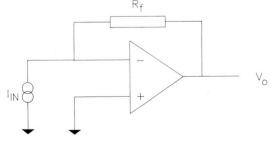

Fig. 8. The current to voltage converter

3.2.2 Sample–and–Hold

Although in practice one would generally use a ready-built sample-and-hold integrated circuit, it is useful to recall its basic operation. Figure 9 shows a very simple sample-and-hold circuit.

An incoming signal is fed into a capacitor when the (electronic) switch is closed. In this mode, the output of the amplifier will continuously follow the input signal. When the switch is opened and isolates the input signal from the capacitor, the output remains at the voltage last seen by the input capacitor.

3.2.3 Analog to Digital Converter

A typical analog to digital converter is illustrated in Fig. 10.

An analog to digital converter changes an analog or continuous voltage into a series of discrete digital values so that a computer can be presented with digital data in a format that it can handle. The conversion starts as soon as a pulse signal is sent to the appropriate input and lasts up to the time that an end of conversion signal is sensed by the analog to digital converter.

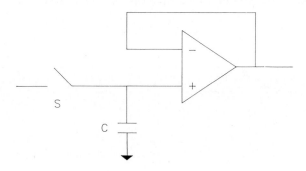

Fig. 9. The sample-and-hold circuit

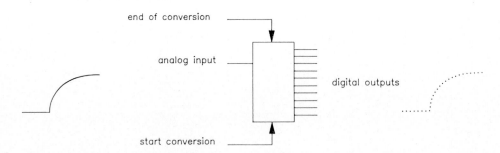

Fig. 10. The analog to digital converter

4 Data Analysis Technique

4.1 Theory

Electrical circuits built exclusively with passive elements such as resistors, capacitors and inductors belong to the class of linear systems. A linear system is one that possesses the property of superposition: if an input consists of the weighted sum of several signals, the output is simply the weighted sum (the superposition) of the responses of the system to each of those signals. Mathematically, this means that if $y_1(t)$ and $y_2(t)$ are the responses of a system to the continuous time functions $x_1(t)$ and $x_2(t)$ respectively, then the system is linear if the response to:

$$ax_1(t) + bx_2(t) \text{ is } ay_1(t) + by_2(t), \tag{18}$$

where a and b are any complex constant, and a system which results in the transformation of signals is defined as a "black box". In plant physiology the black box might be a membrane, a cell, a tissue or even the whole plant. Such a system is usually represented pictorially, as in Fig. 11, where $x(t)$ is the input signal and $y(t)$ is the output signal.

The behaviour of a linear system can be characterized by a linear constant-coefficient differential equation of the form:

$$y(t) + \sum_{i=1}^{n} A_i \frac{d^i y(t)}{dt^2} = B_0 x(t) + \sum_{k=1}^{n} B_k \frac{d^k x(t)}{dt^k}, \tag{19}$$

where t is the time, n is the order of the system and of its equation, A_i and B_k are constant coefficients depending on the circuit structure and on the magnitude of its elements, $x(t)$ is the input signal imposed to the system and $y(t)$ is the output signal. When $x(t)$ is a known and derivable time function, the right hand side of Eq. (19) is a known time function, and the integration of this equation yields the output signal as solution.

In the case of plant cells, the system in consideration is an electric dipole which may be represented by a black box with two conducting wires coming out of it (Fig. 11). There is a potential difference (V) between those wires, and they carry the same current (I), owing to the fact that there is no charge accumulation inside the black box.

4.1.1 Complex Impedance

Let us first consider the case where the steady state input voltage signal has the form:

$$V(t) = V_0 \cos(\omega t) = R\{V_0 \exp(j\omega t)\} = R\{V^*\}, \tag{20}$$

where V_0 is the amplitude of the signal and R is the symbol for real numbers,

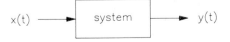

x(t) ⟶ system ⟶ y(t)

Fig. 11. Basic diagram of a system

indicating that we are dealing with the real part of the complex number enclosed in brackets only. This periodic signal has a single frequency component. Then the output current response (I) will have the form:

$$I(t) = I_0 \cos(\omega t + \phi) = R\{I_0 \exp j(\omega t + \phi)\} = R\{I^*\}, \tag{21}$$

where I_0 is the amplitude and ϕ the phase of the current. Both amplitude and phase of the current response depend on ω and on the content of the black box. The quantity Z^*, which is defined in the frequency domain by the relation:

$$Z^* = \frac{V^*}{I^*}, \tag{22}$$

is called the complex impedance of the system. Changing the frequency of the input signal given by Eq. (20) and recording the corresponding current response will permit one to learn the complex impedance spectrum of the system. This procedure is time consuming because the experiment must be repeated for each frequency of interest.

4.1.2 Operational Impedance

Let us now consider the more complicated case where a transient signal formed from a whole set of frequencies is applied to the linear system. Equation (19) will describe the relation between the input and output signals. Applying the unilateral Laplace transform to both sides of Eq. (19) gives:

$$Y(s) \left\{ 1 + \sum_{i=1}^{n} A_i s^i \right\} = X(s) \left\{ \sum_{k=0}^{n} B_k s^k \right\}, \tag{23}$$

where $s = j\omega$, $X(s)$ is the Laplace transform of the input signal (the voltage) and $Y(s)$ is the Laplace transform of the output signal (the current). The quantity $Z_{op}(s)$, which is defined by:

$$Z_{op} = \frac{X(s)}{Y(s)}, \tag{24}$$

is called the operational impedance. It is worth noting that for a given circuit the expression of Z_{op} is identical to that of Z^*, where s is substituted for $j\omega$. Substituting Eq. (23) in (24) we obtain a general analytical expression for the operational impedance:

$$Z_{op} = \frac{\sum_{k=0}^{n} B_k s^k}{1 + \sum_{i=1}^{n} A_i s^i}. \tag{25}$$

As a relevant example we shall consider the operational impedance of the electrical equivalent circuit of Fig. 1. The differential equation relating the current (i) and the voltage (v) is:

$$i(t) = \frac{v(t)}{R} + C\frac{dv(t)}{dt}. \tag{26}$$

Taking the Laplace transform of Eq. (26) we find:

$$I = \frac{V}{R} + CsV, \tag{27}$$

where I and V are the Laplace transform of the current and voltage respectively. Then from Eq. (24):

$$Z_{op} = \frac{R}{1+RCs}. \tag{28}$$

If a step of current is applied to this circuit, then $i(t) = 0$ for $t < 0$ and i_{max} for $t > 0$. The Laplace transform of $i(t)$ is i_{max}/s. The Laplace transform of the voltage (V) is then given by Eq. (27), which, after rearrangement, is written:

$$V = \frac{i_{max}R}{1+RCs}. \tag{29}$$

We may consult a table of Laplace transform (Abramowitz and Stegum 1970) to find that the function of t, which has Eq. (29) as its Laplace transform, is:

$$v(t) = i_{max}R\left(1 - \exp\left(-\frac{t}{RC}\right)\right). \tag{30}$$

In this simple example it is easy to verify that Eq. (30) is the solution of Eq. (26).

4.2 Experimental Method

4.2.1 Data Acquisition

A schematic diagram of the electrical measuring system is shown in Fig. 12. The transient change in membrane potential was recorded between an intracellular axial electrode (Pt/Ir wire) and an external Ag/AgCl electrode. Each electrode was connected to a high impedance (10^{12} ohms) buffer operational amplifier (LH 002 of National semiconductor), the output of which was fed into an instrumentation amplifier (AD 521 of Analog Devices). This amplifier can be viewed as a high performance differential amplifier. The resting membrane potential was memorized by means of a sample-and-hold circuit (LF 398 of National Semiconductor), and only the transient change was recorded in the first channel of a Nicolet digital oscilloscope (model 206). The current was provided by a voltage source in series with a high resistor (four 10^7 ohms) and two Ag/AgCl electrodes. The current was converted into a voltage by means of a LF 157 operational amplifier (National Semiconductor) and was recorded in the second channel of the digital oscilloscope. Data were then transferred into a HP 1000 computer (Hewlett Packard) which was used to treat the signal.

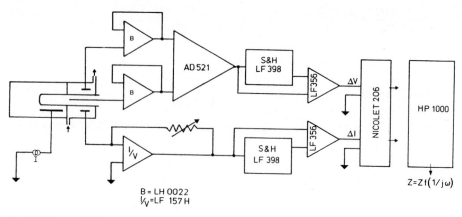

Fig. 12. Schematic diagram of the electrical measuring system used for operational impedance analysis of *Chara* cells ((Homblé and Jenard 1987a)

4.2.2 Computations

Each current and voltage time course was divided in up to twenty unequal segments using an interactive procedure. Each segment was fitted by a polynomial of the third degree using a conventional least square method, except for the final segment, which was assumed to be approximated by an expression of the form:

$$y = A + B \exp(\alpha t). \tag{31}$$

A best fit procedure searched for the parameters A, B and α. The analytical Laplace transform is then calculated over the whole spectrum of frequencies. According to the sampling theorem (Bendat and Piersol 1971) the highest frequency component is half of the sampling rate. The lowest frequency is the reciprocal of half of the sample length. Thus, if a signal is sampled at 1 kHz for 1 s, the highest and lowest frequencies will be 500 Hz and 2 Hz respectively.

The next problem is to find out the values of the elements of an electrical equivalent circuit which has the same operational impedance spectrum as that calculated from the experimental results. One major problem with such modeling arises from the inherent ambiguity of equivalent circuit fitting: an equivalent circuit involving three or more circuit elements can often be rearranged in various ways and still yield exactly the same operational impedance spectrum. Fortunately, an analysis of the structure of the system under investigation often suggests the equivalent circuit which is the most appropriate to describe the system. For instance, the simplest electrical circuit which may be used to describe a plant cell is shown in Fig. 13. The choice of a resistor (R_1) in parallel with a capacitor (C_1) rests on considerations given in Section 1. The series resistor (R_0) arises from the cell wall and from the solution which is between the measuring electrode and the membrane surface. The differential equation of this circuit is given by:

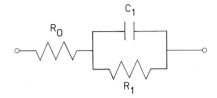

Fig. 13. The simplest electrical equivalent circuit of a plant cell

$$C_1 R_1 \frac{dv}{dt} + v = C_1 R_0 R_1 \frac{di}{dt} + (R_0 + R_1)\, i, \tag{32}$$

and its operational impedance is written as:

$$Z_{op} = R_0 + \frac{1}{\frac{1}{R_1} + sC_1}. \tag{33}$$

Expanding Eq. (33) in power series of $1/s$ we get:

$$Z_{op} = R_0 + \frac{1}{sC_1} - \frac{1}{s^2 R_1 C_1^2} + \dots . \tag{34}$$

When Z_{op} is plotted versus $1/s$, the intercept on the ordinate is R_0, and C_1 is equal to the reciprocal slope at the origin of the operational impedance spectrum. We can then calculate a new operational impedance spectrum (Z_{op}'):

$$Z'_{op} = \frac{Z_{op} - R_0}{1 - sC_1(Z_{op} - R_0)}. \tag{35}$$

If the operational impedance spectrum of the system under investigation can be correctly fitted by the electrical equivalent circuit of Fig. 13, then Z_{op}' is constant (independent of s) and is equal to R_1. If it is not the case, Z_{op}' will be a function of s, which means that the system contains at least one more resistor-capacitor loop. Assuming that R_1 is itself an impedance which has the same structure as that considered at the beginning (Fig. 13), we get the circuit of Fig. 14. Then Z_{op}' can be treated as a new operational impedance spectrum. Using the same procedure as that starting from Eq. (34) one can evaluate R_1 and C_2. If the new operational impedance (now called Z_{op}'') is constant, Z_{op}'' will then be equal to R_2. The analytical expression of the operational impedance of the circuit of Fig. 14 is written:

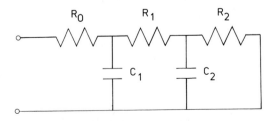

Fig. 14. Electrical equivalent circuit used to fit the operational impedance spectrum of a two time-constant system

$$Z_{op}^o = \frac{R_0 + R_1 + R_2 + (R_1C_2R_2 + R_0(C_1R_1 + C_2R_2 + C_1R_2))s + R_0C_1R_1C_2R_2s^2}{1 + (C_1R_1 + C_2R_2 + C_1R_2)s + R_1C_1R_2C_2s^2}, \quad (36)$$

or using the suitable substitution:

$$Z_{op}^o = \frac{X_1 + X_2s + X_3s^2}{1 + X_4s + X_5s^2}. \quad (37)$$

Our experience has taught us that, starting with 1024 experimental points for both voltage and current signals, the iterative procedure described above may be used to satisfactorily analyze a system consisting of up to three resistor-capacitor loops.

In mature plant cells there is a vacuole which occupies 90% of the cellular volume. Therefore, when electrical signals are recorded between the vacuole and the extracellular solution, at least two membranes are in series: the plasmalemma and the tonoplast. Thus, in this case at least two parallel resistor-capacitor loops must be considered in the electrical equivalent circuit (Fig. 15). The operational impedance of this circuit may be written:

$$Z_{op} = \frac{R_0 + R_1 + R_2 + (R_1C_2R_2 + C_1R_1R_2 + R_0(C_1R_1 + C_2R_2))s + R_0C_1R_1C_2R_2\,s^2}{1 + (C_1R_1 + C_2R_2)s + C_1R_1C_2R_2s^2} \quad (38)$$

Equations (36) and (38) belong to two different electrical equivalent circuits which have the same operational impedance spectrum. Therefore, in order for these two circuits to have an identical impedance spectrum, each coefficient of the different powers of s in the numerator and in the denominator in Eq. (36) must be equal to those in Eq. (38) respectively. Using Eqs. (37) and (38) it is easy to find the equations for the transformation of parameters:

$$R_0 = \frac{X_3}{X_5},$$

$$C_1 = \frac{Y_3 - Y_4}{Y_1 - Y_2Y_3},$$

$$R_1 = \frac{Y_3}{C_1},$$

$$C_2 = \frac{Y_3 - Y_4}{Y_2Y_4 - Y_1},$$

$$R_2 = \frac{Y_4}{C_2},$$

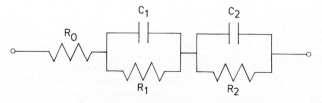

Fig. 15. Electrical equivalent circuit of a plant cell consisting of a plasmalemma and tonoplast in series

where

$$Y_1 = X_1 - \frac{X_3}{X_5},$$

$$Y_2 = \frac{X_2 X_5 - X_3 X_4}{X_5},$$

$$Y_3 = \frac{X_4 + \sqrt{X_4^2 - 4X_5}}{2},$$

$$Y_4 = \frac{2X_5}{X_4 + \sqrt{X_4^2 - 4X_5}}. \tag{39}$$

4.2.3 Example

Chara corallina is a freshwater plant. The large size of its cells (about 1 mm in diameter and up to several cm in length) makes it a convenient material for electrical studies of the plant cell membrane. When the membrane of *Chara corallina* is stimulated with a current step of sufficient magnitude the transient voltage response displays an overshoot, which has been attributed to the time- and voltage-dependent conductance of K^+ channels (Homblé and Jenard 1984; Smith 1984; Homblé 1985) and could be described as inductive behaviour. The use of a low intensity current stimulus (not causing an overshoot) yields a capacitive like curve (Fig. 16).

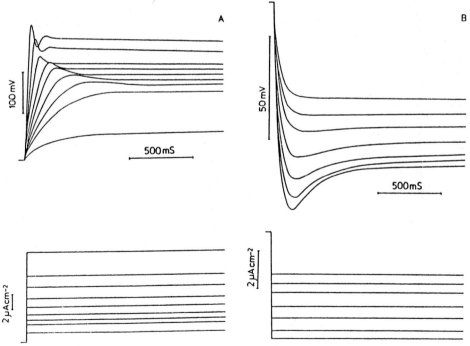

Fig. 16. Response of the membrane potential to increasing (**A**) inward current pulses and (**B**) outward current pulses (Homblé and Jenard 1984)

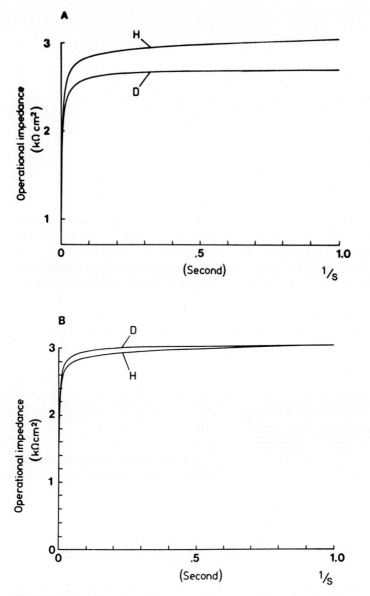

Fig. 17A,B. Operational impedance spectra of *Chara* cell (Homblé and Jenard 1987b)

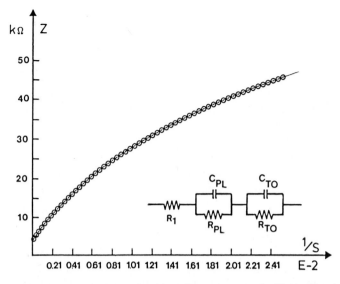

Fig. 18. Simulation of operational impedance spectrum of a *Chara* cell under saturating light at pH 6.0. The calculated values of the components of the electrical equivalent circuit shown in the inlet are: $R_1 = 1.0$ kΩcm^2; $R_{pl} = 26.2$ kΩcm^2; $C_{pl} = 1.9$ μF cm^{-2}; $R_{to} = 4.4$ kΩcm^2 and $C_{to} = 1.1$ μF cm^{-2}

Figure 17 shows the operational impedance of a *Chara* cell stimulated by a hyperpolarizing and depolarizing current step of 0.5 μA cm^{-2}. Since the steady state behaviour of the membrane is not symmetrical, the depolarizing curve is scaled to the hyperpolarizing curve by multiplying it by a constant factor in order to obtain the upper right point of the operational impedance spectrum in common to both curves (Fig. 17B). It then becomes apparent that the spectrum (the change with 1/s) of the membrane operational impedance is also asymmetrical around the resting state. A typical example of the simulation of the membrane impedance spectrum by means of an electrical equivalent circuit is shown in Fig. 18.

5 The Concept of Membrane Capacitance

Strictly speaking, the capacitive current results in a change in the amount of free charges on each side of the membrane. This current flow does not involve the displacement of charges (ions) through the membrane but a redistribution of the charges on the membrane surface. According to the basic laws of electricity, the capacitive current determines the rate at which the voltage across the membrane changes. This is expressed by the equation:

$$i = C \frac{dv}{dt}. \tag{40}$$

The larger the capacity to be charged, the more current is required to reach a given voltage.

However, there are different kinds of biophysical phenomenon related to ion transport across biological membranes, which can produce an electrical response similar to that of a capacitor. Because these phenomenon are not directly related to the dielectric properties of the membrane, they are usually called pseudo-capacitive effects. It is important to be able to separate these pseudo-capacitance components from the true membrane capacitance, because the later can provide information about the structure and properties of the cell membrane itself.

Passive transport occurs through ion channels at a rate of about 10^7 ions per second. In the membrane each channel opens and closes randomly. The small signal capacitance of a population of voltage dependent ion channels is given by (Ferrier et al. 1985):

$$C(\omega) = -\gamma N_o(V-E) \ \frac{(q'[p/q]-p')}{([q+p]^2+\omega^2)}, \tag{41}$$

where γ is the conductance of a single open channel, N_o is the number of open channels, V is the membrane potential differences, E is the equilibrium potential of the ions crossing the channel, q is the frequency of channels changing from closed to open state, p is the frequency of channels changing from open to closed state and q' and p' are the first derivatives of q and p with respect to the membrane potential difference. In case the frequency of opening of channels is sufficiently voltage dependent, so that q'p/q is greater than p', a negative membrane capacitance (or pseudo-inductance) will then be measured. Moreover, from Eq. (41) it may be concluded that an increase in the probability of opening of voltage-dependent ion channels can lead to an apparent (or pseudo-) capacitive effect.

Ion transport and concentration changes occur in the unstirred layers of solution adjacent to each face of the membrane when a current is injected through it. Let us assume that the membrane is surrounded by two solutions made of different concentrations of the same uni-valent binary salt. The current will be carried by negative and positive ions which will, in general, have different transport number (the fraction of current carried by an ion), both in the bathing solutions and in the membrane. As shown by Barry and Hope (1969), the transport number discontinuities at the membrane-solution interfaces will cause solute enhancement or depletion at these interfaces in the unstirred layers width adjacent to the membrane. Electro-osmosis will also cause changes in the solute concentration in the unstirred layers of solution adjacent to the membrane as solute is swept up or away by the electro-osmotic flow which occurs when a current is passed through the membrane. Transport number effects and electro-osmosis effects have opposite consequences on concentration changes at the membrane-solution interfaces. Such effects which change the local concentration at the membrane-solution interface also give rise to an apparent low-frequency (< 0.01 Hz) component of capacitance (Barry 1977; Segal 1967). This is so because the local gradient across the membrane gives rise to a time-dependent diffusion potential, which thus appears as if it were a slowly increasing membrane resistance which mimics a capacitance effect.

Finally, a low-frequency ($<$ 1 Hz) negative capacitance can also result from the proton or hydroxyl diffusion in the unstirred layers adjacent to the membrane (Ferrier et al. 1985; Homblé and Ferrier 1988). This effect is especially significant in plant cells where the membrane conductance is higher for protons (or hydroxyl) than for the other ions. When a steady state ion current crosses the membrane, it is transported away from the membrane in the unstirred layer by means of a concentration gradient-driven diffusion (Ferrier 1981). Because protons (or hydroxyl ions) have a greater diffusion coefficient than the other ions in the unstirred layer, the diffusion-driven component of the current will lag behind the total current, which will give rise to a pseudo-inductive effect.

6 Conclusion

The impedance spectroscopy estimated by Laplace transform analysis permits one to obtain information quickly about the impedance of plant cells, because only a single experimental run is sufficient to obtain the frequency dependence of the membrane impedance. One of the most attractive aspects of impedance spectroscopy as a tool for investigating the electrical properties of plant membranes is the direct connection that often exists between the behaviour of the membrane and that of an idealized model circuit representative of the physical process taking place in the membrane. There are, however, dangers in the indiscriminate use of electrical analogies to describe plant systems. The first point to be made is that electrical equivalent circuits are seldom unique and, in many cases, other experimental data are often necessary to choose the relevant circuit unambiguously. A further limitation is the frequency-dependent electrical properties of interface layers (for instance, the unstirred layer adjacent to the membrane surface), which mimic the electrical response of plant membranes. Finally, an important requirement for a valid impedance spectroscopy analysis is that the system must be linear.

Acknowledgments. This work was supported by a grant from the Belgian Ministry for Scientific Policy (convention ARC 86-91 to Professor R. Lannoye), whose financial support is kindly acknowledged.

References

Abramowitz M, Stegum IA (1970) Handbook of mathematical functions. Dover Publ, NY
Barry PH (1977) Transport number effects in the transverse tubular system and their implications of low frequency impedance measurement of capacitance of skeletal muscle fibers. J Membrane Biol 34:243–292
Barry PH, Hope AB (1969) Electroosmosis in membranes: effects of unstirred layers and transport numbers. I. Theory. Biophys J 9:700–728
Beilby MJ, Coster HGL (1979) The action potential in *Chara corallina*. II. Two activation-inactivation transients in voltage clamps of the plasmalemma. Aust J Plant Physiol 6:323–335
Bendat JS, Piersol AG (1971) Random data: analysis and measurement procedures. Wiley, NY
Boas ML (1966) Mathematical methods in the physical sciences. Wiley, NY

Coster HGL, Smith JR (1977) Low frequency impedance of *Chara corallina:* simultaneous measurements of separate plasmalemma and tonoplast capacitance and conductance. Aust J Plant Physiol 4:667–674

Ferrier JM (1981) Time-dependent extracellular ion transport. J Theor Biol 92:363–368

Ferrier JM, Dainty J, Ross SM (1985) Theory of negative capacitance in membrane impedance measurements. J Membrane Biol 85:245–249

Graeme JG, Tobey GE, Huelsman LP (1971) Operational amplifier. McGraw Hill, NY

Homblé F (1985) Effect of sodium, potassium, calcium, magnesium and tetraethylammonium on the transient voltage response to a galvanostatic step and of the temperature on the steady state membrane conductance of *Chara corallina.* J Exp Bot 36:1603–1611

Homblé F (1988) A fast and high-current voltage clamp device for biophysical investigations. J Phys E: Sci Instrum 21:1100–1102

Homblé F, Ferrier JM (1988) Analysis of the diffusion theory of negative capacitance: the role of K^+ and the unstirred layer thickness. J Theor Biol 131:183–197

Homblé F, Jenard A (1984) Pseudo-inductive behaviour of the membrane potential of *Chara corallina* under galvanostatic conditions: a time-variant property of potassium channels. J Exp Bot 35:1309–1322

Homblé F, Jenard A (1986) Membrane operational impedance spectra in *Chara corallina* estimated by Laplace transform analysis. Plant Physiol 81:919–921

Homblé F, Jenard A (1987) Membrane impedance of internodal cells of *Chara corallina* obtained by analysis of low level transients. Bioelectrochem Bioenerg 17:131–139

Homblé F, Ferrier J, Dainty J (1987) Voltage-dependent K^+-channel in protoplasmic droplets of *Chara corallina.* A single channel patch clamp study. Plant Physiol 83:53–57

Homblé F, Jenard A, Hurwitz HD (1988) Some techniques for investigation of electrochemical and ion transport properties of biological cell membranes: the case of *Chara corallina.* In: Dryhurst G, Niki K (eds) Redox chemistry and interfacial behavior of biological molecules. Plenum Press, NY

Jorgensen PL (1975) Isolation and characterization of the components of the sodium pump. Quart Rev Biophys 7:239–274

Kishimoto U (1961) Current voltage relations in Nitella. Biol Bull 121:370–371

Kishimoto U (1974) Transmembrane impedance of the *Chara* cell. JPN J Physiol 24:403–417

Marmarelis PZ, Marmarelis VZ (1978) Analysis of physiological system: the white-noise approach. Plenum Press, NY

Ross S, Ferrier J, Dainty J (1985) Frequency-dependent membrane impedance in *Chara corallina* estimated by Fourier analysis. J Membrane Biol 85:233–243

Segal JR (1967) Electrical capacitance of ion-exchanger membranes. J Theor Biol 14:11–34

Singer SJ, Nicolson GL (1972) The fluid mosaic model of the structure of cell membranes. Science 175:720–731

Smith PT (1984) Electrical evidence from perfused and intact cells for voltage-dependent K^+ channels in the plasmalemma of *Chara australis.* Aust J Plant Physiol 11:303–318

Vorobiev LN, Musaev NA (1979) Electrical characteristics of the cell wall and plasmalemma of *Nitellopsis obtusa.* Low frequency impedance. Sov Plant Physiol 26:570–577

Image Instrumentation Methods of Plant Analysis

K. Omasa

1 Introduction

Knowledge and understanding of the biological world result from information about organisms and their interactions with their surroundings. Although information can come from many sources, tremendous advances in science have occurred with advances in instrumentation and technology, e.g. microscopes. The recent advances in electronics have greatly increased the amount of information that can be obtained. The development of image instrumentation technologies, which gather two- or three-dimensional information about the organism in a non-destructive manner, has been particularly remarkable. In the field of medicine, instrumentation technologies for surface, microscopic, X-ray, RI (radio isotope) and ultrasonic images have been put to practical use for patient diagnoses. In conjunction with advances in computed tomography (CT), tomographic images that give not only morphological information but also functional and physiological information have also been obtained (Herman 1979; Onoe 1982; Mansfield and Morris 1982).

In the field of plant research, a variety of image instrumentation technologies, similar to those in the field of medical science, have been developed for cells, individual plants, and small-scale plant communities (Omasa and Aiga 1987; Omasa et al. 1988). These technologies are used to elucidate the reaction mechanisms of living plants in fields such as physiology and physiological ecology. Based on the information obtained from these fundamental studies, diagnosis of plant growing conditions, injury by disease and pests, nutrition disorders, and environmental pollution injury, etc., have become possible. In addition, in the field of biotechnology these technologies can also be used for screening plants obtained by such methods as breeding, cell fusion, and gene recombination.

Development of wide-area remote sensing for plant communities by artificial satellites or airplanes has also been quite remarkable (Colwell 1983). The sensing is used to survey cropping acreage of field crops, to estimate yields, and also to investigate changes in vegetation and ecosystems. Since a TM (Thematic Mapper) with many spectral bands and about 30 m resolution has been mounted in Landsat, the expansion in applications of remote sensing to agriculture or vegetation investigation are expected. However, wide-area remote sensing for large-scale plant communities are limited in terms of the physiological information obtained from living plants. Therefore, in order to extend the application fields, image instrumentation technologies for individual plants and small-scale plant communities, which will combine physiology at the cell level with information obtained from the wide-area remote sensing, are important.

The application fields for image instrumentation of living plants include those of the future, listed in Table 1. The instrumentations are expected to be used in a great many fields, ranging from the fundamental to the applied, such as botany, agriculture, environmental science, space science, and pedagogy.

Table 1. Major application fields for image instrumentation of living plants

* Diagnoses of injury by diseases and pests, nutrition disorders, environmental pollution injury, etc.
* Plant growth monitoring and culture control
* Biotechnology (mainly screening)
* Automatization of farm work (robotics)
* Diagnoses of environmental purification capacity
* Educational systems for plant diagnoses and culture control
* Auxiliaries for wide-area remote sensing
* Other research in fields such as physiology, ecology, agriculture, environmental science and space science

2 Image Sensor Selection and Processing System

2.1 Image Sensor and Plant Information

Human eyes perceive only a very narrow band (visible rays) of the electromagnetic spectrum and limit the information obtained. In image instrumentation and wide-area remote sensing, however, information that cannot be obtained simply through human eyes can be obtained by using sensors capable of detecting electromagnetic (or sound) waves of various bands, and high-sensitivity sensors. Figure 1 shows the wavelength of electromagnetic (or sound) waves and typical sensors used in the respective bands.

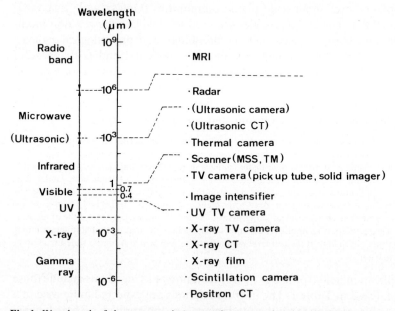

Fig. 1. Wavelength of electromagnetic (or sound) waves and typical image sensors

One requirement for actual instrumentation using these sensors is the presense of electromagnetic waves for the active or passive bearing of information about objects. Active instrumentation methods, in which electromagnetic waves are irradiated to obtain information from living plants, have been developed recently, as an addition to the passive method, which uses only electromagnetic waves from the environment. Active instrumentation method features selective acquisition of information on specific physiological and biochemical reactions, but it has the disadvantage of exerting influence on the plant environment. The information such as minute area, internal conditions, and metabolic processes of the living plants is obtained by combining such techniques as RI, microscopes, and CT. Plant information and to appropriate image instrumentation apparatuses are listed in Table 2. The list includes techniques developed for medical or industrial use, which are applicable to the plant field, in addition to those developed specifically for plant research. Typical image instrumentation methods for obtaining plant information are described in detail in Section 3 and thereafter.

Table 2. Plant information and to appropriate image instrumentation apparatuses

Plant information	Image instrumentation apparatus
* Growth, shape, community structure	Multispectral camera (TV, scanner), stereo or moire camera, ultrasonic camera, CT (X-ray, NMR, ultrasonic)
* Plant temperature, transpiration, gas absorption	Thermal camera (including scanner)
* Stomatal response	Thermal camera, remote-control light microscope system
* Photosynthetic system, activity	High-sensitivity spectral camera
* Visible injury, plant pigment, color and luster	Multispectral camera (TV, scanner)
* Cell, organelle	Light microscope system
* Structure, physiological function inside organism (e.g., annual ring, rot, water content, transfer and metabolism of organism components)	CT (X-ray, NMR, ultrasonic, positron), X-ray TV camera, scintillation camera

2.2 Image Processing System

The amount of data obtained by image instrumentation is huge. For example, think about displaying a color photograph as digital images of red (R), green (G), and blue

(B). In order to obtain an image quality as good as that of the original seen only with the eyes, the amount of information required is above 512 × 512 pixels and 256 gradations in each digital image. If this amount is expressed by bytes, a unit used in computers, it is about 786 KB, corresponding to almost the total capacity of one floppy disc, as used for personal computers. If the object is a three-dimensional and moving image, the amount of information handled greatly increases. Recent advances in computer technologies, represented by 32-bit super personal computers, array processors, and optical discs with large-capacity memory are outstanding and complicated calculations that until now have only been possible using large-scale computers can now be carried out using portable image processing systems. Packages of image processing software with good compatibiliy, such as SPIDER (Kyodo System Kaihatsu, about 700 subroutines of FORTRAN 77 if SPIDER-II is included) have also been sold on the market.

As an example of a portable high-speed image processing system, our system is shown in Fig. 2. This system consists of a super personal computer (MASSCOMP MC 5400) with an array processor for floating point operations, a color graphic display (KRC, nexus 6800) with simple functions for image processing, and an optical disc unit of the write-once type (Matsushita Denso CU-15, DU-15), with a large capacity for image recording.

The MC5400, used as the host computer, is a small light-weight computer with a 32-bit virtual CPU (central processing unit) based on Motorola's MC 68020 (16.7 MHz) and uses UNIX (AT&T System V + Berkeley 4.2BSD), with real-time operating functions as the OS (operating system). Many languages such as C, FORTRAN 77, Franz LISP, Common LISP, and Prolog are applicable, and the

Fig. 2. Portable high-speed image processing system

software compatibility is good. For high-speed processing of floating-point operations, required for image processing, the system has a double precision high-speed processor by pipeline processing (FPA-1; 0.6 Linpack MFLOPS; 3.5 MWhets/s) and a single precision array processor of the built-in type (VA-1; 14 MFLOPS), in addition to a standard MC 68881 (0.1 Linpack MFLOPS; 1.1 MWhets/s). In particular, the array processor has approximately 250 libraries and can carry out, for example, addition and multiplication of 1024 points in 160 μs and complex FFT in 4.5 ms. Due to this capacity, the system realizes a processing speed from several times to some tens times faster than that of a system with only an MC 68881, and can process complicated operations such as CT reconstruction.

The color graphic display with eight frame memories (512 × 512 × 8-bits) is capable of displaying any three frame memories as a color image (approximately 16.7 million colors; RGB each 8 bits), and also has functions such as zooming, and plotting. The graphic display can also attain various processings required for display, such as image smoothing, space filter, and addition/subtraction/multiplication between frame memories and segmentation, in 1/30 to several seconds using an 8- or 16-bit integer image processor of the built-in type. Since this processor has a processing speed from several to several hundred times faster than that of general purpose personal computers, it is able to carry out simple image processing on its own. The host computer and the graphic display are connected by a GP-IB bus (100KB/s) in order to send and receive data and control commands. Image data is digitized by a high-speed image A/D converter connected to the display, if the image signal is an RGB or NTSC type. If it is another type of image signal, it is digitized by a multipurpose high-speed A/D converter connected to the host computer. For example, a TV image can be digitalized in 1/30 of a second, and preprocessing such as noise removal, shading correction, and density level conversion can be executed in real time by the processor built into the display.

The optical disc unit connected to the graphic display is used to store the image data. One optical disc (1.2GB) can store approximately 5000 digital images (512 × 480 × 8 bits). It can write (2.7 s) and read (1.4 s) between the optical disc and the display at high speed. With its tree structure, the directory makes it easy to edit the files for an image data base. Due to remarkable advances in optical disc technologies in recent years, practical and erasable devices will be marketed in the near future. A combination of technologies for optical read-only discs will allow the spread of inexpensive image data bases.

The high-speed image processing system described above is still costly, although computers and graphic displays have become inexpensive. However, 32-bit work stations have recently become popular and inexpensive, and can be purchased for about one million yen. The work station transplanted SPIDER may serve almost all purposes if one is not overly concerned with processing speed and the number of colors displayed. Due to advances in VLSI technologies, single-chip real time signal processors with capacities equivalent to that of array processors have become commercially available. In the near future, high-speed image processing systems will also be available cheaply and will become personal systems in fact as well as in name.

3 TV Spectral Image Instrumentation

Since spectral reflection, transmission, and absorption properties of plants in wavelength from near ultraviolet to near infrared (0.3 to 2 μm) are influenced by such factors as surface or internal structure, type and amounts of plant pigments, and water conditions, they are important as plant information (Gates et al. 1965; Myers 1983). Particularly in the band below 0.8 μm, plants are known to absorb light and emit fluorescence in relation to physiological reactions such as photosynthesis, photomorphogenesis, and stomatal reactions (Trebst and Avron 1977; Kendrick and Kronenberg 1986; Sharkey and Ogawa 1987). This band sensed by human eyes is important for obtaining information on characteristics of plant growth such as shape, community structure, and visible injury. The TV spectral image in-strumentation method is used to obtain plant physiological information or growth characteristics using TV spectral cameras with various optical filters (a special one is a color camera by RGB synthesis), and by using cameras according to the stereo or moire method. Movement of a stoma and morphological or physiological information at the cell level can be obtained by a light microscope with the high-sensitivity camera (See Sect. 4). Techniques for measuring chlorophyll fluorescence transients with information of photosynsetic system will be described in Section 5.

3.1 Types and Features of TV Cameras

TV cameras with image pickup tubes and solid imagers for detecting bands from near ultraviolet to near infrared have been placed on the market. Pickup tube cameras have a variety of spectral sensitivities in the range of 0.2 to 2 μm, resolution, dark current, and after-image. The image distortion, image stability, and shading are automatically corrected by electric circuits in the camera. The solid camera has also been developed as a next generation camera and a substitute for the pickup tube camera. Cameras with solid imagers (sensitivity range; 0.4 to 1.1 μm) such as CCD (charge coupled device) and MOS (metal-oxide-silicon) have recently been mar-keted. Color cameras of broadcasting standard are now becoming available as well, and these solid cameras are expected to be small in size and light in weight, to work at low voltage and low power consumption, to be highly reliable, and to have a long service life. The resolution, however, is usually worse than that of pickup tube cameras. They are also characterized by the lack of image distortion and burning, and the decrease of after-images. For instrumentation under very weak light, the use of a SIT (silicon intensifier target) camera and the mounting of an image intensifier are required.

The household VTR is often used to record color or black-and-white images for image analysis. However, if high image quality and time code accuracy are required, VTRs of broadcasting standard with time base correctors and time code recording functions are more suitable, although they are expensive. Optical disc devices for recording TV images are commercially available, although they are also expensive.

3.2 Spectroradioanalyzer for Field Measurement

The selection of TV camera and optical filters requires a knowledge of the spectral properties of the light environment and subject. A portable spectroradioanalyzer (Omasa et al. 1982) and a standard light source for calibration are shown in Fig. 3.

The analyzer consists of two diffraction grating spectroscopes for scanning a range of 0.250 to 0.900 μm (Ch 1: photomultiplier R636 is used as a detector) in 1 s and a range of 0.850 to 2.50 μm (Ch 2: PbS cell) in 5 s, and a signal processor to control the spectroscope and to analyze the measured data. The detector is electronically cooled to improve the sensitivity and S/N. The wavelength resolution and the stray light are ± 1 nm at Ch 1, ± 3 nm at Ch 2, and 1×10^{-4}. Light is introduced from a condenser or an integrating sphere attachment (Fig. 4) to the spectroscopes via optical fibers. The condenser is used to measure the spectral reflection properties of a subject far away (0.8 to infinity m), and the attachment is used to measure the special reflectance and transmittance by the built-in light source and integrating sphere.

The signal processor is an interactive system with a CRT (cathode ray tube) display for data collection and analysis. The measured data are automatically calibrated based on the verification value of the standard light source and are stored in a magnetic cassette tape. Since the signal processor has functions of continuous addition and collection of data when a subject is measured under very weak light, it can expand its dynamic range for the quantity of light in combination with the selection of slit width for the spectroscope. In addition, the processor has additional

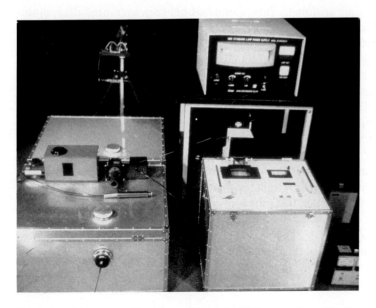

Fig. 3. Portable spectroradioanalyzer and standard light source for calibration

Fig. 4. Attachment with light source and integrating sphere for introducing light to spectroscopes via optical fibers

functions such as four arithmetic rule operations between measured data, multiplication of coefficient, and calculation of energy in a specified band.

Spectral reflectances of the healthy part and the visibly injured part of plant leaves measured using the integrating sphere attachment are shown in Fig. 5. In the healthy part, the reflectance below 0.7 μm becomes smaller due to absorption by plant pigments, and in 1.4 μm (1.9 μm absorption band is impossible to measure) it becomes smaller due to water absorption. However, in the visibly injured part of the leaf (in a dry state), the reflectances of these bands are large. In the non-dried state, the absorption by water in the near infrared band remains. An optical fiber of UV grade is used for Ch 1, and one with no absorption by water in the near infrared is used for Ch 2. Measurements using this attachment are limited to the band between 0.4 and 1.8 μm because of the spectral properties of the light source energy (tungsten halogen lamp), optical fiber transmission, and integrating sphere reflection in addition to sensitivity of detectors. Spectral reflectance and transmittance can be measured within an error margin of about 3%. On the other hand, the condenser allowed the measurement of the band from 0.25 to 2.3 μm.

Reflectance

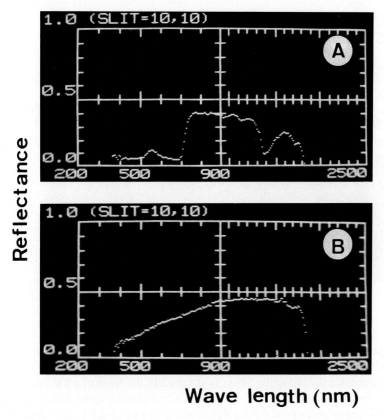

Wave length (nm)

Fig. 5. Spectral reflectances of leaves measured using the integrating sphere attachment. **A** Healthy part; **B** visibly injured part

3.3 Image Instrumentation of Plant Growth and Shape

Plants grow because of cell division and cell elongation. The growth speed differs by location and changes at each growing stage. Therefore, it is important to investigate spatial differences and changes over time in the growth rate of plant organs. The important problem in growth instrumentation is distinguishing the objective organs from the background. Plant reflectance is larger in the near infrared band of 0.8 to 1.3 µm in comparison with that of the background soil (Myers 1983). Therefore, the plant community growing in the soil is easily extracted by near infrared spectral images, although some parts remain hidden because of surface instrumentation. The community growth characteristics such as leaf area, leaf area index, dry weight, and plant height are estimated from calculations of matrix elements of binary images extracted by measuring a community from multiple directions (Matsui and Eguchi 1978). Also, it is possible to estimate, although not very accurate, the characteristics of community growth by measuring the intensity of reflected light from plant community.

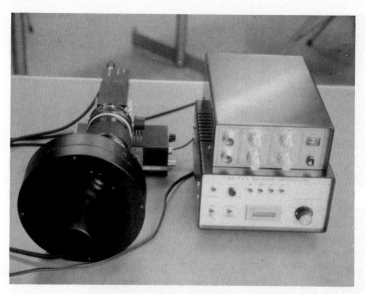

Fig. 6. Multispectral TV camera with 4-band rotary attachment

A multispectral TV camera developed by the author for image instrumentation of plant growth and visible injury is shown in Fig. 6. A rotary attachment with an automatic function for the changing filters is attached to the front of a TV camera, thus allowing spectral images to be made at four different wave bands. The combination of image pickup tube (or solid imager) and optical filters is selected depending on the purpose. For plant instrumentation, a typical combination is a silicon vidicon tube (or a CCD imager) camera with sensitivity in the band of 0.4 to 1.1 μm and interference filters whose central wavelengths are 0.45, 0.55, 0.67, and 0.90 μm (half-bandwidth: 0.01 to 0.03 μm).

Figure 7A shows a spectral image in the near infrared band (0.90 μm) of a sweet potato plant community growing in the field. The use of spectral image allows the separation of the plant community from the background by thresholding (Fig. 7B). The spectral image instrumentation is an effective method for extraction or growth analyses of the leaf, petal, fruit, and other organs.

Although stereo and scanning moire methods using TV cameras permit the measuring of simple three-dimensional shapes, it is impossible to obtain information such as the intricately shaped inside structure of plants and roots in the soil. These subjects must be measured by other methods such as CT.

3.4 Image Instrumentation of Visible Leaf Injury

Necrotic and chlorotic visible leaf injury is caused by the destruction of leaf tissue and the loss of plant pigments. The reflectance of a healthy leaf at wavelengths below 0.7 μm is small, due to absorption by plant pigments such as chlorophylls and

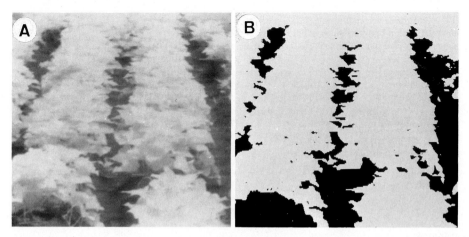

Fig. 7A,B. Spectral image of sweet potato plants growing in the field and their areas extracted by thresholding. **A** Spectral image (0.9 μm); **B** two-valued image; *white* is plant area and *black* is soil area

carotenoids (see Fig. 5). However, in the visibly injured leaf, the reflectance becomes greater, due to the loss of plant pigments. Figure 8 shows two types of typical visible injuries. One features numerous, discrete and unclear injuries, and the other features broad and clear injuries.

Spectral image analysis provides two objective methods for evaluating these visible injuries (Omasa et al. 1983a, 1984). One method uses the average gray level or band ratio obtained from the spectral image as an index for the evaluation. Table 3 shows the correlation coefficients and standard errors which indicate the relationship between the total chlorophyll content, which is a major component of the lost pigments, and the average gray level or band ratio of their spectral images measured through various interference filters (central wavelength 0.45, 0.55, 0.67, 0.78, 0.90 μm with respective half-bandwidth 0.03, 0.01, 0.01, 0.01, 0.01 μm) under constant lighting with a silicon vidicon camera. The correlation is the highest in the case of a band ratio of 0.55/0.90, where the correlation coefficient is –0.95. The standard error in evaluating the total chlorophyll content under this band ratio is 4.2

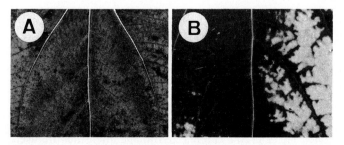

Fig. 8A,B. Two types of typical visible injuries. **A** O_3 injury; **B** SO_2 injury (Omasa et al. 1984)

Table 3. Correlation coefficients and standard errors which indicate the relationship between the total chlorophyll content and the average gray level or band ratio of various spectral images (Omasa et al. 1983a)

Wavelength of image or band ratio	Correlation coefficient	Standard error
0.45 (μm)	-0.91	5.6 (μg cm^{-2})
0.55	-0.89	6.1
0.67	-0.87	6.7
0.78	-0.09	
0.90	0.03	
0.45/0.90	-0.93	5.1
0.55/0.90	-0.95	4.2
0.67/0.90	-0.88	6.4
0.78/0.90	-0.20	
0.45/0.78	-0.91	5.7
0.55/0.78	-0.91	5.8
0.67/0.78	-0.86	6.9
0.45/0.67	0.16	
0.55/0.67	0.60	10.9
0.45/0.55	-0.78	8.5

μ gcm^{-2}. This method is especially effective for the evaluation of numerous, discrete, and unclear injuries, such as O_3 injury.

The other method uses the ratio of the injured area to the leaf area. The most effective wavelength of the interference filter for separating the injured part from the healthy part is 0.67 μm, because the difference between the average gray levels of these parts in the spectral leaf image is the greatest. This method is effective for the evaluation of broad and clear injuries, such as SO_2 injury. The complex injury caused by mixed pollutants, pathogenic fungi, noxious insects, etc., can be evaluated by the combined use of both methods.

Also, since the reflectance in 1.4 μm and 1.9 μm depends on the water content of the plant tissue, the spectral image of these bands provides information relating to the severe drying out of visibly injured leaves (Knipling 1970; Myers 1983). However, a slight change in water content or the water potential of a healthy leaf is not detected by the spectral image.

4 Remote-Control Light Microscope System

In many research fields of biology, the optical light microscope is widely used to observe plant tissues and cells. The use of a TV camera and monitor instead of the naked eye facilitates observation. The use of an image processor for analyzing signals from the camera also makes it possible to evaluate cell growth, the shape of organelles, their color tones, etc. The SIT camera and image intensifier are used effectively to observe them under weak light.

Direct observation of the stomatal movement of attached leaves had been very difficult under the plant's actual growing conditions (Meidner and Mansfield 1968; Meidner 1981). Although the scanning electron microscope (Shiraishi et al. 1978) and the light microscope, in which a piece of leaf is immersed in water or liquid paraffin (Monzi 1939; Stålfelt 1959; Meidner 1981), can provide a clear image at high magnification, observation of intact stomata under their growing conditions is impossible. Observation with an ordinary light microscope under a plant's growing conditions poses some problems (Heath 1959; Meidner and Mansfield 1968); first, visual observation under weak light is very difficult; second, the environment of the lower side of the leaf cannot be controlled, because the leaf is directly held on the microscope stage, and the environment is also affected by human manipulation of the microscope; and third, the working distance, that is, the distance between the leaf and the objective during the observation, is very small at high magnification, thus subjecting the leaf to the danger of sticking to the objective during focusing and destroying the environment between the leaf and the objective. Omasa et al. (1983b) and Kappen et al. (1987) have solved the above problems of the ordinary light microscope. As an example of a light microscope system coupled to an image processor, our system for the measurement of stomatal movements of attached leaves under actual growing conditions is introduced in this section.

4.1 Outline of the System and Its Performance

Figure 9 shows our remote-control light microscope system. This system has a light microscope (Bausch & Lomb, MicroZoom) with a wide working distance (ca. 13 mm) at high magnification (a 50x objective, 1.5x and 2x amplifiers and a TV adapter lens; ca. 1600-fold magnification on a TV monitor), an SIT video camera (Hamamatsu TV, Model C1000-12) with high sensitivity S20 type spectral response, image resolution of over 600 TV lines, distortion within 2% and shading within 20% as a detector of the microscope image, a monochromatic TV monitor (Chuomusen, Model MD2002A) with image resolution of ca. 1000 TV lines and distortion within 3% and remote controllers for adjusting camera sensitivity, a microscope focus, and movement of the microscope stage in a separate room. The microscope images are projected on the TV monitor in a separate room and recorded photographically on black-and-white (B & W) film or by a VTR (SONY, BVU820) (horizontal resolution, 340 TV lines at B & W mode; S/N, ca. 50dB) with a digital time base corrector (SONY, BVT800) and a time code generator/reader. The image processing for evaluating stomatal aperture and cell injury is carried out by a system composed of a high-speed video processor, graphic displays, and a host computer (See Sect. 2.2).

Figure 10 shows a schematic cross-sectional view of the microscope stage designed to hold the leaf of an intact plant. The leaf (C) is held on a ring (F, 30 mm in inner diameter, 10 mm wide and 10 mm high) fixed to a remote-control movable stage (G) by a holding ring (E, the same diameter and width as F, 3 mm in height), in order to have the conditioned air pass under the surface of the leaf. Furthermore, since the center of the movable stage is cut to a circle 30 mm in diameter, and the distance between the movable stage and plate (H) fixed on base (K) is kept at 10 mm,

Environment-controlled chamber

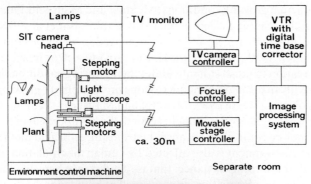

Fig. 9. Remote-control light microscope system (After Omasa et al. 1983b)

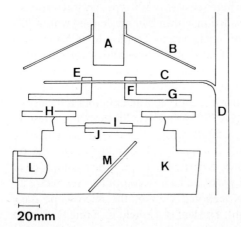

20mm

Fig. 10. Schematic cross-sectional view of the microscope stage for holding an intact leaf. *A* Objective; *B* shade cover; *C* leaf; *D* stem; *E* holding ring; *F* ring fixed to remote-control movable stage; *G* remote-control movable stage; *H* plate; *I* heat-absorbing glass filter; *J* diffusing filter; *K* base; *L* halogen lamp; *M* mirror (Omasa et al. 1983b)

the same temperature and humidity can be maintained on both sides of the leaf. The movable stage and the plate are made of transparent acrylic resin, to allow light from the environment to enter. The shade cover (B) is not used except for observation with transmitted light. Although observation is usually carried out with light from the environment, a halogen lamp (L) is sometimes used as a supplementary light source for observation with transmitted light.

Figure 11 shows photomicrographs of an intact stoma observed with reflected and transmitted light using the remote-control light microscope system. The stomatal image was clear at high magnification (ca. 1600–fold magnification on the TV monitor). The stoma was observed with reflected light and then rapidly observed with transmitted light, and the stomatal aperture was found to be the same for both. Although this system could provide stomatal images with a mixture of reflected and transmitted lights, the clearest images were obtained with either reflected or transmitted light alone. The stomata could be observed with either reflected or transmitted light above ca. 0.1 mWcm^{-2} [environment illumination; ca. 2 klx (0.5 mWcm^{-2}) with reflected light, ca. 0.5 klx (0.5 mWcm^{-2}) with transmitted light]. If the observation with a single light of 0.1 mWcm^{-2} is done with the naked eye through the eyepiece instead of the SIT camera, the eye must be sufficiently acclimatized to the dark room. The attachment of an image intensifier to the SIT camera produces a further increase in sensitivity, although the image quality becomes poor. Although the resolution of the microscope image was within 1 μm, it is improved by a digital image processing technique. Omasa and Onoe (1984) were able to evaluate the stomatal aperture within 0.3 μm standard error using this technique, even when the microscope image was of poor quality.

4.2 Continuous Observation of Guard and Epidermal Cells

When the light microscope system was used for continuous observation of the stomatal movement of an intact growing plant, we obtained photomicrographs

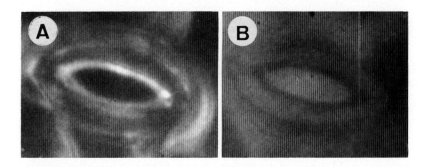

20 µm

Fig. 11A,B. Photomicrographs of an intact sunflower stoma observed with reflected or transmitted light using the light microscope system. **A** Reflection image; **B** transmission image (Omasa et al. 1983b)

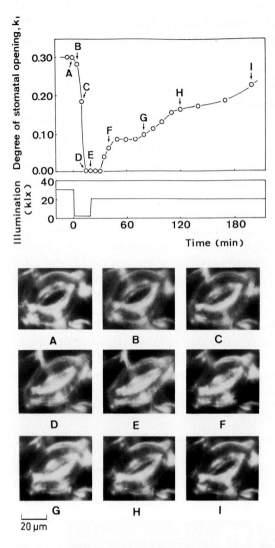

Fig. 12. Responses of an intact stoma of the adaxial epidermis of a broad bean plant to illumination changes; k_1 is the degree of stomatal opening expressed by the ratio l_a/l_{bmax}, where l_a is the width of the stomatal pore and l_{bmax} is the maximum length of the fully opened stomatal porem Photomicrographs (**A–I**) correspond to time points (**A–I**) in the k_1 change. The illumination was changed from 30 to 2 klx at 0 min (**A**) and from 2 to 20 klx at 20 min (**E**). Other environmental conditions: air temperature, 20.0°C; RH, 70% (Omasa et al. 1983b)

similar to those in Fig. 12, which show the response of an intact stoma of the adaxial epidermis of a broad bean plant to an illumination change. The illumination was changed from 30 klx (11.9 mWcm⁻²) to 2 klx (0.5 mWcm⁻²) at 0 min (A) and from 2 to 20 klx (7.7 mWcm⁻²) at 20 min (E). The movement of the central pore of the stoma could be continuously observed; the degree of opening (k_1) of the stomatal

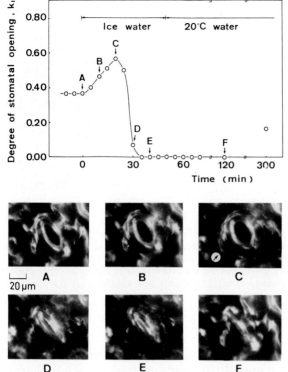

Fig. 13. Responses of an intact sunflower stoma to water deficit. The water deficit was caused by ice water perfusion to roots. Photomicrographs (**A-F**) correspond to time points (**A-F**) in k_1 change. The *arrow* (↗) in **C** shows first hollow of subsidiary cell. Environmental conditions: air temperature, 25.0°C; RH, 60%; light intensity, 600 μmol photons m^{-2}s^{-1} (Omasa and Maruyama 1990)

pore is expressed by the ratio l_a/l_{bmax}, where l_a is the width of the stomatal pore and l_{bmax} is the length of the fully opened stomatal pore. The stoma began to close within 5 min (B) of lowering the illumination (30 to 2 klx) and became completely closed after ca. 15 min (D). It began to reopen within 15 min of raising the illumination (2 to 20 klx), and after 180 min (I), had recovered to ca. 75% of aperture before the illumination change.

Figure 13 shows changes in an intact sunflower stoma as a result of water deficit caused by ice water perfusion to its roots. The stoma increased its aperture within a few minutes of perfusion and reached a maximum opening at 20 min (C). Thereafter, it began to close and reached complete closure at 35 min (E). Subsidiary cells observed in the photomicrographs began to become hollow at ca. 20 min (C) and then the hollow expanded from the epidermal cell to the guard cell (C to E). The transient opening from "A" to "C", therefore, may be caused by the rapid decrease in subsidiary cell turgor in comparison to guard cell turgor, due to a decrease in water uptake from the root. After 20°C water perfusion, the cell form slowly recovered (F) and the stoma began to reopen ca. 3 h later.

Figure 14 shows the varying responses of the stomata to 1.5 μ 1 l^{-1} SO$_2$ at the border of the injured region near a veinlet. These stomata continued to maintain constant apertures until about 15 min after the start of the exposure, when a wide variety of stomatal responses began. Stomata in area I contiguous to the veinlet

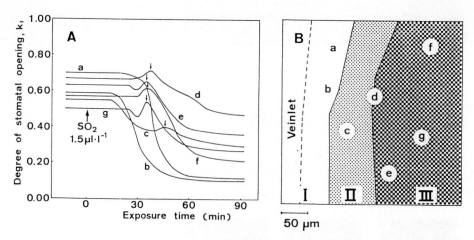

Fig. 14. Responses of neighboring stomata to 1.5 μl l^{-1} SO_2 at the border of an injured region near a veinlet. **A** Stomatal responses; *small arrows* (↓) show when water soaking and cell collapse began to appear. **B** development of injury at the region; *I* uninjured region; *II* injured region where only cell collapse appeared, without water soaking; *III* injured region where both water soaking and cell collapse appeared. Environmental conditions: air temperature, 25.0°C; RH, 60%; light intensity, 600 μ mol m^{-2} s^{-1} (Omasa et al. 1985a)

showed rapid and continuous closure, and the guard and epidermal cells maintained normal turgor. Stomata in area III, at a distance from the veinlet, had a transient opening after either the closing or the keeping of a constant aperture, because of the rapid decrease in subsidiary cell turgor in comparison to guard cell turgor, caused by the appearance of water-soaking and cell collapse in the epidermal cells. The stomata then closed as the injury expanded to all cells. In area II, cell collapse occurred gradually without water-soaking from the cells near area III by the turgor loss of cells in area III. Therefore, the appearance of a transient stomatal opening in area II was later than that in area III.

Porometry or gravimetric measurements of transpiration do not provide the real aperture of stomata and the information for cell form and surface conditions (Omasa et al. 1985a). They also involve the risk of losing important information about the stomatal response, because they average the behavior of many stomata. Therefore, it is important to observe the individual stomata and their neighboring cells directly in order to examine the stomatal movement. Also, since this system is effective for observing many intact stomata because of its easy and rapid operation, it can be used to analyze the relationship between stomatal aperture and conductance under the plant's growing conditions.

5 Image Instrumentation of Chlorophyll Fluorescence Transients

Abiotic and biotic stresses such as air pollutants, water deficit, high or low temperature, and virus infection cause a spatially heterogeneous impairment in attached leaves. The invisible impairment was indicated in stomatal responses and photosynthetic activity. For example, recent investigations employing thermal imaging methods (See Sect. 6) and DLE (delayed light emmission) imaging methods (Ellenson and Raba 1983; Ellenson 1985) have shown evidence of spatially different responses of stomata in situ to various stresses. The DLE imaging method, furthermore, clarified localized changes in photosynthetic activity induced by the stresses (Björn and Forsberg 1979; Ellenson and Amundson 1982). However, these techniques did not provide any information about the site of inhibition in the photosynthetic apparatus.

Rapid changes in intensity of chlorophyll *a* fluorescence during dark-light transition (CFI) reflect the various reactions of photosynthesis (Kautsky et al. 1960; Murata et al. 1966; Papageorgiou 1975), especially the photosynthetic electron transport system. Therefore, the measurement and analysis of CFI in plant leaves in situ has been developed as a sensitive and nondestructive assay for the functional state of the photosynthetic apparatus (Smillie and Hetherington 1983; Shimazaki et al. 1984; Sivak and Walker 1985). Recently, Omasa et al. (1987) developed a new instrumentation system using a CCD image sensor for a quantitative analysis of CFI, the system of which would give information not only about localized differences in photynthetic activity on the whole leaf in situ but also about the inhibition site in the photosynthetic system. In this section, the system is introduced.

5.1 Outline of the System and Its Performance

Ordinary TV cameras and recording systems are not suitable for a quantitative analysis of CFI in attached leaves, because of their low sensitivity, large after-image, bad image quality, AGC (automatic gain control) function, and the indistinctness in timing of the playback image. Common tungsten and fluorescent lamps also cannot be used as light sources for CFI imaging, because of the unevenness and fluctuation in light intensity. The new image instrumentation system (Fig. 15) was designed to overcome these problems. A highly sensitive CCD imager with uniformity in sensitivity and an after-image suppression was selected for a TV camera (SONY XC-47, improved type). The image quality was improved by the use of a computer-control VTR (SONY BVU-820) with a digital time base corrector (SONY BVT-800) and the preprocessing by using a high-speed TV image processor (KCR nexus 6800). The AGC function, which changes a relationship between fluorescence intensity and the gray level of the recorded image, was removed from the TV camera and the VTR. The timing of the playback image was exactly defined by the use of a shutter synchronized to the TV signal and by the time code recorded in each image. The unevenness in light source intensity was improved by the use of two xenon lamp (CERMAX LX-300F) projectors, and by attaching a special ND filter with concentric circles of different density.

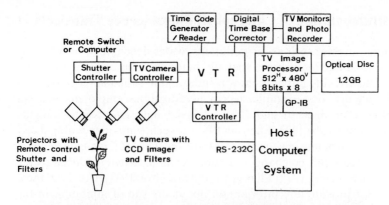

Fig. 15. Image instrumentation system for quantitative analysis of CFI in attached leaves (Omasa et al. 1987)

After the plant had adapted to the dark for 30 min, the CFI was provoked by irradiation of the whole leaf with two beams of blue-green light (380–620 nm) from the projectors, with band-pass filters via the shutter opening. The fluorescence image was continuously measured at a TV field interval of 1/60 s by the TV camera equipped with an interference filter (683 nm; half-band width, 10 nm) and a red cut-off filter (> 650 nm), and recorded, with time code, by the VTR. The VTR image was digitized by a video A/D converter after it was played back to a still image without guard band noise through the time base corrector. A series of the digitized images (512H × 480V 8 bits) was stored on an optical disc (National DU-15). A host computer system (MITSUBISHI MELCOM 70/40 and NEC PC9801 vm4) was used to control the VTR and TV image processor.

The CFI curves, intensity images of characteristic transient levels (I,D,P,S,M,T), and amplitude images of major transient characteristics (ID, DP, PS, MT) were calculated by the TV image processor on the basis of a series of images after preprocessing for shading correction and noise removal. These were indicated by scales which correspond to the A/D conversion level.

Figure 16 shows the relationship between the intensity of artificial light in the likeness of fluorescence and the A/D conversion level of VTR image. The data indicated a linear correlation between the light intensity and the A/D conversion level. Since the artificial light was sufficiently diffused, and the unevenness of the light intensity in the visual field of the TV camera was maintained within 0.1%, the standard deviation of 6.3% (maximum) in the A/D conversion level was due to image shading caused by unevenness in sensitivity of the TV camera and any noise other than VTR guard band noise. The image shading was corrected by calculating the ratio of an original image to a specific image (shading master), obtained by measuring a uniform light of definite intensity, because the shading was mainly caused by the lens and optical filters of the TV camera. The noise was removed by the use of a spatial smoothing filter, and the averaging of images digitized from a still VTR image. For example, the image quality could be improved to within 1% standard deviation by the shading correction and the smoothing of 3 × 3 pixels after an averaging of 10 images, when the image resolution was 280 lines. The after-image for the TV camera was about 4% at 30 ms after shutter opening, and decreased to 0.3% at 50 ms.

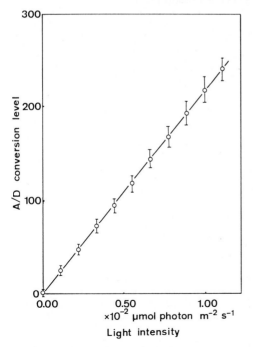

Fig. 16. Relationship between the intensity of artificial light in the likeness of fluorescence and the A/D conversion level of VTR image. Symbol ○ represents the mean value of A/D conversion levels of an image (512 × 480 pixels) measured at a light intensity and the *vertical bar* indicates ±SD. The artificial light was sufficiently diffused and the unevenness of the light intensity in the visual field of the TV camera was maintained within 0.1%. The light intensity was measured by a quantum sensor through two filters placed in front of the TV camera (Omasa et al. 1987)

The CFI from a defined part of a cucumber leaf in situ was measured by our system under different intensities of actinic blue-green light. The CFI curves were calculated on the basis of a series of fluorescence images (Fig. 17). Those clearly revealed the typical IDPSMT transients (Papageorgiou 1975) under light intensities from 50 to 200 μ mol photons m^{-2}s^{-1} at leaf surface. In these intensity ranges of light, we could resolve IDPSMT transients from any part of the leaf surface which has an area of at least 1 mm^2. Fluorescence intensities of I,D,P,S,M, and T, and rates in transients of DP, PS, and MT increased as the actinic light intensity increased. The appearance of peak P became more rapid with the increase of actinic light intensity.

Because the CFI is light intensity dependent, as described above, it is important to illuminate the whole leaf with uniform light when we want to compare CFI curves derived from different areas (each about 1 mm^2) of the leaf to the entire leaf. The combination of special ND filters and two projector lamps, which were placed at angles of 60° and 120° to the leaf surface, overcame this problem. The combination kept the spatial deviation of the intensity of actinic light within 5% over a flat surface of 20 cm in diameter.

Ellenson (1985) has reported a video recording system for DLE analysis, but its performance was not characterized. In general, the quality of DLE image is worse than that of chlorophyll fluorescence image, because the image intensifier was used to image DLE. The irradiation of a leaf with a tungsten projector lamp placed at an oblique angle to the leaf surface (Ellenson and Raba 1983) causes spatial differences in the intensity of actinic light. Moreover, it should be noted that the AGC function of the ordinary TV camera and VTR system changes a linear correlation between the intensity of DLE or CFI and the A/D conversion level.

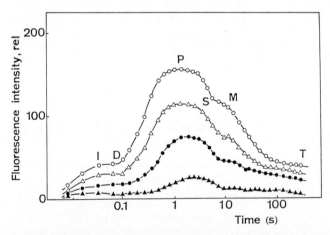

Fig. 17. The CFI curves of a small area (about 1 mm^2) of a healthy cucumber leaf in situ measured under different intensities of actinic blue-green light. Before measurement, the leaf was dark-adapted for 30 min. Intensity of actinic light (unit = μ mol photons m^{-2} s^{-1}) : (○) 200, (△) 150, (●) 100, and (▲) 50. Light intensities were varied using ND filters. Environmental conditions: air temperature, 25.0°C; RH, 70% (After Omasa et al. 1987)

5.2 Diagnosis of Photosynthetic System

The CFI image instrumentation system was used to diagnose the SO_2 effect on the photosynthetic apparatus (Omasa et al. 1987). Figure 18 shows the CFI curves and the images with a gray scale. In the unfumigated leaf area (UF), the CFI clearly showed the typical IDPSMT transients (Papageorgiou 1975) and almost identical transients at any location on the leaf area. Since the CFI observed upon dark-light transition of the leaf reflects the partial reactions of photosynthesis, we can detect the alteration in photosynthetic apparatus by SO_2 from the changes in CFI curves. As shown in Fig. 18, B and C, both the intensity images of characteristic transient levels (I, P, M, T) and amplitude images of major transient characteristics (ID, DP, PS, MT) in the fumigated counterpart (F) strikingly differed from those in the un-fumigated area. Fluorescence intensity at I was raised and at P reduced markedly, and that at T was increased in the fumigated area. Amplitude of fluorescence transients of DP-rise and PS- and MT-decline, indicating photosynthetic activity, was reduced in the fumigated leaf. The changes in the intensity and amplitude varied with the location on the leaf surface; the effect of SO_2 was more severe in locations of interveins and veinlets than in those near large veins. Contrary to the perturbation in photosynthetic apparatus shown above, there was no visible injury on the whole surface of leaf at the end of SO_2 treatment and 2 days later.

The significance of the changes in CFI induced by SO_2 fumigation was as follows: Because fluorescence intensity in the early induction phenomena is regulated by the redox state of Q, a primary electron acceptor of PSII, the elevated I level suggest that some portion of Q was brought to a reduced state by the SO_2-fumigation. Since the DP rise in CFI reflects the photoreduction of Q through reductant from H_2O, a diminished rise of DP was consistent with the inactivation of the water-splitting enzyme system. Since PS decline involves energy-dependent quenching, the suppression of PS decline suggested the depression of formation of trans-thylakoid proton gradient, probably due to the inactivation of the water-splitting enzyme system. However, the possibility that the PS decline was affected by the inhibition of electron flow Q to PSI cannot be excluded, because PS decline partly reflects the oxidation of Q by PSI. Suppression of MT decline was probably due to the inhibition of the trans-thylakoid proton gradient formation in addition to unidentified reactions in chloroplasts. Although the extent of the SO_2-effect on CFI differed from area to area on a single leaf, the mode of SO_2 action was essentially the same.

The recovery from perturbation by SO_2 became evident by monitoring the CFI when the fumigated plants were transferred to SO_2-free air (Fig. 19). Adjacent to the large midvein, where the alteration of CFI was relatively small, the CFI recovered completely and showed the IDPSMT transient identical to that of the unfumigated area. In most of the area near veinlets, the CFI which was affected strongly was also almost completely restored. However, the fluorescence emitted from the interveinal area was still affected during this period; the peak P reappeared but its intensity was still low, and the I level was elevated further. The elevated I level suggests that the irreversible injury to the PSII reaction center took place in chloroplasts. In contrast, the elevated T level in the quasi-stationary state became normal in this leaf area.

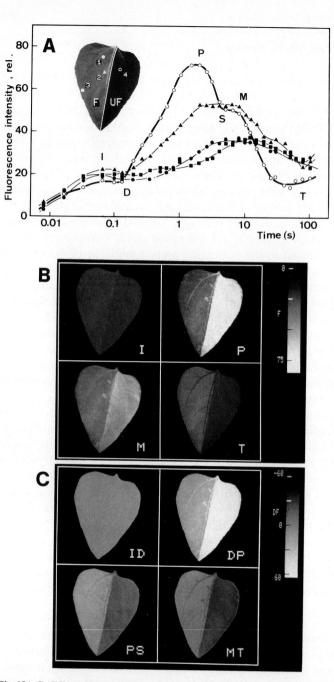

Fig. 18A-C. Effect of SO_2 on CFI in an attached sunflower leaf. Sunflower plant was fumigated with 62.4 μ mol SO_2 m^{-3} (1.5 μ l l^{-1}) at 25.0 °C air temperature, 70% RH, and 350 μ mol photons m^{-2}s^{-1} light intensity for 30 min. After dark-adaptation for 30 min, CFI of an attached leaf was measured under 125 μmol photons m^{-2}s^{-1} actinic light. **A** CFI curves at different sites in fumigated area (F: ● interveinal site 1; ▲ site 2 near a large vein; ■ site 3 near a veinlet) and unfumigated area (UF: ○ interveinal site 4). The vein and corresponding sites ($1–4$) are denoted in a photograph of the whole leaf. **B** Intensity images of characteristic transient levels (I, P, M, T). **C** Amplitude images of major transient characteristics (ID, DP, PS, MT) (Omasa et al. 1987)

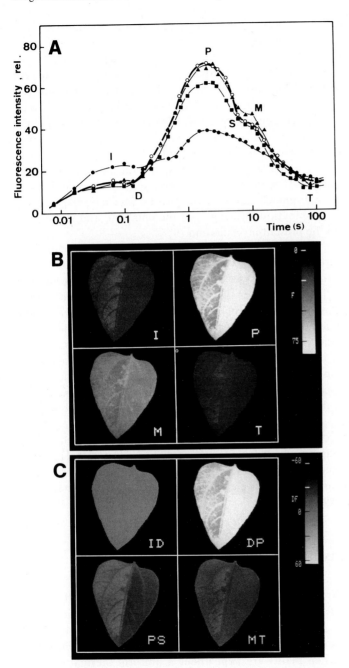

Fig. 19A-C. Recovery of CFI from disturbances by SO_2. After the measurement described in Fig. 18, the same plant was kept at 25.0°C air temperature and 70% RH in darkness and SO_2-free air. Measurement of CFI was then repeated for the same leaf every 60 min. The CFI measured at 6 h after the initiation of SO_2 fumigation are presented. **A** CFI curves at leaf sites marked in Fig. 18A; **B** intensity images of characteristic transient levels; **C** amplitude images of major transient characteristics. *Symbols* are the same as those in Fig. 18 (Omasa et al. 1987)

In general, the leaf area close to the veinal region was resistant to SO_2-fumigation. The resistance is probably due to the smaller absorption of SO_2 in the proximity of veins, because of the small stomatal conductance in these area and/or to the fast dilution of absorbed SO_2 by the larger provision of water from the vein (Omasa et al. 1981b; 1985a). Therefore, we observed that SO_2 injured the leaf heterogeneously, and the recovery from the SO_2 injury also proceeded heterogeneously. Although the heterogeneous response of a plant leaf to environmental stress is a common process microscopically, the conventional method (Schreiber et al. 1975; Schreiber 1983), which measures the 'averaged' CFI derived from a defined leaf area, could not resolve a spatially different response in the same leaf. For example, Shimazaki et al. (1984) described the partial recovery of O_2 evolution and CFI from the SO_2 injury in a spinach leaf macroscopically. Such a partial recovery is probably due to the combination ('average') of the complete and incomplete recovery, as indicated in the present investigation.

As shown above, our image instrumentation system demonstrated not only the affected location in photosynthetic activity on the leaf but also the inhibition site in the photosynthetic electron transport system. The phytoluminographic technique, which measures the DLE spatially over the single whole leaf, was developed to detect injury to the photosynthetic activity (Björn and Forsberg 1979; Ellenson and Amundson 1982). This method detects the injured area on the leaf in photosynthetic activity but does not reveal the inhibition site in the photosynthetic system, because the intensity of the DLE is very similarly reduced by various factors; these include inhibition of the oxidizing and reducing sides of PSII, uncoupling of photophosphorylation, anaerobic conditions, and high CO_2 conditions (Satoh and Katoh 1983). On the contrary, the CFI imaging points to SO_2 fumigation, inhibiting the oxidizing side of PSII.

Together with the techniques of computerized image processing, such as DLE imaging and thermal imaging, which provide information about stomatal response and gas exchange, the image analysis of CFI provides an integrative approach for early warning diagnosis and for functional analysis of disorders during stress, as well as for the plant's capacity to recover.

6 Thermal Image Instrumentation

The rate of physiological reactions in living plants is dependent upon cell temperature (Sutcliffe 1977, Berry and Björkman 1980, Sakai and Larcher 1987), which, in turn, is determined by the conditions of the thermal environment (i.e., air temperature, humidity, radiation, and air currents) (Monteith 1973, Jones 1983). As such, when plants are kept under a uniform thermal environment, leaf temperature can indirectly provide physiological information on stomatal movement, transpiration, CO_2 uptake, and air pollution absorption (Hashimoto et al. 1984, Omasa and Aiga 1987). Thermal image instrumentation is a method for obtaining plant temperatures and the physiological information involved in it, by measuring the electromagnetic waves of the thermal infrared band (Omasa et al. 1980, 1981a,b,c; Horler et al. 1980; Hashimoto et al. 1984).

6.1 Method for Measuring Plant Temperature and Its Accuracy

Thermal infrared radiation from the environment (approximately 10 μm) is almost all absorbed by the plant. However, since thermal infrared radiation is emitted from the plant according to Plank's law of radiation, the plant temperature can be obtained by measuring the radiation in this band.

For a perfectly diffuse and opaque plant surface the spectral energy of infrared radiation from the surface at temperature T is $R(\lambda,T,T_s)$, given by:

$$R(\lambda,T,T_s) = \varepsilon(\lambda,T)W(\lambda,T) + [(1-\varepsilon(\lambda,T)] E(\lambda,T_s), \tag{1}$$

where λ is the wavelength, $\varepsilon(\lambda,T)$ is the spectral emissivity of the plant surface, $W(\lambda,T)$ is the spectral radiant energy of a black body at temperature T, and $E(\lambda,T_s)$ is the spectral radiant energy from the environment at temperature T_s to the plant surface.

When $R(\lambda,T,T_s)$ is measured by an infrared detector with spectral sensitivity in the band from λ_1 to λ_2, the output voltage $V_T(T,T_s)$ of the detector is:

$$V_T(T,T_s) = \int_{\lambda_1}^{\lambda_2} f(\lambda)R(\lambda,T,T_s)d\lambda$$
$$\simeq \bar{\varepsilon}(T)V_w(T) + [1-\bar{\varepsilon}(T)] V_E(T_s), \tag{2}$$

where

$$\bar{\varepsilon}(T) = [\int_{\lambda_1}^{\lambda_2} \varepsilon(\lambda,T)f(\lambda)W(\lambda,T)d\lambda]/[\int_{\lambda_1}^{\lambda_2} f(\lambda)W(\lambda,T)d\lambda],$$
$$V_w(T) = \int_{\lambda_1}^{\lambda_2} f(\lambda)W(\lambda,T)d\lambda,$$

and

$$V_E(T_s) = \int_{\lambda_1}^{\lambda_2} f(\lambda)E(\lambda,T_s)d\lambda,$$

since $f(\lambda)$ is the function describing the radiation-electricity conversion of the detector, the amplification with the amplifier, and the transmission and reflection of the air, lens, filter, etc.; $\bar{\varepsilon}(T)$ is the average emissivity in the band from λ_1 to λ_2; $V_w(T)$ is the output voltage in the measurement of $W(\lambda,T)$, and $V_E(T_s)$ is the output voltage in the measurement of $E(\lambda,T_s)$.

When $\bar{\varepsilon}(T)$ and $V_E(T_s)$ are given, $V_w(T)$ can be obtained from $V_T(T,T_s)$:

$$V_w(T) = [V_T(T,T_s)-V_E(T_s)]/\bar{\varepsilon}(T) + V_E(T_s). \tag{3}$$

Furthermore, the plant temperature T is evaluated from $V_w(T)$ using the characteristic relation between T and $V_w(T)$.

The emissivity $\bar{\varepsilon}(T)$ of the plant in the spectral sensitivity range (8–13 μm) of the thermal camera is 0.95 to 0.99 (Fuchs and Tanner 1966; Omasa et al. 1980). The influence of radiation from the environment is corrected by adjusting $V_E(T_s)$ in Eq. (3). The influence of the change in $f(\lambda)$ is also corrected by continuously monitoring a standard blackbody source within the camera. As a result, it is possible to measure the plant temperature with an accuracy of $\pm 0.1°C$ (Omasa et al. 1980).

A thermal camera of the optical-mechanical scanning type, called a thermo-graphy, with a InSb (3.5–5.5 μm) or HgCdTe (8–13 μm) detector, is now on the market. To use for plant measurements, the scanning type of mirror vibration, which has good temperature sensitivity and resolution, is better than that of prism rotation. Since the detector is cooled by liquid nitrogen (77K), it is necessary to supplement it every few hours. At present, the detector with an electronic cooler (200K) and the new image sensor of the electrical scanning type cannot be used to accurately measure plant temperature because of insufficient sensitivity and resolution.

Figure 20 shows our thermal image instrumentation system and blackbody source for calibration. The thermal camera (JEOL JTG-IBL) is an optical-mech-anical scanning type, and its detector is an HgCdTe (8–13 μm, cooled by liquid nitrogen). The detected signals from the thermal camera are converted into 12-bit digital signals (256H×240V, quantization error 0.0125°C) by a thermal image processor and analyzed by a host computer. The detected image is enhanced by integrating, using the thermal image processor. The temperature resolving power, the uniformity of image, and the drift is within 0.05°C, ± 0.1°C and ± 0.05°C/ 4 h respectively. The image resolution is ca. 50H × 40V in measuring at 4% error (0.2°C) (Omasa et al. 1981a). Emissivity of the blackbody source (Electro Op-tical Industries PD1401X) is 0.99, and the surface temperature is automatically controlled at 0.05 °C accuracy.

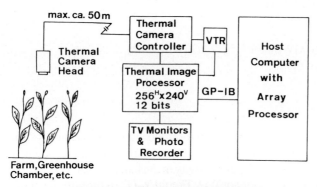

Fig. 20. Thermal image instrumentation system and blackbody source for calibration

6.2 Evaluation of Stomatal Response, Transpiration, and Gas Sorption

,Plants and the surrounding atmosphere exchange CO_2 and water vapor, which are related to photosynthesis, respiration, and transpiration, through the stomata. In atmospherically polluted areas, toxic air pollutants also enter the plants through the stomata, and have various effects.

By analyzing relationships between leaf temperature and thermal environment factors such as air temperature, humidity, radiation, and air currents, from the standpoint of the plant-environment system, the spatial distributions of the transpiration rate, stomatal resistance to water vapor-diffusion (which is an indicator of stomatal aperture), and gas absorption are exactly evaluated from the leaf temperature (Omasa et al. 1981a,b,c). In the case of evaluating them at local sites over a thin flat leaf, the model is simplified as follows:

The transpiration rate W_x $(gcm^{-2}s^{-1})$ at local site x on the leaf is:

$$W_x = [\alpha_p E_{sx} + \varepsilon \{E_{wx} - 2\sigma (273.15 + Tl_x)^4\} + 2\rho C_p (T_a - T_{1x})/r_{kax}]/L, \qquad (4)$$

where E_s is shortwave radiation from the environment (wavelength $< 3\mu m$, $calcm^{-2}s^{-1}$), E_w is longwave radiation from the environment (wavelength $> 3 \mu m$, $calcm^{-2}s^{-1}$), α_p is absorption coefficient of shortwave radiation of the leaf, ε is emissivity of longwave radiation of the leaf, T_1 is leaf temperature $(°C)$; T_a is air temperature $(°C)$, σ is Stefan-Boltzmann constant $(calcm^{-2}s^{-1}K^{-4})$, ρC_p is volumetric heat capacity of air $(calcm^{-3} °C^{-1})$, r_{ka} is boundary layer resistance to heat transfer, and L is latent heat by evaporation $(calg^{-1})$. The subscript x denotes the values at local site x on the leaf. Assuming that thermal environment factors such as air temperature, humidity, radiation, and air current are kept constant all over the leaf, the only variable in the right side of Eq. (4) is leaf temperature T_{1x}. Therefore, by previously determining the parameters other than T_{1x}, the transpiration rate W_x is evaluated from T_{1x}, measured with the thermal instrumentation system. The stomatal resistance to water vapor diffusion, r_{wax} (scm^{-1}), is furthermore expressed by:

$$r_{wax} = 2\{X_{sx}(T_{1x}) - \phi X_s(T_a)\}/W_x - (\kappa/D_w)^{2/3}r_{kax}, \qquad (5)$$

where $X_s(T)$ is saturated water vapor density at $T °C$, ϕ is relative humidity, κ is thermal diffusivity of air, and D_w is air-water vapor diffusivity. The absorption rate Q_x $(gcm^{-2}s^{-1})$ of CO_2 or air pollutants at a local site x is evaluated by the following equation:

$$Q_x = 2(P_a - P_{1x})/(r_{gax} + r_{gsx}), \qquad (6)$$

where

$$r_{gax} = (\kappa/D_g)^{2/3}r_{kax}$$
$$r_{gsx} = (D_w/D_g)r_{wsx},$$

since P_a is atmospheric gas concentration (gcm^{-3}); P_1 is gas concentration at the gas-liquid interface in the substomatal cavity (gcm^{-3}), r_{ga} is boundary layer resistance to gas diffusion (scm^{-1}), r_{gs} is stomatal resistance to gas diffusion (scm^{-1}), and

D_g is air-gas diffusivity (cm^2s^{-1}). The gas concentration P_{1x} of SO_2, NO_2, O_3 and PAN (major air pollutants) at the gas-liquid interface in the sub-stomatal cavity has been assumed to be zero μ 11^{-1}, because physiological functions such as metabolism and transfer which reduce the gas concentration, are considered sufficient (Omasa 1979; Omasa et al. 1979). However, the CO_2 concentration concerned with photosynthesis and respiration is changed by conditions of environment and growth (Omasa 1979; Jones 1983).

Figure 21 shows changes in the spatial distributions of stomatal resistance to water vapor diffusion, transpiration rate, and O_3 absorption rate evaluated from leaf temperature images of a sunflower plant during exposure to 1.2 μ 11^{-1}. During the O_3 exposure, the stomatal resistance increased, and the transpiration rate decreased, because of stomatal closure. However, their behaviors varied randomly at different sites on the leaf. This result means that stomatal sensitivity to O_3 varies at the local site of the leaf. By this method, the transpiration rate, stomatal resistance, and gas absorption rate are evaluated with errors of $0.02 \times 10^{-5} gcm^{-2}s^{-1}$, 0.3 scm^{-1}, and ca. 10% respectively. (Omasa et al. 1981a,b,c).

It is very difficult to evaluate exactly the stomatal resistance and transpiration rate of plants growing in the field. However, since stomatal closure occurs before the appearance of visible leaf injury, thermal image instrumentation can be used for the early detection of plant stress under steady-state thermal environments. Figure 22 demonstrates the diagnosis of some zelkova trees on Aoba street in Sendai city. A tree on the left was high in leaf temperature because of stomatal closure caused by gasoline and water stress, although damage was not visible. The use of both thermal image instrumentation and portable porometer (e.g., LI-COR, Model LI-1600) makes a precise diagnosis of field plants possible.

The leaf temperature image is also used for screening plants. In Fig. 23, the leaf temperature of a poplar tree was lower than that of a white oak and a spindle tree. This indicates that the poplar tree has large transpiration and gas absorption rates, and potential for growth and air purification.

7 Computed Tomography

Computed tomography (CT), first developed in the medical field and currently undergoing remarkable further development, is an effective method for investigating conditions in a living organism without destroying it (Herman 1979; Onoe 1982; Mansfield and Morris 1982). Cormack and Hounsfield were awarded the Nobel Prize in Physiology and Medicine in 1979 for developing the X-ray CT (Hounsfield 1973) which is now popularly used in clinical medicine and in industry. In addition to X-ray CT, CTs using radioisotopes (RI), including positron CT, and CTs using an energy medium such as ultrasonic waves, nuclear magnetic resonance (NMR or MR), heavy particle beams or microwaves are currently being studied, partly for practical use. Methods for investigating objects at the cell level are now being developed. In the plant field, investigations of annual rings and internal rot in living trees by X-ray CT, and studies of root systems and water absorption of

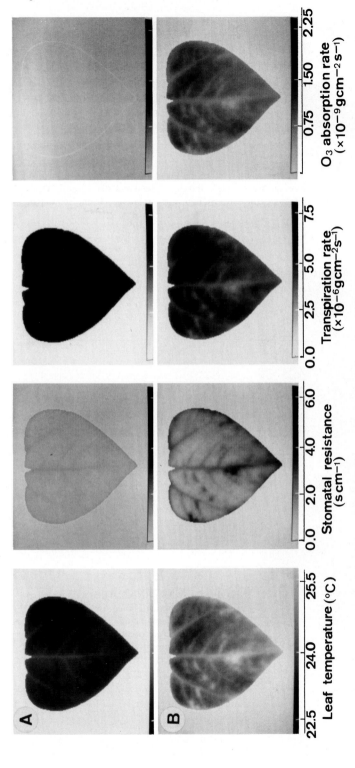

Fig. 21A-B. Changes in spatial distributions of stomatal resistance to water vapor diffusion, transpiration rate, and O_3 absorption rate evaluated from leaf temperature images of a sunflower plant during exposure to 1.2 μ 1 1^{-1}. **A** Just before exposure; **B** 30 min after exposure start. The distributions of shortwave radiation, longwave radiation, illumination, and boundary layer resistance to heat transfer on the leaf surface were maintained at 2.37 ± 0.05 × 10^{-3} cal cm^{-2} s^{-1}, 2.23 ± 0.01 × 10^{-2} cal cm^{-2} s^{-1}, ca. 25 klx, and 1.5 ± 0.1 s cm^{-1}, except at the leaf edges. Other environmental conditions: air temperature, 25.0 ± 0.1 °C; RH, 62 ±1% (After Omasa et al. 1981c)

Fig. 22A,B. Diagnosis of zelkova trees on Aoba street in the city of Sendai. **A** Photograph; **B** thermal image. Environmental conditions: air temperature 26.5°C; light intensity, about 500 μmol photons m^{-2}s^{-1} (Omasa et al. 1990)

Fig. 23A,B. Difference in leaf temperature of various healthy plants. **A** Photograph (*a* poplar tree; *b* white oak tree; *c* spindle tree); **B** thermal image. Environmental conditions: air temperature, 31.0°C; RH, 70%; light intensity, 500 μmol photons m^{-2}s^{-1} (unpublished data)

plants by MRI (magnetic resonance imaging), are now being studied, and their future development is expected to continue.

7.1 Instrumentation of Living Trees by X-Ray CT

X-ray CTs, widely used in clinical medicine and industry, are suitable for obtaining information about the internal structure of a living organism. Portable equipment, suitable for field instrumentation, has been developed recently, and is powerful enough for investigation and diagnoses of annular rings and internal rot in living trees (Onoe et al. 1983). Although various methods for obtaining a reconstructed image from X-ray projection images have been proposed, an easily understood method based on Fourier's projection theorem is shown in Fig. 24.

Projection $p(s,\theta)$ of a subject with a density distribution of $f(x,y)$ to the s axis, with rotation angle θ to the x axis is given by the following equation, by assuming a t axis perpendicular to the s axis:

$$p(s,\theta) = \int_{-\infty}^{\infty} f(x,y)dt. \tag{7}$$

The two-dimensional Fourier transformation $F(u,v)$ of $f(x,y)$ is given by:

$$F(u,v) = \int\int_{-\infty}^{\infty} f(x,y)\exp[-j2\pi(ux+vy)]\,dxdy. \tag{8}$$

Recognizing that $s = x\cos\theta + y\sin\theta$ and $t = -x\sin\theta + y\cos\theta$, the one-dimensional Fourier transformation $P(w,\theta)$ of $p(s,\theta)$ relating to the s is obtained and rearranged:

$$
\begin{aligned}
P(w,\theta) &= \int_{-\infty}^{\infty} p(s,\theta)\exp(-j2\pi ws)ds \\
&= \int\int_{-\infty}^{\infty} f(x,y)\exp[-j2\pi w(x\cos\theta + y\sin\theta)]\,dxdy \\
&= F(w\cos\theta, w\sin\theta).
\end{aligned} \tag{9}
$$

This relationship, called projection theorem, means that the one-dimensional Fourier transformation of projection is equal to a center-cross-section gained by cutting the two-dimensional Fourier transformation of the original distribution to the angle. Therefore, projections from various directions allow the production of

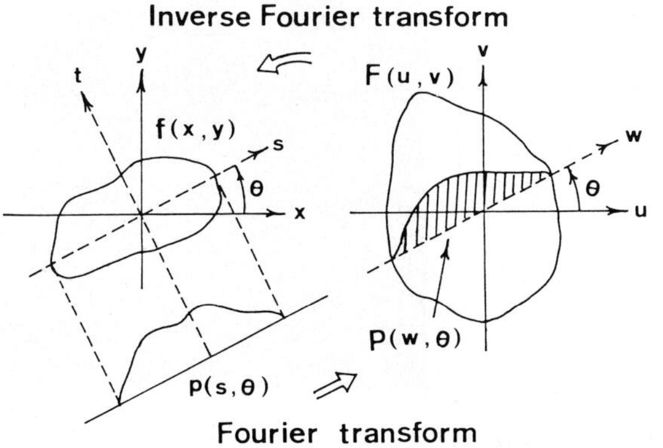

Fig. 24. The concept of Fourier's projection theorem

F(u,v), and an inverse Fourier transformation of F(u,v) reconstructs the original image:

$$f(x,y) = \iint_{-\infty}^{\infty} F(u,v)\exp\left[j2\pi(ux+vy)\right]dudv \qquad (10)$$

Changes of variables, $u = w\cos\theta$ and $v = w\sin\theta$, are performed and the integral limit relating to θ is halved to obtain Eq. (11):

$$f(x,y) = \int_{v}^{\infty} |w|\,dw \int_{0}^{\pi} d\theta\, P(w,\theta)\exp(j2\pi ws). \qquad (11)$$

If the integral order is changed, the integral relating to w is the inverse Fourier transformation of $|w|\,P(w,\theta)$. This is set as $q(s,\theta)$, and Eq. (12) is obtained:

$$q(s,\theta) = \int_{-\infty}^{\infty} |w|\,P(w,\theta)\exp(j2\pi ws)dw. \qquad (12)$$

As a result, Eq. (11) becomes the following:

$$f(x,y) = \int_{0}^{\pi} q(s,\theta)d\theta. \qquad (13)$$

In Eq. (12), q is an inverse transformation of the product of $|w|$ and P. The product of the Fourier transformation in the frequency domain is a convolution in the space domain. Therefore, q is calculated by convoluting the inverse transformation of filter $|w|$ to eliminate blur in the original projection p. Actual processing differs depending on the equipment, and various methods have been proposed.

Portable X-ray CT equipment developed by Onoe et al. (1983) is shown in Fig. 25. In the equipment, an X-ray tube of 40–120 kV and three NaI scintilation counters are used. The tube and the detectors are relatively fixed in an assembly, which rotates around the target focus of the X-ray tube, so that three collimated X-ray beams, 8° apart, scan across the object under test. The target focus is eccentric from the center of the tree, hence a fan-beam algorithm is used for reconstruction. Typically, 1200 samples of 16 bits of projection data are taken in 2° intervals.

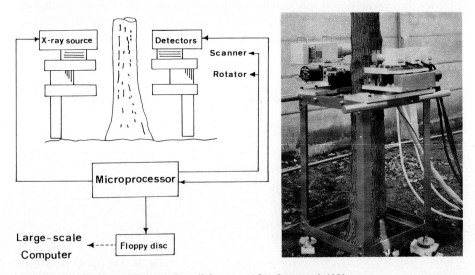

Fig. 25. Portable X-ray CT measuring a living tree (After Onoe et al. 1983)

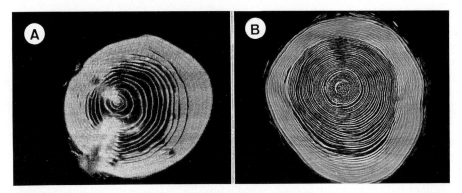

Fig. 26A,B. Reconstructed images of living Japanese cedar and hinoki trees. **A** Japanese cedar tree; **B** hinoki tree. Large X-ray absorption parts are *white* and small parts are *black*. The annual rings in **B** are enhanced by unsharp masking (Onoe et al. 1983)

Scanning, rotation, and data collection are controlled by a microprocessor. The projection data is stored in a floppy disc, and calculations for the reconstruction are carried out by a large-scale computer.

Figure 26 shows reconstructed images of living Japanese cedar and hinoki trees. Large X-ray absorption parts are white and small parts are black, although the annual rings in B are enhanced by unsharp masking. In the reconstructed images, annual rings, differences in water content, and knots (Japanese cedar) are observed. Since cut and dried woods differ in X-ray absorption, because of a difference in growth density of springwood and summerwood, the annual rings are easily distinguishable. However, it is difficult to observe the annual rings of a standing tree containing water. This tendency is particularly strong in the sapwood, which is a passageway for the transpiration flow, and which has more water content than the heart wood. Therefore, enhancement of the annual rings is required.

The formation of annual rings is greatly affected by changes in the environment. Therefore, environmental changes at each growth stage are estimated by investigating the growth in annual rings of standing trees using the portable X-ray CT. Also, since the CT gives information about rot and water content in addition to that about annual rings, it can be used to investigate the growth conditions of trees needing protection.

7.2 Instrumentation of Root Systems and Soil Moisture by MRI

Nuclear magnetic resonance (NMR or MR) instrumentation methods are widely used in the fields of biological and medical research as an analytical means for obtaining information about the chemical composition and reaction process in organisms, from the resonance phenomena between the magnetic moment of the nuclear spin system and the impressed magnetic field (Linskens and Jackson 1986). Since Lauterbur (1973) announced it in 1973, an MR imaging method

(Zeugmatography) using a linear magnetic field gradient has attracted considerable attention as a diagnostic method for clinical medicine. It is characterized as a useful method for providing not only morphological information about organisms but also physiological and biochemical information, and also for easily providing images of any section, by electrical control of the impressed magnetic field. Superconductive magnet MRIs with a high resolution as good as that obtained by X-ray CT have been developed recently, and success in imaging at cell level has been announced (Aguayo et al. 1986). At present, commercially available MRIs can give information relating to density, T_1 (spin-lattice relaxation time), and T_2 (spin-spin relaxation time), etc., for proton (hydrogen atomic nucleus).

Figure 27 shows the concept of the MRI. The nucleus of the atoms (e.g., hydrogen) which make up organisms has a positive charge and spins like the earth. When the nucleus is placed in a magnetic field, it oscillates like a top (Larmor precession) at a frequency proportional to the field intensity. Here, an impression of a high-frequency magnetic field at the same frequency as that of the precession causes a resonance, and the nucleus absorbs its energy and becomes excited-state. The high-frequency field is then cut, and the nucleus returns to the original state, while releasing the absorbed energy (relaxation phenomenon). The energy released from the nucleus at this time is measured as FID (free induction decay) or SE (spin echo) signals, to make a tomographic image. In concrete terms, a linear gradient magnetic field, in which the change in intensity depends linearly on the position, is produced by the addition of an inclined magnetic field to a static magnetic field. Under this condition, a specific high-frequency magnetic field is impressed, so that just the nuclei located in a certain section are excited (selective irradiative process). After this, the inclined magnetic field is turned off, and a newly inclined magnetic field is put on the excited section to produce a linear gradient magnetic field. In this way, the output signal is obtained as a synthetic signal of frequencies corresponding to the respective magnetic field intensity. The Fourier transformation of this signal

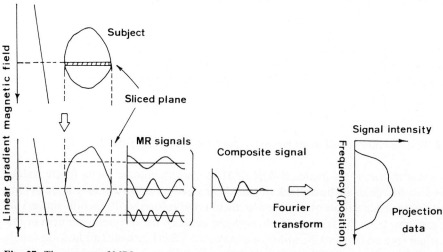

Fig. 27. The concept of MRI

gives the same projection data as that from X-ray CT. The direction of the linear gradient magnetic field on the excited section is changed to obtain projection data in each direction. A reconstructed image is calculated from a series of the projection data by the same method as X-ray CT. Actual processing differs depending on the equipment, and various other imaging methods have been proposed.

Resonance frequencies of high-frequency magnetic fields are in RF (radio-frequency) wave; for protons ($'H$), 42.6 MHz at 1 T (Tesla). The magnetic field is impressed by the SR (saturation recovery) method, the IR (inversion recovery) method, and the SE method, etc. The SE method is frequently used because it detects signals easily. In this method, first a $90°$ pulse is impressed to excite the system and, after τ (echo time) elapses, a $180°$ pulse is applied. The FID signal after the $90°$ pulse and the SE signal, which has a peak at τ elapses after the $180°$ pulse, are then detected. The intensity I_{SE} of the SE image is proportional to the xy component of magnetization at occurrence time of echo and an approximation is given by the following equation:

$$I_{SE} = k\rho_p \exp(-2\tau/T_2) [1-\exp(-T_r/T_1)], \tag{14}$$

where k, ρ_p, and T_r are constant, proton density, and repetition time of the pulse series respectively. I_{SE}, if τ is sufficiently small, is expressed by:

$$I_{SE} = k\rho_p [1-\exp(-T_r/T_1)], \tag{15}$$

and agrees with the image (SR image) obtained from the SR method. Here, if T_r is made sufficiently larger than T_1, the image shows proton density. If τ is made large, $\exp(-T_r/T_2)$ has an effect and the influence of T_2 appears. In addition, it is possible to obtain T_1 and T_2 images with information related to phase or viscosity, but this is not advisable from the points of view of S/N and calculation time.

As an example of MR imaging of plants, distributions of root systems and soil moisture were measured by Omasa et al. (1985b) and Bottomley et al. (1986). Figure 28 shows $'H$ images of dry and wet soils with broad bean roots in horizontal sections of pots. The gray level indicated by the numerals under the gray scale shows the intensity of the MR signal. The MR signal from air-dried soil (ca. pF 5.5) in A was very weak except for that from the root in the center, and the background noise was observed as black-and-white spots. The wet soil (ca. pF 2) in B showed high gray level, although the level differed from region to region in the pot. Since the bulk of protons in soil is water, the gray level represents its water content, T_1 and T_2. Care needs to be taken concerning image distortion caused by diamagnetism and ferromagnetism of soils (Bottomley et al. 1986).

Figure 29 shows an $'H$ image of a broad bean and its main root in a vertical section of a pot with wet soil. Since the gray level of the bean and root was higher than that of the soil, we could easily discriminate the bean and its main root from the soil. However, roots finer than 2 mm could not be detected by our CT system. The lack of uniformity in the soil-water content was probably influenced by the water uptake of roots, as well as by the spatial distribution of the soil structure. Recent developments in MRI technology have increased the spatial resolution to 0.05 mm, making it possible to observe differences in the water content of various tissues of plant organs (Fig. 30) (Brown et al. 1986).

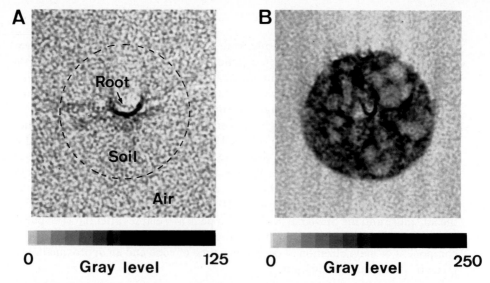

Fig. 28A,B. 'H images of dry and wet soils with broad bean roots in horizontal sections of pots. **A** Air-dried soil; **B** wet soil. The gray level indicated by the numerals under the gray scale shows the intensity of MR signal. Parameters: $T_r = 525$ ms, $\tau = 40$ ms (Omasa et al. 1985b)

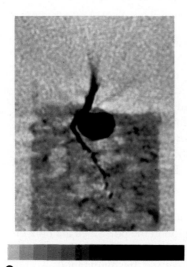

Fig. 29. 'H image of a broad bean and its main root in a vertical section of a pot with wet soil. Parameters: $T_r = 525$ ms, $\tau = 40$ ms (Omasa et al. 1985b)

The results suggest that the MR imaging is an effective method for measuring spatial distributions of water in soil and roots. By analyzing the image, information about the growth in seedling and root, and the water uptake of the root, may be provided without destroying the plant itself and the soil environment. Since an MR signal, especially T_2, relates to water potential as well as to water content in plants (Van As 1982), the CT system also may used to measure spatial distributions of water potential. The MRI is more suitable for making three dimensional images, in comparison with X-ray CT.

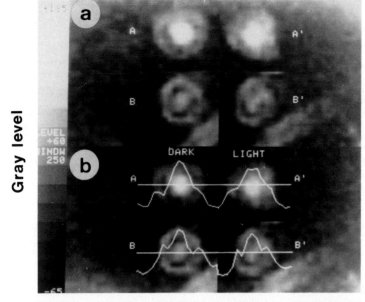

Fig. 30. Microscopic ¹H images of changing water content in *Pelargonium hortorum* roots. **a** Images of cross-sections of adventitious roots showing apparent changes in distribution of water after a period of transpiration. Roots *A* and *B* represent the plant imaged in the dark, and *A'* and *B'* represent the roots imaged after the plant was exposed to light and allowed to transpire for 8 h. **b** Graphs superimposed on the corresponding image represent the relative MR signal intensity for each pixel on its respective transecting line. These graphs better illustrate changes in signal intensity caused by changes in proton concentration within the root during transpiration. Parameters: $T_r = 400$ ms; $\tau = 20$ ms (After Brown et al. 1986)

Another application of MRI has been to detect the presence of postharvest internal disorders in stored fruit. In apples, avocados, and pears, MRI clearly revealed the extent of breakdown in the core tissue. (Omasa et al. 1989, Wang and Wang 1989).

References

Aguayo JB, Blackband SJ, Schoenniger J, Mattingly MA, Hintermann M (1986) Nuclear magnetic resonance imaging of a single cell. Nature (Lond) 322:190–191

Berry J, Björkman O (1980) Photosynthetic response and adaptation to temperature in higher plants. Annu Rev Plant Physiol 31:491–543

Björn LO, Forsberg AS (1979) Imaging by delayed light emission (photoluminography) as a method for detecting damage to the photosynthetic system. Physiol Plant 47:215–222

Bottomley PA, Rogers HH, Foster TH (1986) NMR imaging shows water distribution and transport in plant root systems in situ. Proc Natl Acad Sci USA 83:87–89

Brown JM, Johnson GA, Kramer PJ (1986) In vivo magnetic resonance microscopy of changing water content in *Pelargonium hortorum* roots. Plant Physiol 82:1158–1160

Colwell RN (ed) (1983) Manual of remote sensing, 2nd edn, vols 1 & 2. Am Soc Photogrammetry, Falls Church

Ellenson JL (1985) Phytoluminographic detection of dynamic variations in leaf gaseous conductivity. Plant Physiol 78:904–908

Ellenson JL, Amundson RG (1982) Delayed light imaging for the early detection of plant stress. Science 215:1104–1106

Ellenson JL, Raba RM (1983) Gas exchange and phytoluminography of single red kidney bean leaves during periods of induced stomatal oscillations. Plant Physiol 72:90–95

Fuchs M, Tanner CB (1966) Infrared thermometry of vegetation. Agron J 58:597–601
Gates DM, Keegan HJ, Schleter JC, Weidner VR (1965) Spectral properties of plants. Appl Opt 4:11–20
Hashimoto Y, Ino T, Kramer PJ, Naylor AW, Strain BR (1984) Dynamic analysis of water stress of
    sunflower leaves by means of a thermal image processing system. Plant Physiol 76:266–269
Heath OVS (1959) The water relations of stomatal cells and the mechanisms of stomatal movement. In:
    Steward FC (ed) Plant physiology vol 2. Academic Press, Lond NY, pp 193–250
Herman GT (ed) (1979) Image reconstruction from projection. Springer, Berlin Heidelberg New York
Horler DNH, Barber J, Barringer AR (1980) Effects of cadmium and copper treatments and water stress
    on thermal emission from peas (*Pisum sativum* L): Controlled environment experiments. Remote
    Sens Environ 10:191–199
Hounsfield GN (1973) Computerized transverse axial scanning (tomography): Part I. Description of
    system. Brit J Radiol 46:1016–1022
Jones HG (1983) Plants and microclimate. Cambridge Univ Press, Cambridge
Kappen L, Andresen G, Lösch R (1987) In situ observations of stomatal movements. J Exp Bot 38:
    126–141
Kautsky H, Appel W, Amann H (1960) Die Fluoreszenzkurve und die Photochemie der Pflanze.
    Biochem Z 332:277–292
Kendrick RE, Kronenberg GHM (eds) (1986) Photomorphogenesis in plants. Nijhoff Publ,
    Dordrecht
Knipling EB (1970) Physical and physiological basis for the reflectance of visible and near-infrared
    radiation from vegetation. Remote Sens Environ 1:155–159
Lauterbur PC (1973) Image formation by induced local interactions – examples employing nuclear
    magnetic resonance. Nature (Lond) 242:190–191
Linskens HF, Jackson JF (eds) (1986) Modern methods of plant analysis. New series vol 2. Nuclear
    magnetic resonance. Springer, Berlin Heidelberg New York Tokyo
Mansfield P, Morris PG (1982) NMR imaging in biomedicine. Academic Press, Lond NY
Matsui T, Eguchi H (1978) Image processing of plants for evaluation of growth in relation to environment
    control. Acta Hortic 87:283–290
Meidner H (1981) Measurements of stomatal aperture and responses to stimuli. In: Jarvis PG, Mansfield
    TA (eds) Stomatal physiology. Cambridge Univ Press, Cambridge, pp 25–49
Meidner H, Mansfield TA (1968) Physiology of stomata. McGraw-Hill, Lond
Monteith JL (1973) Principles of environmental physics. Arnold, Lond
Monzi M (1939) Die Mitwirkung der Stomata-Nebenzellen auf die Spaltöffnungsbewegung. Jpn J Bot
    9:373–394
Murata N, Nishimura M, Takamiya A (1966) Fluorescence of chlorophyll in photosynthetic systems II.
    Induction of fluorescence in isolated spinach chloroplasts. Biochim Biophys Acta 120:23–33
Myers VI (ed) (1983) Remote sensing applications in agriculture. In: Colwell RN (ed) Manual of remote
    sensing, 2nd ed, vol 2. Am Soc Photogrammetry, Falls Church, pp 2111–2228
Omasa K (1979) Sorption of air pollutants by plant communities – Analysis and modelling of
    phenomena –. Res Rep Natl Inst Environ Stud JPN 10:367–385 (in Japanese)
Omasa K, Onoe M (1984) Measurement of stomatal aperture by digital image processing. Plant Cell
    Physiol 25:1379–1388
Omasa K, Aiga I (1987) Environmental measurement: Image instrumentation for evaluating pollution
    effects on plants. In: Singh MG (ed) Systems and control encyclopedia. Pergamon Press, Oxford,
    pp 1516–1522
Omasa K, Maruyama S (1990) Study on changes in stomata and their surrounding cells using a
    nondestructive light microscope system: Responses to changes in water absorption through roots. J.
    Agr. Meteorol. 45:265–270 (in Japanese and English summary)
Omasa K, Abo F, Natori T, Totsuka T (1979) Studies of air pollutant sorption by plants. (II) Sorption
    under fumigation with NO_2, O_3 or $NO_2 + O_3$. J Agric Meteorol 35:77–83 (in Japanese and English
    summary); Res Rep Natl Inst Environ Stud JPN 11:213–224 (1980) (in English translation)
Omasa K, Abo F, Hashimoto Y, Aiga I (1980) Measurement of the thermal pattern of plant leaves under
    fumigation with air pollutant. Res Rep Natl Inst Environ Stud Jpn 11:239–247
Omasa K, Abo F, Aiga I, Hashimoto Y (1981a) Image instrumentation of plants exposed to air pollutants
    – quantification of physiological information included in thermal infrared images. Trans Soc
    Instrum Control Eng 17:657–663 (in Japanese and English summary); Res Rep Natl Inst Environ
    Stud Jpn 66:69–79 (1984) (in English translation)
Omasa K, Hashimoto Y, Aiga I (1981b) A quantitative analysis of the relationships between SO_2 or NO_2

sorption and their acute effects on plant leaves using image instrumentation. Environ Control Biol 19:59–67

Omasa K, Hashimoto Y, Aiga I (1981c) A quantitative analysis of the relationships between O_3 sorption and its acute effects on plant leaves using image instrumentation. Environ Control Biol 19:85–92

Omasa K, Matsumoto S, Aiga I (1982) Trial manufacture of a portable spectroradioanalyzer. Proc ann meeting (Sendai), Environ Control Biol 16–17 (in Japanese)

Omasa K, Aiga I, Hashimoto Y (1983a) Image instrumentation for evaluating the effects of air pollutants on plants. In: Striker G, Havrilla K, Solt J, Kemeny T (eds) Technological and methodological advances in measurement vol 3. North Holland Publ Co, Amst, pp 303–312

Omasa K, Hashimoto Y, Aiga I (1983b) Observation of stomatal movements of intact plants by using an image instrumentation system with a light microscope. Plant Cell Physiol 24:281–288

Omasa K, Hashimoto Y, Aiga I (1984) Image instrumentation of plants exposed to air pollutants (4) Methods for automatic evaluation of the degree of necrotic and chlorotic visible injury. Res Rep Natl Inst Environ Stud 66:99–105

Omasa K, Hashimoto Y, Kramer PJ, Strain BR, Aiga I, Kondo J (1985a) Direct observation of reversible and irreversible stomatal responses of attached sunflower leaves to SO_2. Plant Physiol 79:153–158

Omasa K, Onoe M, Yamada H (1985b) NMR imaging for measuring root system and soil water content. Environ Control Biol 23:99–102

Omasa K, Shimazaki K, Aiga I, Larcher W, Onoe M (1987) Image analysis of chlorophyll fluorescence transients for diagnosing the photosynthetic system of attached leaves. Plant Physiol 84:748–752

Omasa K, Kondo N, Inoue Y (eds) (1988) Instrumentation and diagnosis of plants. Asakura, Tokyo (in Japanese)

Omasa K, Unno K, Yamada H, Nagano T, Funada S (1989) Diagnosis of fruits and rootcrops by MRI. Proc joint ann meeting (Tsukuba), Environ Control Biol and Agr Meteorol JPN 262:263 (in Japanese)

Omasa K, Tajima A, Miyasaka K (1990) Diagnosis of street trees by thermography: Zelkova trees in Sendai city. J Agr Meteorol 45:271–275 (in Japanese and English summary)

Onoe M (ed) (1982) Medical image processing. Asakura, Tokyo (in Japanese)

Onoe M, Tsao JW, Yamada H, Nakamura H, Kogure J, Kawamura H, Yoshimatsu M (1983) Computed tomography for measuring annual rings of a live tree. Proc IEEE 71:907–908; Nuclear Inst Methods Phys Res 221:213–220

Papageorgiou G (1975) Chlorophyll fluorescence: An intrinsic probe of photosynthesis. In: Govindjee (ed) Bioenergetics of photosynthesis. Academic Press, Lond N Y, pp 319–371

Sakai A, Larcher W (1987) Frost survival of plants. Springer, Berlin Heidelberg New York Tokyo

Satoh K, Katoh S (1983) Induction kinetics of millisecond-delayed luminescence intact *Bryopsis* chloroplasts. Plant Cell Physiol 24:953–962

Schreiber U (1983) Chlorophyll fluorescence yield changes as a tool in plant physiology. I. The measuring system. Photosynth Res 4:361–373

Schreiber U, Groberman L, Vidaver W (1975) Portable, solid-state fluorometer for the measurement of chlorophyll fluorescence induction in plants. Rev Sci Instrum 46:538–542

Sharkey TD, Ogawa T (1987) Stomatal responses to light. In: Zeiger E, Farquhar GD, Cowan IR (eds) Stomatal function. Stanford Univ Press, Stanford, pp 195–227

Shimazaki K, Ito K, Kondo N, Sugahara K (1984) Reversible inhibition of the photosynthetic water-splitting enzyme system by SO_2-fumigation assayed by chlorophyll fluorescence and EPR signal in vivo. Plant Cell Physiol 25:795–803

Shiraishi M, Hashimoto Y, Kuraishi S (1978) Cyclic variations of stomatal aperture observed under the scanning electron microscope. Plant Cell Physiol 19:637–645

Sivak MN, Walker DA (1985) Chlorophyll *a* fluorescence: can it shed light on fundamental questions in photosynthetic carbon dioxide fixation? Plant Cell Environ 8:439–448

Smillie RM, Hetherington SE (1983) Stress tolerance and stress-induced injury in crop plants measured by chlorophyll fluorescence in vivo. Plant Physiol 72:1043–1050

Stålfelt MG (1959) The effect of carbon dioxide on hydroactive closure of the stomatal cells. Physiol Plant 12:691–705

Sutcliffe J (1977) Plants and temperature. Edward Arnold, London

Trebst A, Avron M (eds) (1977) Encyclopedia of plant physiology, New series vol 5, Photosynthesis I. Springer, Berlin Heidelberg New York

Van As H (1982) NMR, water and plants. PhD thesis, Agricult Univ, Wageningen

Wang CY, Wang PC (1989) Nondestructive detection of core breakdown in 'Bartlett' pears with nuclear magnetic resonance imaging. HortSci 24:106–109

Energy Dispersive X-Ray Analysis

J.S. HESLOP-HARRISON

1 Introduction

1.1 Capabilities

Energy dispersive X-ray (EDX) microprobe analysis detects the presence and amount of chemical elements in plant tissues while they are being viewed under an electron microscope (EM). The detection depends on the atomic number of an element, and cannot distinguish between ionized, bonded or free atoms. The standard EDX analyzers can detect elements with an atomic number from 11 (sodium) upward. The windowless EDX analyzer can detect elements from boron (atomic number 5) upwards, including the major elements present in tissues (carbon, nitrogen and oxygen). Depending on the instrument capabilities (scanning EM, SEM, or transmission EM, TEM), tissue type and preparation (sectioned or bulk, fixed or frozen), concentrations of particular elements can be localized with a spatial resolution better than 0.01 μm, and 1 μm is easily achieved. Quantification of relative and absolute amounts of elements is possible with element concentrations from a few hundred parts per million (typically millimolar concentrations in hydrated plant tissues).

In plant tissues, elements such as sodium, magnesium, potassium, aluminium, silicon, phosphorus, sulphur, chlorine and calcium are often present in high enough concentrations to be located and quantified using EDX analysis. Elements in trace concentrations, such as those associated with proteins at enzyme active sites or some signalling functions, are normally not detectable.

Results from qualitative EDX analyses indicate whether a particular element is present or is below the minimum detectable limit in the specimen prepared in a particular manner. Semi-quantitative analyses use the comparison of spectra, or numbers of X-rays emitted from particular elements, to make limited conclusions about differences in absolute or relative amounts of elements between specimens or regions of a specimen. Quantitative analysis can provide accurate and absolute measurements of element concentrations and distributions. Extreme care is required at all stages in quantitative analysis, because artefacts can be introduced and may obscure results.

In common with many other modern, advanced techniques, EDX analysis is very straightforward to use and for many elemental analyses is the preferred or most practical method to use. Many research workers are involved in the continuing development of EDX analysis and associated techniques to increase spatial resolution, reduce the possibility of artefacts, detect trace elements and accurately quantify compositions. The highly demanding and rigorous requirements for the highest accuracy and resolution should not, however, restrict the application of the technique by the non-specialist, who can obtain useful results easily and quickly.

The present chapter provides information about the basic principles and applications of EDX analysis, with a guide to some of the technical difficulties experienced with plant tissues. I have emphasized the use of the technique at low levels of resolution for qualitative or semi-quantitative analysis of plant specimens. In particular, descriptions of methods for accurate quantitative analysis are outside the scope of the chapter. Discussion is also restricted to the type of equipment available commercially, which might be found in a well-equipped, general purpose biological electron microscopy facility.

There are several books written about EDX analysis and many mention biological applications. The reader is recommended to refer to Zierold and Hagler (1989), Newbury (1988), Bald (1987), Joy (1987), Newbury et al. (1986), Morgan (1985), Revel et al. (1984), Goldstein et al. (1981), and Echlin (1978) who review aspects of the subject and give details of specimen preparation. Journals such as Scanning Microscopy (Scanning Electron Microscopy until 1987), Micron and Microscopia Acta (formerly Micron) and the Journal of Microscopy regularly publish new techniques for EDX analysis.

The preparation of the specimen for analysis in the EM-EDX instrument is extremely critical. Normally, plant tissue is dehydrated or frozen before analysis, and is not alive, although some fresh tissues can be analyzed quickly, while they are dehydrating in the vacuum. The techniques of preparation for EDX analysis must be specifically designed to avoid artefacts which are common because of element mobility or loss during preparation and sampling. Conventional dehydration procedures will lead to unacceptable element mobility, and the usual heavy element staining for EM cannot be used. For quantitative analysis, extreme care and control is required to maintain the chemical integrity of the specimen. The preparation techniques required restrict the resolution of the electron microscope image.

Figure 1a shows an example of an energy dispersive X-ray energy spectrum. The data were acquired from a particularly rugged, hydrated plant tissue, the stigma of a sub-tropical grass, *Pennisetum americanum* (Fig. 1b; Heslop-Harrison and Reger 1985). The stigma was not pretreated; it was placed in the SEM chamber immediately after excision from the plant. The spectrum was acquired in 200 s, before the tissue had time to dehydrate, and shows large peaks ('lines') generated by the potassium and chlorine atoms in the stigma, as well as peaks for aluminium (from instrument contamination) and other minor elements in the stigma.

1.2 History of EDX Analysis

Moseley (1913) first reported that characteristic X-rays were produced from elements which had been excited by electrons. The concept was very important for the understanding of atomic orbitals and later, the development of quantum theory. By the late 1950s, Cosslett and Duncumb (1956) and Oatley et al. (1965) were developing scanning electron microprobes and scanning electron microscopes in Cambridge, England. Laeuchli and Schwander (1966) first used EDX analysis on

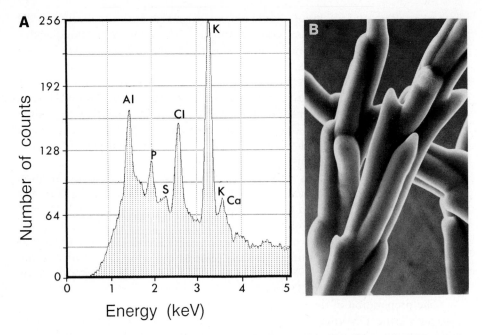

Fig. 1A. An energy dispersive X-ray (EDX) emission spectrum collected from the trichomes near the tip of a fresh stigma of *Pennisetum americanum*; **B** in the scanning electron microscope (SEM) final mag. × 800. The stigma was examined at a magnification of 350 x and 20kV beam voltage in a Philips 505 SEM with a Tracor-Northern 5500 X-ray analyzer. X-rays occurring between 0 and 5.12 keV (X-axis) were collected for 200 s live-time from a region approximately 0.02 mm² and the number occurring in each 0.01 keV wide energy band was plotted (Y-axis). The curve was then smoothed. Peaks arising from characteristic X-rays emitted by the major detectable elements are labelled

plant specimens to look at the distribution of phosphorus and potassium. The introduction of the solid state EDX detector (Fitzgerald et al. 1968), made the equipment more reliable, rugged and cheaper. Material science techniques rapidly developed, and EDX analysis was used with geological specimens, in forensic science and, particularly, for the analysis of the composition of scrap metal! In the 1980s, the increasing understanding of the theoretical basis of the technique and improvements in the computer analysis of data, in scanning TEM (STEM), and particularly in cryopreparation techniques, have enabled biological EDX analysis to become more widespread (Hall 1986).

1.3 Applications of the Technique

EDX analysis has diverse applications in plant science. It can identify the presence of an element in a tissue, and can generate qualitative maps of the distribution of major elements within a tissue, cell or organelle, in combination with morphological or physiological studies. It can be used to examine the composition of surface particles, cell inclusions, and exudates or saps. The data allow the comparison of tissues, cells and organs at various times or in different physiological states which can contribute to a fundamental understanding of plant function and development.

Because no single EM technique gives all the information available about a specimen, many instruments include several analytical and imaging modes, of which EDX analysis is one. It is almost always necessary to use EDX analysis in conjuction with the normal electron image, at least to examine the area of the specimen under analysis. Other data available from the instrument increase the information available and hence help the interpretation and the accuracy of analysis.

1.4 The EDX Equipment

Equipment is available from many manufacturers to fit most current SEMs and STEMs. The EDX detector probe is fitted into the vacuum chamber of the instrument, where X-rays released from the specimen can enter the detector and be measured (Fig. 2). The electronic circuits associated with the detector produce output signals which are proportional to the energy of the X-rays entering the detector. The signals are fed to a computer which carries out operations ranging from plotting the energy data as a frequency histogram, to semi- or completely automatic, quantitative analysis.

EDX analysis instruments are also available which do not include EM imaging functions, but it is difficult to identify the area under analysis and hence large specimens tend to be used. EDX analyzers can also be attached to conventional TEMs. The principles of specimen preparation and analysis in dedicated instruments or TEMs are similar to the SEM and STEM methods discussed in the present chapter, but STEM and SEM-EDX analysis systems are more frequently used in plant biology.

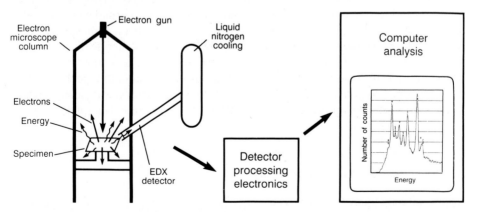

Fig. 2. A diagram of a typical SEM-EDX analysis system. The *EDX* detector is in the column, and connected to the processing electronics and a computer which analyses and displays the energy spectrum

EM-EDX analysis systems usually have highly automated modes, and are designed to be simple in use. Most instruments are designed primarily for imaging of stable material science specimens which include high proportions of elements with high atomic numbers. Plant specimens often consist largely of hydrated hydrocarbons, oxygen and nitrogen, with low concentrations of other, mostly low atomic number, elements in a labile matrix. The examination of such tissues must avoid the analysis or imaging of preparative artefacts, and the misquantification, and even the misidentification, of elements which are present.

For readers who have an EDX analysis facility, the information in the present chapter and elsewhere should be read in conjunction with the manuals for the particular EDX analysis system available. Although the principle of all instruments is the same, and the computer programs have similar capabilities, there are differences in their use. This chapter avoids descriptions of instrument-specific hardware and software. Graphs and distribution maps were produced using the Tracor-Northern TN-5500 analyzer on a Philips 505 SEM or Jeol STEM.

2 Principles – Application and Physics

2.1 Specimen – Electron Interaction

The quantum mechanical processes leading to the production of characteristic X-rays and the properties of the detector are not directly relevant to plant science. An explanation of the source and detection of X-rays, in non-quantum mechanical terms, is helpful to enable interpretation of the spectra and to show the limitations on the use of the instrument arising from physical considerations.

Figure 3 shows the major interactions that occur when an electron beam enters a specimen (Goldstein et al. 1981). Most of the electrons and electromagnetic waves produced in the interaction can be imaged or analyzed to give information about the specimen. Every interaction depends to a greater or lesser degree on the composition of the specimen and its topology, density and structure. In the normal imaging mode in the TEM, the unscattered electrons are visualized directly on a phosphorescent plate. In the SEM, the secondary electrons are collected and generate the image. Of the different electron beam-specimen interactions shown in Fig. 3, three types allow direct and specific characterization of elements or molecules in the specimen, and hence can be used analytically. The three are Auger electron spectroscopy, measurement of cathodoluminescence, and, the subject of the present chapter, analysis of the X-rays generated.

2.2 Excitation and Emission of X-Rays

When an electron from an electron beam passes near or through an atom in the specimen, one of the possible interactions (Fig. 3) results in the incident electron transferring some of its energy to an electron belonging to the atom. The atom's

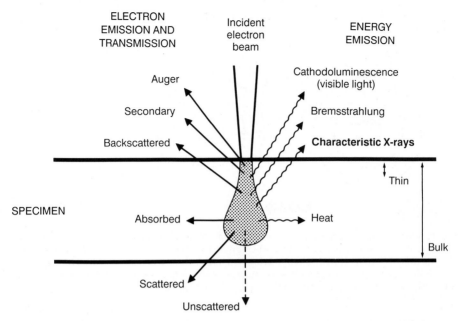

Fig. 3. Interactions between a specimen and an electron beam in an electron microscope. Electrons which are emitted are shown on the *left-hand side* and energy which is emitted (as X-rays, photons or heat) on the *right-hand side*. The volume of the specimen from which the characteristic X-rays originate is *shaded*. In thick or bulk specimens, few electrons from the incident beam penetrate the section, and the volume from which X-rays emanate is large. In thin sections, many electrons penetrate (*broken line*), and the interaction volume is small

electron is excited and moves to a higher energy state (Fig. 4a). The original incident electron loses a specific amount of energy in causing the excitation, and then continues through the specimen, usually in a different direction. Some short time after the interaction, the electron which has been excited loses energy and returns to its original energy level or ground state (Fig. 4b). The energy may be emitted as an X-ray.

2.3 Characteristic X-Rays

The amount of energy emitted when an electron changes its energy level is the difference in energy between the two energy states. The energy in a change between two levels depends on the atomic number of the atom involved. The energy of an X-ray is usually measured in kilo electron volts (keV), but can also be given as a wavelength. X-rays, which can be called X-ray photons, or described as X-ray fluorescence, are a form of radiation with a very short wavelength. Most EDX analyzers detect energies between about 0.5 and 10 keV (wavelength between 240 and 12 nm, compared to 400 nm for red light). All elements have at least one transition which gives X-rays below 10 keV in energy, although some transitions in

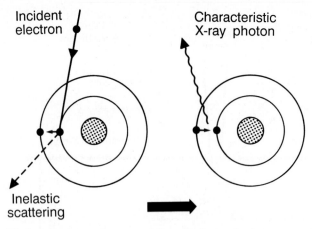

Fig. 4. A diagrammatic representation of an interaction between an electron in an electron beam, and the electrons in an atom from a specimen, which gives rise to a characteristic X-ray. *Left* The incident electron from the beam excites an electron in an atom, which causes it to change to a higher energy level. The incident electron is scattered, and loses a characteristic amount of energy. *Right* The excited electron relaxes to a lower energy level, emitting the energy it loses as a characteristic X-ray

heavy elements involve changes of more than 100 keV. For biological work, energies from 0.5 to 5 keV are normally measured, unless heavy metal stains or other heavy elements are present in the tissue, when a greater range may be measured. Table 1 gives the characteristic energies for selected transitions in a range of low atomic number elements and some others of biological importance.

The energy levels of electrons are conventionally named from the lowest, as K, L, M, N, etc., and the transitions from successively higher energy levels as alpha, beta, gamma, etc. Figure 5 shows the transitions which give rise to some of the X-ray energy peaks observed in spectra; the nomenclature becomes more complex when the initial vacancy occurs at higher energy levels. Hydrogen and helium, with electrons in only the K-energy level, cannot emit characteristic X-rays. Between lithium and neon, only K-alpha X-rays are emitted. For sodium and elements of higher atomic number, three or more characteristic X-rays with different energies can be emitted from the element. Sodium has an M-energy level electron, and hence shows K-alpha, K-β and L-alpha transitions, although it is only in elements heavier than phosphorus that the energies of the two K peaks are different enough to separate with a typical EDX detector. The energies of the major X-rays are characteristic of each element, and are related mathematically to the atomic number of the element and the transition (such as K-alpha) involved.

2.4 Spatial Resolution of the X-Rays

The various interactions between an electron beam and a thick or bulk specimen (where no incident electrons penetrate directly) are shown diagrammatically in Fig. 3. The different interactions occur within various volumes in the specimen

Table 1. X-ray energies in keV from selected transitions for various elements of interest in biological EDX analysis. Energies are taken from the calculator provided for the Tracor Northern 5500 EDX system

Atomic Number	Element	K-alpha	K-β	L-α
5	B	0.18		
6	C	0.28		
7	N	0.39		
8	O	0.52		
11	Na	1.04	1.07	
12	Mg	1.25	1.30	
13	Al	1.49	1.55	
14	Si	1.74	1.84	
15	P	2.02	2.14	
16	S	2.31	2.47	
17	Cl	2.62	2.82	
19	K	3.31	3.59	
20	Ca	3.69	4.01	
26	Fe	6.40	7.06	
29	Cu	8.04	8.91	
82	Pb	74.23	85.36	10.55
92	U	97.14	111.79	13.61

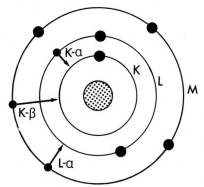

Fig. 5. The nomenclature of characteristic X-ray emissions (giving a peak or line on the energy spectrum). Each different electron energy transition in an element gives rise to a characteristic X-ray of a different energy

relative to the point where the electron beam enters. Some interactions lead to reabsorbance of electrons and X-rays. The secondary electrons are those that produce the image which is normally viewed in the SEM. They originate from near the surface, and hence the surface topology is imaged clearly. The characteristic X-rays emanate from a relatively larger, pear-shaped volume of the specimen, which is shaded in Fig. 3. The actual size and shape of the volume are dependent on the characteristics of the electron beam (size and accelerating voltage) and the atomic number of the elements in the specimen; the diagram shows a typical shape for light (low atomic number) elements. The large volume from which X-rays emanate (shaded) means that the signal has poor spatial resolution compared to the secondary electron signal.

2.5 Detection

As an electron changes energy level, an X-ray is emitted. If the X-ray subsequently enters the window of the EDX analyzer, the amount of energy it contains is measured. The energy data are used to generate a frequency histogram in a multi-channel analyzer. The range of energies of X-rays which arrive at the detector is divided into energy classes, conventionally shown on the horizontal axis (in many papers without the information of class width).

The solid state detector consists of a crystal of silicon diffused with a small amount of lithium which is typically 3-mm-thick and 7-mm in diameter. Opposite ends are plated with gold contacts and are maintained with an electrical potential difference of about 1000 V. The crystal is placed in a chamber, normally behind a thin beryllium window. The chamber protects the crystal from light arising from cathodoluminescence (the specimen glowing because of heat or charge), vacuum changes in the microscope chamber, scattered electrons from the specimen, and from contamination. When an X-ray enters the crystal, its energy is dissipated by collisions within the crystal. Each collision separates an electron from a positive charge, called a 'hole' in the silicon crystal, which requires an energy of about 3.85 eV. The separated pair then moves to opposite contacts on the crystal, causing a short pulse of current to flow. The number of pairs generated (for the potassium K-alpha transition, about 850), and hence the size of the pulse, are directly proportional to the energy of the X-ray entering the detector. Hence its energy can be measured by suitable amplifiers and pulse processors.

2.6 Specimen Independent Spectral Features

The background, white or continuum radiation, often known by the German *Bremsstrahlung* ("braking rays") is a source of X-rays (Fig. 3) which are not characteristic of the elements in the specimen. They arise from the inelastic scattering of incident electrons by atoms in the specimen, during which the incident electrons lose energy as X-rays. Since any amount of energy may be lost, the X-rays can have any energy up to that of the incident electron, so the *Bremsstrahlung* is a continuous spectrum. Most *Bremsstrahlung* X-rays have very low energies and are absorbed in the specimen, detector window or contacts. Thus, the *Bremsstrahlung* (shown in Fig. 6a, with a system peak for aluminium) is observed as a curve reaching a maximum at about 1.5 keV which underlies the sharp peaks from the characteristic X-rays.

The detector system is not free from random noise which is generated in the electronics. The detector is maintained at a low temperature (with liquid nitrogen) to limit random electron-hole pair formation. Some of the effects of random noise can be mitigated by smoothing the spectrum (as in Fig. 1; at the expense of some resolution) or by increasing the time of acquisition of a spectrum. The number of electron-hole pairs formed by a particular energy of X-ray shows random variation. Randomness leads to a broadening of the sharp line at the energy of each characteristic X-ray energy in a spectrum, so spread peaks are seen on the energy

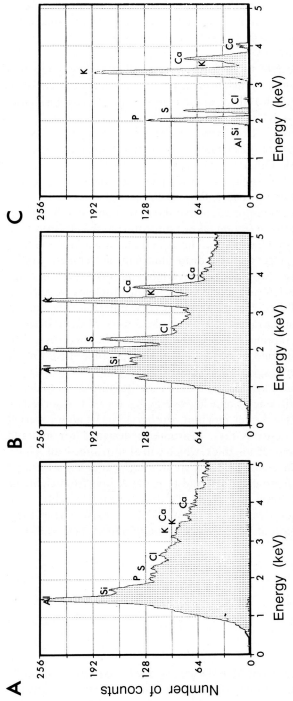

Fig. 6A-C. EDX spectra from the plumose region of a fresh stigma of rice, *Oryza sativa*, examined under the conditions of Fig. 1. **A** Background spectrum acquired from a region of the carbon support stub next to the stigma showing the *Bremsstrahlung* and system peaks. **B** Spectrum acquired from the stigma. **C** Plot of **B** after subtraction of **A** showing the positive differences

distribution graph (Fig. 1). Broadening of peaks can cause X-rays of similar energy to overlap. One of the most common problems in analysis of biological specimens is the overlap of the K-β peak (energy 3.59 keV) from potassium with the K-alpha peak from calcium (energy 3.69 keV).

When an X-ray enters the detector, it normally loses all its energy by creating electron-hole pairs. If the X-ray has sufficient energy, however, it may transfer some of its energy to excite an electron in a silicon atom in the detector crystal. The X-ray then has its original energy less that of the silicon K-alpha excitation (1.74 keV) so it is the energy of the remaining X-ray which is detected. The resulting X-ray spectrum will have a main peak for the element and a second 'silicon escape peak' 1.74 keV below the main peak. Other X-ray energies may be wrongly measured because they are absorbed in the contacts or dead-space near the contacts in the detector crystal, or part of the energy passes through the crystal.

The description of the detection process above shows that only one X-ray can be detected at a time, and that the detection takes a finite time as the electron-hole pairs are produced and move to the contacts. Three problems arise from the time taken to detect the energy of a pulse: there is a dead-time in the detector electronics while it recovers from one pulse; two pulses can overlap; and two pulses can enter at almost the same time and be detected as one X-ray. All the problems are reduced by keeping the counting rate of the detector (the number of events counted) at under $3000 \, s^{-1}$ (see Sect. 4.3). At a counting rate of $3000 \, s^{-1}$, the dead-time (when the detector would not be able to detect another pulse) is about 30% of the total time. The live-time, when X-rays are detected, is hence 70% of the real time. Live-time is normally quoted when an energy spectrum is shown.

Most elements in plants have low atomic numbers, so it is important to detect low energy X-rays with high efficiency. Some low energy X-rays are reabsorbed by the specimen, and many are lost in the detector unit. The beryllium window covering the detector crystal absorbs many X-rays. Instruments designed for material sciences may have windows up to 25 μm thick, which will absorb almost all X-rays below 1.5 keV. Windows which are 7.5 μm thick are still thick enough to withstand pressures changes in the specimen chamber, and transmit X-rays from elements with the atomic number of sodium and above. To detect elements lighter than sodium, thin window or windowless detectors (WEDX) are used. The WEDX instrument is more delicate and sensitive to contamination and cathodo-luminescence.

3 The Specimen

3.1 Introduction

The aims of specimen preparation for EM-EDX are to make the specimen amenable to analysis and stable in the EM chamber, while minimizing loss of structure and the movement of elements between the living state and that in the EM.

The short distance which electrons and X-rays travel in air or through tissues makes analysis of the specimen under vacuum conditions, in practice, essential. Specimen preparation for analysis is therefore required. Specialized instruments have been designed to image tissues in liquid water (Danilatos 1981); the arrangement is impractical for routine use, and the results are those expected from imaging specimens prepared for vacuum examination.

In the present chapter, the techniques of specimen preparation and EDX analysis are not discussed in detail. I have given a brief description of the major methods which have been used for specimen preparation and an example of the use of many of them.

In SEM-EDX analysis, bulk specimens (such as the stigma in Fig. 1) give a high signal strength, and allow the use of various simple preparation techniques. Quantitation is difficult in bulk specimens (see Sect. 6), and preparation for quantitative analysis requires much greater care and the use of more rigorous techniques. STEM-EDX analysis is more demanding than SEM analysis because specimens are normally sections (Sigee 1988, reviews STEM-EDX analysis of non-sectioned material) on more delicate supports, and give very low numbers of characteristic X-rays per unit area because the specimen is thin. STEM generally gives greater resolution and the possibility of much more accurate quantification of either relative or absolute amounts of elements present in a specimen. The basic methods described here may be used in all EDX analyses, whether qualitative or quantitative, and whether the specimen is analyzed in the SEM or STEM.

3.2 Support Material and Adhesives

The specimen to be examined must be supported in the EM. The support should not give a significant signal which will interfere with the spectrum from the specimen. Supports made from low atomic number elements are preferred as they make little contribution to the background X-ray signal.

For SEM-EDX analysis, EM supply companies sell ultra-pure carbon stubs, or carbon discs to stick on aluminium stubs. In normal (non-windowless) detectors, carbon stubs give no detectable characteristic X-rays, and hence are a good support to use. Pieces of pure carbon rods used in carbon arc coating equipment are also suitable. Beryllium discs on stubs may be used as they give less background (particularly in WEDX); the element is toxic so great care must be taken in handling. To stick specimens to stubs, pure colloidal graphite or conducting carbon cements are sold specifically for EDX analysis. They are rapid drying, conductive and give no background, although each new bottle must be tested.

For STEM-EDX analysis, the specimen is normally supported on an EM grid. Beryllium or carbon-coated nylon grids are often used. Hydrocarbon films may be used on grids if greater support of the specimen is required. Normal EM slot grids (made of copper, gold etc.) may also be used with hydrocarbon support films.

3.3 Fresh Specimens

The spectrum acquired from the *Pennisetum* stigma shown in Fig. 1 was from an untreated, fresh stigma which had approximately 85% water content when placed in the microscope. The stigma contains potassium chloride in a high concentration solution (approximately 100 mM), which reduces specimen charging as the solution is quite conductive. The stigma is adapted to be exposed in dry air to the sun, so the tissue is unusually resistant to damage by the EM vacuum and electron beam, and highly amenable to fresh analysis by SEM-EDX analysis. The fresh stigma was stuck to a carbon stub by squashing the ovary at the base so that the fluids produced a conductive path to the stub, and held the stigma in place. The specimen could be examined in the SEM chamber for some 10 min before substantial dehydration and charging effects were noticeable.

The use of fresh tissues in the SEM reduces the quality of the electron image, since the water vapour and other gasses from the specimen reduce the vacuum, and a much shorter filament life may be expected. Normally, a fairly low accelerating voltage is used, so that charging and burning of the specimen is reduced. The specimen tends to be unstable, and may move under the electron beam during analysis.

SEM-EDX analysis of fresh specimens provides useful results with many tough plant tissues, e.g. seeds, pollen, spores and lignified or dead tissue. Such tissues are fresh, but their water content may be below 10%. Kahn et al. (1985) examined the elemental composition of seeds of two halophytes by cutting mature seeds in half and mounting them on carbon stubs with graphite before EDX analysis; heavy charging of the specimen did not interfere with analysis. A similar technique was used by Stewart et al. (1988) for examination of minerals in barley seeds.

Drier specimens are more stable in a vacuum although their surfaces usually need to be coated with a conductive layer to reduce charging. In hydrated specimens, the water and dissolved ions may be conductive enough that charging is not a major problem. Dry specimens or those which charge in the EM are often coated in a vacuum coating unit. As with the support, carbon is suitable for coating as it produces only low energy characteristic X-rays; a thin coat is preferable to avoid reabsorbance of X-rays from the specimen.

3.4 Heat or Air Drying

Seeds and various other tissues such as those containing lignin are naturally dry. Other plant tissues can be air dried before analysis. Lanning and Eleuterius (1983) examined silica in the leaves of coastal plants by fixing dry leaves onto stubs with graphite paste, coating them with carbon, and using SEM-EDX analysis to identify and localize silica.

Drying is a quick technique which avoids the need to freeze or otherwise prepare tissue. As well as examining non-labile elements on or in cell surfaces, it has been used for the study of more mobile elements, which are present in the cytosol, in both plant and animal specimens. It is particularly suitable for isolated cells. For example, Elbers (1983) used EDX analysis to examine dried snail egg cells. We have applied the method to pollen tubes (Heslop-Harrison et al. 1985). Fresh pollen grains of *Pennisetum* were allowed to germinate in a medium of 20% sucrose and 1 mM calcium nitrate and boric acid. When tubes had reached a suitable length, they were gently centrifuged and quickly rinsed twice in distilled water before being spread and dried under a lamp onto the surface of a carbon stub. Spectra were then acquired from different regions of undamaged tubes between the tip and the grain aperture. Since the tubes were fairly uniform in area and topology, qualitative comparisons were possible and showed the differences in calcium (which was present in approximately mM concentrations) in different tube regions. Here, cytochemical reactions were used to confirm the results because the drying process might have led to major redistribution of the calcium within the tube.

3.5 Dehydration

Conventional TEM preparation techniques involve dehydration of a specimen after fixation, using an alcohol or acetone series to replace water with another solvent. The fixation and dehydration are often followed by critical point drying for SEM and embedding and sectioning for TEM. Such 'wet', often room temperature, preparative methods are generally unsuitable for use with EDX analysis. The processes involve the loss of elements from the specimen into solutions, the gain of elements from solutions (e.g. sodium in buffers) and the redistribution of elements within the specimen. Loss and redistribution occurs at all stages, e.g. during fixation, dehydration, drying, embedding and staining (see Morgan 1979, 1980, for a review of wet-chemical preparative techniques used in EDX analysis).

There are a limited number of specimens where particular characteristics of the tissue make wet-chemical preparation suitable for EDX analysis. For example, insoluble silica may be deposited in a plant tissue, and be unaffected by the preparation method, although even here, silica can be leached out or incorporated from glassware.

3.6 'Chelation' and Ion Precipitation

Various precipitation procedures have been used to immobilize elements in tissues, sometimes also coupling them with heavier elements so that X-rays can be more easily detected. Van Steveninck and van Steveninck (1978) reviewed some of the techniques of ion precipitation for localization of elements such as chlorine, sodium and potassium. More recently, the improvement and wider use of cryopreservation techniques has limited the use of ion precipitation: Morgan (1985) stated that precipitation reactions do not represent satisfactory alternatives to cryoprocedures for ion-localization studies.

Although the use of ion precipitation for examination of mobile ions in tissue is declining, EDX analysis is used increasingly to detect stains at the electron microscope level. Westermark et al. (1988b) and Westermark et al. (1988a) linked mercury to lignin in spruce and birch wood and used SEM-EDX analysis to determine the distribution of the lignin between the middle lamella and cell wall.

In the future, SEM or STEM-EDX analysis might be applied more widely to determine both the organ distribution and intracellular location of immuno cytochemical reactions (for proteins or other antigens), of DNA or RNA in situ hybridizations, of specific cytochemical stains or of reaction products from enzymes. Current procedures for EM localization often use gold or silver molecules linked to the antibodies or nucleic acids; if EDX analysis were used to detect the site of precipitation, any element not present in the tissue could be used.

An example of the use of EDX analysis for localization of stains is the work of Sumner (1981). He used EDX analysis to examine the mechanism of staining of chromosomes. Chromosomes were prepared with standard methods before transferring to an STEM support and EDX analysis to examine stain distribution. Kausch et al. (1983) examined the location of cerium precipitated by glycolate oxidase in leaves and roots.

3.7 Formless Specimens

EDX analysis can be used for elemental analysis of bulk tissue, potentially using very small amounts. Dried liquids from tissues, or dried solvents which have been used to extract tissues, can be analyzed to determine whether particular elements are present.

Chong et al. (1983) correlated the cooking quality of field peas with the ratio of magnesium, phosphorus and potassium in protein bodies using EDX spectra acquired from protein bodies in freeze-dried powder. Analysis allows the cooking quality to be measured from a small portion of the cotyledon without destroying the seed. In a plant breeding program, selection could be based on the results of such an analysis.

The elements included in exudate volumes of 0.1 μl (0.1 mm^3) or less can be measured by placing a droplet on a suitable stub in the SEM-EDX and collecting a spectrum after the droplet has dried in the EM vacuum. Hyatt and Marshall (1985) discussed the quantitative analysis of fluid microdrops obtained from insects. Their techniques could be applied to plant tissue exudates.

3.8 Cryopreparation

3.8.1 Applicability

The ultimate aim of fixation, i.e. introducing no artefacts, can be approached by using quick-frozen specimens. By maintaining the tissue in a state close to that in vivo, with little redistribution or introduction of elements, cryofixation is often the preferable specimen preparation technique for EDX analysis, and increasingly, preferable for any other EM analysis. Cryopreservation can be used with both SEM and STEM-EDX analysis on both bulk and sectioned or fractured specimens.

The use of frozen-hydrated bulk specimens for SEM-EDX analysis provides useful data on element presence and distribution within a tissue or organ (Echlin et al. 1982). Cryopreserved and sectioned specimens analyzed in the STEM can give excellent quantitative data on element amounts and locations. Robards (1985) has given a short review of cryopreparation methods, and Bald (1987) has reviewed the subject in depth.

Artefacts may be introduced at all stages of cryopreservation for EDX analysis. During freezing and tissue processing, ice can recrystallize and elements can be grossly redistributed. During analysis, an uneven and distorted specimen surface, as well as the vaporization and redistribution of elements, can affect results.

3.8.2 Preparation for Freezing

Specimens can be frozen with no preparation, although often a support material is used. The support will give more strength to the specimen, and may be required so the specimen can be freeze-fractured or cryosectioned. Small or soft specimens, such as pollen grains, may be suspended in glutinous carriers such as polyvinylpyrrolidone (PVP), dextran sulphate or proteins. Substances such as agar or gelatine should only be used if the non-hydrocarbon elements they may contain (such as chlorine) will not affect the final analysis, and there is no chance of elements diffusing into or out of the specimen.

Cryoprotection can be used to reduce damage caused by ice crystal formation during freezing (Wolf and Schwinde 1988). Penetrating protectants cannot be used because they can damage cells and allow ion movement. Non-penetrating substances such as PVP can give effective cryoprotection without causing ion movement artefacts (Goldstein et al. 1981).

3.8.3 Freezing

Good cryopreservation of a specimen depends on very rapid cooling. The most commonly used technique is the plunging of the specimen into a cryogen, such as liquid nitrogen. Alternatively, the specimen can be pressed onto or between cold metal blocks, or sprayed with a liquid cryogen. If cooling is not rapid, ice crystals will grow within the specimen and physically disrupt the tissue, concentrating ions in the part of the cell or tissue which is frozen last. Vitrifaction of the water in cells, where the freezing occurs so rapidly that no ice crystals form, can be achieved near a specimen surface and in single cells (Bald 1987).

Rapid cooling is usually achieved by plunging the specimen into a liquid cryogen. If a gas layer is allowed to form between the specimen and the cryogen, heat transfer is slowed substantially because the gas acts as a thermal insulator. The phenomenon is clear to anyone who has spilt liquid nitrogen onto their hand; the fingers do not freeze quickly. Spring loaded catapults which shoot the specimen into the cryogen can give very rapid rates of cooling and avoid formation of a gas barrier.

Liquid nitrogen is the coldest easily available substance (boiling point 77 K or $-196\,^\circ$C), and can be cooled a further 14 K to a melting slush by placing a container of the liquid in low pressure. Liquid nitrogen can give excellent cryopreservation if the specimen is rapidly plunged into it (Bald 1987). Other widely used plunge coolants are propane (obtained from gas blow-torch cylinders; as it is inflammable, the liquefied gas is a hazard, particularly in basement SEM laboratories) and chlorofluorocarbons (obtained from dust-blowing cans in most EM laboratories). Although they are liquefied in liquid nitrogen, and boil at a higher temperature, they do not produce as large a gas barrier as nitrogen and under most conditions give higher cooling rates.

Once the specimen has been frozen, it can be freeze-dried, freeze-substituted, critical point dried or else maintained frozen and hydrated for EDX analysis.

3.8.4 SEM-EDX Analysis of Frozen Hydrated Specimens

If the SEM is fitted with a cooled stage, the specimen can be analyzed without further preparation. The water is kept frozen throughout the analysis, and little redistribution of elements occurs. The surface may be prepared by fracturing or even polishing, although ice crystals may grow unless the specimen temperature remains below the recrystallization temperature of water (143 K for pure water; 50 K or more higher in biological materials). Surface drying and condensation or contamination must be avoided (normally by analyzing the specimen quickly). An example of the use of frozen-hydrated tissue in a semi-quantitative EDX analysis is given by Robards (1985) who analyzed potassium frozen-fractured hydrated *Abutilon* nectary hairs.

3.8.5 Freeze Drying and Freeze-Dried Embedding

When a specimen is freeze-dried, the ice in the frozen specimen is sublimed under vacuum. It is a widely used technique in EDX analysis. Freeze drying should be carried out at a temperature below the recrystallization point of water, or ice crystals may grow during drying and cause disruption similar to those formed during slow freezing. Sakai and Sandford (1984) used freeze-dried leaves of sugar-cane to investigate the deposition of silica semi-quantitatively at different stages of development in different cell types.

Following freeze drying, many tissues are fragile, and embedding materials can be infiltrated. Embedding may not disturb structure or ion distribution, and the material can then be sectioned for EDX analysis (see Edelmann 1986, for review).

3.8.6 Cryosectioning

For quantitative analysis of specimens in the STEM or SEM, thin or semi-thin cryosectioning of hydrated or freeze-dried specimens is an important and accurate technique. Sections give much higher resolution of element distribution than bulk specimens. A critical strategy for cryosectioning has been advanced by Saubermann (1988a). In 1- to 3-μm-thick sections, a lateral resolution of about 1 μm can be achieved. Under ideal conditions, a resolution of 25 nm can be approached; extreme rigour in both specimen preparation and data analysis is required to obtain high resolutions (Goldstein et al. 1981). Frozen sections between 2 and 0.05 μm thick can be cut, although the process may be technically difficult. Most manufacturers of ultramicrotomes have cryochambers which can be fitted and allow control of specimen, knife and chamber temperatures. Single sections are cut on a dry knife and picked up on an EM grid or stub. The temperature of the cryochamber used for cutting depends on the specimen, and must be optimized to obtain a suitable section or cut surface for analysis; it must be smooth, but not show evidence of melting or of fracturing.

3.8.7 Use of Cryosections

After sections are cut, they may be freeze-dried, or transferred to the EM in a fully hydrated state. Saubermann (1988a,b) and Morgan (1985) have reviewed the advantages of analysis of hydrated over freeze-dried sections. Hydrated sections might be expected to give the most undisturbed picture of element amounts and locations, but analysis has 'formidable' problems (Saubermann 1988b), including the low height of EDX energy peaks over the background, poor image contrast in the section and specimen damage under the electron beam. Examples of the use of cryosectioning for quantitative analysis of sections are given in Section 6. Freeze-dried sections provide an alternative, and careful drying may not introduce many artefacts.

Figure 7 shows a freeze-dried cryosection of a stigma of *Zea mays*. The fresh tissue was quench frozen in liquid Freon ('Dust-off'), mounted on a block holder with a drop of water which was then frozen, and sectioned at about −150°C with a glass knife. Sections were placed on a liquid nitrogen cooled carbon SEM stub. Since a freeze-drying apparatus was not available, an insulating block (made from Teflon or a similar material was cooled in liquid nitrogen, and the frozen section and stub were mounted on the block. The assembly was placed under vacuum in a coating

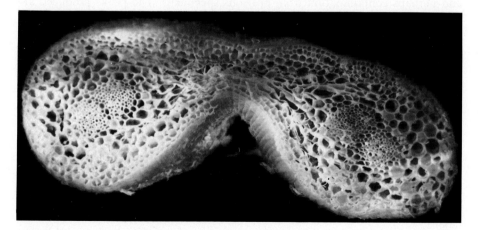

Fig. 7. A freeze-dried cryosection of a *Zea mays* stigma. The stigma was quench cooled in liquid Freon, sectioned on a Reichert Ultramicrotome at -150°C, and dried under vacuum on a carbon stub. The section was photographed in the SEM before EDX analysis × 170

unit, while it dried out and warmed up over 72 h. Because the vacuum and block are good insulators, much of the ice in the specimen sublimed before the tissue reached the water recrystallization temperature.

4 The Spectrum

4.1 Calibration of the Detector

Before the X-ray detector can be used, it must be calibrated. Many operators leave the EDX detector system on all the time because the high gain amplifiers required to detect the pulses in the detector crystal tend to drift if they change in temperature. Most systems should be allowed to stabilize for at least 1 h after switching on before calibration. After moving (or other disturbance, such as filling with liquid nitrogen), the detector may be unstable for many minutes. The actual method of calibration depends on the system. Aluminium or copper, available in stubs or grids, or part of the metal of the specimen chamber are usually the most convenient standards. After acquiring a spectrum, the analyzer is programmed to standardize itself on the major peaks detected. Calibration should be checked every session if the detector is left on.

4.2 System Peaks and *Bremsstrahlung*

Before analyzing a spectrum from a specimen, it is important to acquire a spectrum from an area of the support next to the specimen for a relatively long period, under the same conditions used for the specimen. Various system peaks will normally be identified, including aluminium and iron from the construction of the instrument, and perhaps contamination from previous specimens. The peaks produced are referred to as system peaks and, when the spectrum from the specimen is analyzed, such peaks should be noted and ignored. Figure 6a shows a substantial system peak for aluminium. The *Bremsstrahlung* will also be seen in the background spectrum.

4.3 Adjustment of the Analyzer

4.3.1 Electron Image

To use the EDX analyzer, a specimen is inserted into the SEM or STEM, and a normal secondary electron or transmitted image obtained. Parameters within the normal ranges which produce good secondary or transmitted electron images are normally suitable for the initial EDX analysis of biological specimens. The acquisition of a spectrum can be started when a suitable image is obtained, and the spectrum watched as it builds up on the computer screen. The background X-rays and a number of peaks from the elements in the specimen should be displayed.

The system can then be optimized for the particular specimen under analysis. As well as changing the count rate, the surface topology and orientation of the area where the X-rays originate can make a large difference to the spectrum acquired. Smooth surfaces generally give the best results. Data from rough surfaces must be interpreted much more cautiously.

4.3.2 Voltage of Instrument

The accelerating voltage of the EM beam needs to be higher than the energy of the highest energy peak required for analysis. Normally, some 5 kV or three times the voltage of the highest energy peak (the 'overvoltage') is used to give an adequate number of X-rays (counts). Thus, for unstained plant specimens with elements up to calcium, 10 kV is a minimum suitable beam voltage. Higher voltages increase the number of X-rays produced, but may damage the specimen by heating, charging or causing mass loss (vaporization); in bulk specimens, the depth of penetration of the beam increases, which will reduce spatial resolution of the X-ray signal.

As was discussed above in Section 2.5, the count rate from the detector must normally be limited to around 3000 counts/s, giving a dead-time of less than 30%. Count rate is affected by the accelerating voltage of the EM and the detector to specimen distance, and both may be adjusted.

4.3.3 Detector-Specimen Distance and Angle

The detector is usually mounted on a screw which allows it to be moved closer or further away from the specimen, and sometimes to control its angle to the specimen or electron beam. In SEM-EDX, it is usual to place the detector as close to the specimen as possible without interfering with opening the chamber to maximize the count rate. Although the detector can normally be moved while in operation, it is important not to push it into the specimen. To find the specimen-detector distance, air can be admitted to the column, one of the observation ports into the chamber removed, and the detector moved in and out.

The detector angle to the specimen is important ('take-off angle') to maximize collection of X-rays from the specimen and to minimize the effect of specimen topology. It is usually more convenient to tilt the specimen to the optimum angle for the detector to maximize the count rate, and then reduce the electron beam voltage or increase the detector distance to the optimum count rate. In quantitative and semi-quantitative analysis, it is important to match the angles for successive analyses (Robards 1985).

4.3.4 Electron Beam Diameter

The actual diameter (spot size) of the electron beam is not of major importance in SEM imaging of bulk specimens, because of the increase in volume imaged below the surface (Fig. 3). Therefore, for qualitative and some quantitative SEM-EDX analyses, the beam diameter giving the optimum secondary electron image, and minimum specimen damage, is usually satisfactory. In STEM-EDX analysis of thin sections, the beam diameter is strongly correlated with the area of X-ray production, and hence must be reduced at the highest levels of resolution.

4.3.5 Acquisition Time

The time of acquisition of a spectrum is the live-time plus the dead-time. A spectrum from a bulk plant specimen acquired over 200 s live-time usually gives satisfactory data. To reduce statistical variation in the background, to assist with detecting trace elements, in thin sections, or if a low overvoltage is required to minimize specimen damage, much longer live-times may be required. Times for acquisition of X-ray maps are discussed in Section 5.4.

4.3.6 Selection of Magnification and Area of Acquisition

Spectra can be collected from any area which is scanned by the electron beam in the EM. The area can range from a point up to the full area which can be imaged. It is usually preferable to collect a spectrum from a relatively small area on the specimen chosen for uniformity of appearance and not including any obvious boundaries. Point imaging, where the electron beam is focussed on a point for the time required for acquisition, may lead to extreme specimen damage. After collecting a spectrum the area should be checked for damage.

Hall and Gupta (1979) consider that 30% mass loss may occur in a typical specimen, and not only carbon and other light elements but also the heavier elements under analysis may be lost. Results from EDX analyses should be interpreted cautiously because selective loss of elements (by vaporization) from the specimen can occur. Reducing the accelerating voltage or the acquisition time will limit specimen damage, although it will also reduce the count rate. Cooling the specimen on a cold stage in the microscope reduces mass loss without changing the count rate, even for dried specimens. The acquisition of several consecutive spectra from one area of a specimen can give an indication whether serious mass loss is occurring; the heights of peaks change and the *Bremsstrahlung* radiation will decrease if mass loss is occurring.

4.4 Analysis of the Spectrum

4.4.1 Aims

The qualitative microanalysis of the elements contained in a specimen is a very important function of the EDX analyzer. Identification of the major detectable constituents of a specimen takes only a few seconds or at most minutes. Minor elements (typically less than 0.1% of the specimen, and usually those of greatest biological interest) may be much more difficult, particularly near the minimum detectable level.

The identification of elements is the first stage in any analysis, quantitative or qualitative, and in many cases no further analysis is required. For example, if one has observed crystals in a particular organ which are suspected to be calcium oxalate, EDX will very quickly indicate whether they contain calcium. Okoli and McEuen (1986) showed such crystals in leaves of *Telfairia* which had been dried and carbon coated.

Elements with low atomic numbers are most common in plant tissues. For minor and trace elements, errors in identification are possible because of spectral artefacts, i.e. poor distinction from the background and overlapping spectral peaks from different elements.

4.4.2 Peak Identification

The starting point for the qualitative analysis of a spectrum is a table of X-ray energies (part of which is Table 1). In plant specimens, peaks may be expected from sodium, phosphorus, sulphur, chlorine, potassium and calcium. After smoothing to reduce some random variation, the X-ray energy spectrum, such as Fig. 1, is inspected with the table of X-ray energies and the elements giving the 'peaks' are identified. Identification can normally be done with the graph displayed on the computer screen, with the aid of a program which suggests the identity of any peaks. Many EDX systems have automatic peak identification programs although they are often inadequate for identifying peaks in spectra from plant specimens.

After acquiring the spectrum from the specimen, it should be compared with a spectrum acquired under the same conditions from a control region (Sect. 4.2). Common peaks which are due to instrument contamination are marked and discounted. Next, the major peaks with the highest energy on the specimen spectrum are identified. Following the identification of a major peak, the graph is checked for the presence of secondary peaks for the element (e.g. K-alpha or L-alpha). If a K-β peak of 10 to 20% of the height of the K-alpha peak occurs, then the first identification was probably correct. Otherwise, a peak from another element might overlap with the secondary or primary peak, or the original identification was incorrect.

Once major peaks and their related secondary peaks are identified, minor peaks can be examined. The minimum peak height above the *Bremsstrahlung* which allows identification of an element (i.e. a statistical minimum level of detection; Heinrich 1981, p. 199) is not defined. Experience with the machine and a specimen type is important, but a general consensus is that a peak which is more than three times the background level (i.e. the fluctuation seen above the *Bremsstrahlung*) is significant. Peak-to-background ratio is an important factor in accurate element identification and also quantification.

The identification process can be followed by reference to Fig. 6b. The large peaks for calcium, potassium, sulphur, phosphorus and aluminium (system peak) were identified first. The calcium K-β peak at 4.01 keV was then found (about the correct height), but the potassium K-β peak overlaps the calcium K-alpha peak. Following their identification, the smaller peak for silicon (a system peak) was identified. A chlorine peak might be present but is not set off from the background.

4.4.3 Background Subtraction

Background subtraction is possible using the spectrum from a specimen (e.g. Fig. 6b) and a spectrum acquired from a blank, or area of the specimen support next to the specimen (Fig. 6a). The resultant graph has both the *Bremsstrahlung* and system peaks removed (Fig. 6c). Subtraction may lead to artefacts because the region of low energy X-rays from elements in plants overlaps with the region where the *Bremsstrahlung* is greatest, and where it changes most rapidly. If the *Bremsstrahlung* height is not identical between the specimen and the background spectra (as in Fig. 6a and b where specimen topology probably caused a difference), real peaks may be removed (perhaps the chlorine peak in Fig. 6b).

In plant specimens, linear interpolation of the *Bremsstrahlung* (using lines connecting areas adjacent to peaks), with separate identification of system peaks, may be more efficient than background spectrum subtraction. Interpolation involves subjective judgement which can make the data more operator-dependent, but is usually reliable. Several complex and rigorous algorithms are available which calculate a theoretical model of the background shape (see Table 6 in Morgan 1985) which is then subtracted from the specimen graph. Such techniques are used widely in quantitative analyses, and are often implemented on the computer analysis system.

4.4.4 Optimization of the Spectrum

Following primary peak identification, optimization of the X-ray detection might be possible following the guidelines in Section 4.3. If some peaks are not well resolved above the *Bremsstrahlung* or background noise, a longer acquisition time might be required. Changing the overvoltage will alter the *Bremsstrahlung*. The detector to specimen distance and take-off angle might also be adjusted; in Fig. 6b, the reduction of the distance would have reduced the height of the aluminium peak and increased the relative size of any magnesium peak. Checking of parameters, such as the percentage dead-time and count rate, is also useful after an initial spectrum has been acquired and analyzed.

5 Maps and Line Scans

5.1 Application

EDX analysis can be used to generate a qualitative compositional map of the distribution of an element within a tissue or cell. It can also show the distribution of an element along a scanned line. As an example, Fig. 9 shows maps and line scans of the distribution of potassium, calcium, and phosphorus together with sulphur, in a sagittal section of an ovule of *Zea mays* through the embryo sac.

5.2 The Specimen

Maps can be generated from thin sections in the STEM or from any tissue in the SEM. Quick-frozen specimens which are maintained frozen during analysis are particularly suitable for semi-quantitative analysis, although techniques involving freeze-drying or heat-drying are also usable.

The highest resolution maps are obtained from thin-sectioned specimens, and may be quantitative (Saubermann 1988a,b; Saubermann and Heyman 1987). The narrow 'neck' to the pear-shaped volume from which X-rays emanate (Fig. 3) means that specimens which are no thicker than the neck region can be mapped with very high resolution. The resolution from relatively thick sections (50–100 μm) is often adequate to show element distribution semi-quantitatively.

A surface which is as smooth and uniform as possible is required to minimize topographic effects because X-rays are collected with different efficiencies depending on the specimen surface. The surface is usually sectioned; hand sectioning of a quick-frozen specimen is adequate (Heslop-Harrison and Reger 1986), although some effects of specimen topology may be seen in maps. For example, the detector used to making the maps in Fig. 9 was to the top right of the specimen, and few background X-rays were detected from the lower left of the stub where the base of the ovary was between the detector and stub. Similarly, some shadowing effect is visible between the ovary wall and top of the nucellus at the top right of the specimen.

5.3 Selection of X-Rays for Line Scanning or Mapping

Line scans of element concentrations in a specimen can be made by collecting a spectrum from a large number of points along a line on the specimen. The resulting numbers of counts from each element is displayed as a histogram (Fig. 9d). In generating a map of element distribution, a spectrum is collected at a large number of points over a series of scan lines (raster) covering an area of the specimen. The number of X-rays from each element at each point is counted and used to modulate the brightness of the point at the corresponding position on the map. Maps for several different elements are usually acquired simultaneously.

To produce an element distribution map, a spectrum of the X-rays produced from the complete specimen or representative portion is first acquired. Within the spectrum, a number of regions of interest (ROIs) are defined, which are the energy bands where X-rays from each element are collected. The number of X-rays in the ROI is then used during mapping or line scanning for measuring the presence or amount of an element at each point. Each ROI included one or more peaks of interest and a minimum area on each side. Although rigorous methods may be used to define the peak positions and widths, a subjective choice of ROI width and position is generally adequate for semi-quantitative mapping. It is important also to define an ROI in the spectrum where no peak occurs, to check the uniformity of the background. Figure 8 shows the spectrum and ROIs used to generate the maps and line scan in Fig. 9. A preliminary map made with another specimen showed that

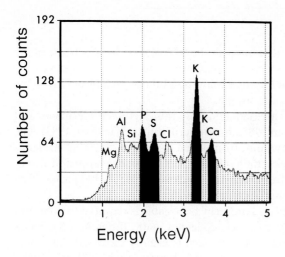

Fig. 8. An EDX spectrum from a thick section of a *Zea mays* ovule (conditions as in Fig. 1). The energy ranges (regions of interest, ROIs) for calcium (*Ca*), potassium (*K*) and phosphorus (*P*) with sulphur (*S*) are *black* and X-rays in the three regions were used to generate the maps shown in Fig. 9

phosphorus and sulphur were distributed reasonably uniformly throughout the specimen. As their peaks are close to each other, they were selected in one ROI, and the two elements were mapped together. Potassium gave a strong signal, and hence was mapped with the least noise. The K-alpha peak for calcium is very close to the K-β peak for potassium, so that the calcium ROI was set to the upper part of the peak to minimize the overlap.

5.4 Map or Line Acquisition

Once the ROIs have been chosen, the EM is programmed to acquire a spectrum at a defined number of points along the line or over the raster for mapping. The time of spectrum acquisition at each point is then chosen. Element mapping with even low resolution is slow, as an X-ray spectrum must be acquired at each point; however, very much shorter acquisition times are used in mapping than are used to obtain spectra from areas. To obtain a map with a resolution of 128×128 picture elements, 16 384 points must be imaged. At only 0.3 s live-time per point (a typical time which was used in Fig. 9) the map takes about 90 min to build up. Although the results at each point are subject to large statistical errors because the acquisition time is so short, the visual nature of the final map allows semi-quantitative interpretation.

→

Fig. 9A-E. Scanning electron micrograph, three EDX maps and a line scan of an ovule of *Zea mays*. **A** Micrograph of the secondary electron image of the ovule showing the stigma base (*s*), nucellus (*n*), ovary wall (*o*) and embryo sac (*e*) × 25. The line used in the scan **E** is shown. **B** Map of the distribution of calcium made by counting the number of X-rays detected in the calcium region of interest (Fig. 8). **C** Similar map to **B** made for phosphorus and sulphur together. **D** Similar map to **B** made for potassium. **E** Line scan showing, as a histogram, the number of counts (*vertical axis*) in the calcium ROI at points along the line in **A**

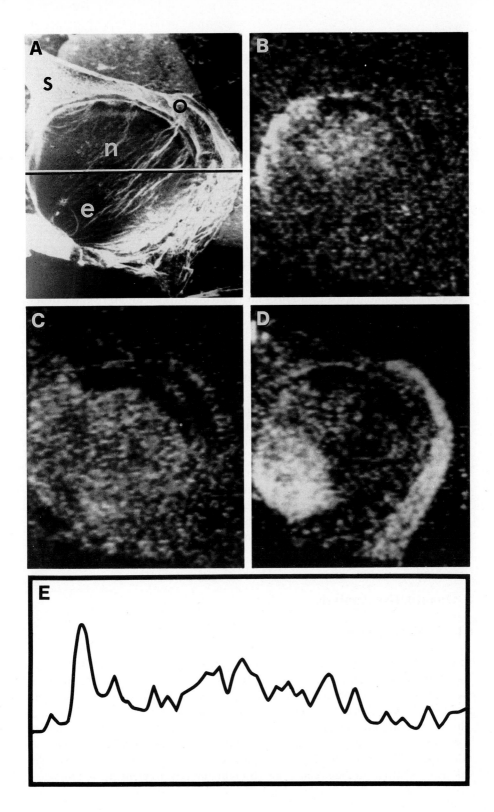

Beam damage to a specimen may be substantial over long acquisition times, particularly with frozen-hydrated specimens. It may be advantageous to build up a map by scanning a specimen several times with very short times at each point. The resulting counts for each ROI at each point are then added. Specimen damage is minimized in a fast scan because points on the specimen are under the beam for many short periods rather than a single long period.

After map acquisition, the specimen should be checked for beam damage which may have caused selective loss of elements, or changed the specimen topology. In the specimen in Fig. 9, some damage was visible, but a second set of maps acquired after the first showed a similar distribution of elements so selective elemental loss was not distorting the results.

5.5 Interpretation of the Map or Line

Figure 9 shows that the distribution of the elements mapped differs over the ovary. Examination of the maps indicates that potassium is heavily concentrated near and in the embryo sac. Calcium is mainly located in the upper part of the nucellus and below the stigma insertion point, which is shown in both the map (Fig. 9b) and line scan (Fig. 9d). Phosphorus and sulphur are more evenly distributed throughout the tissue. Since the maps are only semi-quantitative, more detailed conclusions would be of limited value.

Maps can be enhanced using image processing techniques, which emphasizes their qualitative nature. A threshold number of counts is usually set, below which no point is displayed on the map. The threshold compensates for the number of counts expected from the *Bremsstrahlung*. In Fig. 9, the counts above the threshold have been used to give a grey scale, where white is a high number of counts. The contrast slope (relationship between point brightness and number of counts) has been altered subjectively to give a visually clear image. Previously, we have published the same data with the maps as high contrast images (Heslop-Harrison and Reger 1986).

6 Quantitative Analysis

6.1 The Measurements

From a physical and mathematical point of view, the quantitative analysis of X-rays is very well defined (Goldstein et al. 1981). It involves counting the number of X-rays from each element, by setting ROIs (Sect. 5.3). The relative or absolute amount of each element present in the specimen is calculated from the data. The relative amount of an element can be measured because the number of X-rays produced (as counts or intensities) is directly proportional to its concentration (Russ 1974) in thin sections. A range of quantitative methods have been developed for measuring the relative or absolute amounts of elements in biological tissues (reviewed by Goldstein et al. 1981; Morgan 1985; and Saubermann 1988a,b) but relatively little use of quantitative EDX-analysis procedures have been made with plant material (Echlin et al. 1982, 1983; Cameron et al. 1984; Moore et al. 1987; Baldi et al. 1987), perhaps because of the difficulty and uncertainties of analysis and the complicating factors (Hall and Gupta 1979) discussed briefly below. Semi-quantitative and qualitative data is often adequate.

6.2 The Specimen

Quantitative analysis is, in theory, straightforward from thin or semi-thin sections (less than approximately 2 μm) which can be used in both the SEM and STEM. Most analyzers have a processing program which assists with the analysis of the spectrum and with quantification. Such programs must be used cautiously because they give spurious accuracy on biological specimens and may make many assumptions which are incorrect. Quantitative analysis of bulk specimens in the SEM is possible, but various factors including the undefined interaction volume (Fig. 3) and the variable X-ray reabsorption by the specimen can give large errors.

The analysis of elements present in plant specimens is both difficult and rarely more accurate than $\pm 10\%$, even for elements present in biologically high concentrations of 1–25%. The accuracy is many times worse than that achievable in material specimens because of difficulties with both specimen preparation and spectrum acquisition and analysis (Goldstein et al. 1981).

Material specimens have high proportions of heavy elements, which produce strong X-ray signals with high count rates. Even the undetected elements are often present as known compounds (e.g. the oxygen in SiO_2), and standards similar in composition to the material under investigation can be manufactured. The elements to be analyzed in biological specimens are usually present in low concentrations in a hydrated organic matrix, where the major constituents (carbon, nitrogen, oxygen) do not produce easily detectable X-rays. Material specimens are usually stable and undamaged by the X-ray bombardment under the EM. They often conduct heat, and sometimes electric charges, and do not often show significant mass loss (which can lead to column contamination), dehydration or other effects. Particular elements are often present in only a few known compounds. In biological systems, one element may be present as ions and in compounds which differ between the cytosol, nucleoplasm, in enzymes or in membranes, and elements may move and change over short times. Material surfaces can be polished smooth so the surface geometry and roughness can be defined, and finally few artefacts are introduced by preparation techniques. Such factors, as well as the "essential inhomogeneity of living tissue" (Morgan 1985) make biological tissue impossible to analyze as accurately as material specimens.

6.3 The Analysis

6.3.1 Counts per Channel

As in qualitative analysis, quantitative analysis starts with the identification of peaks as regions of interest (ROIs; Sect. 5.3). The background is then removed (using one of the methods discussed in Sect. 4.4.3) and peak overlaps are unfolded (deconvolved), usually using available software. The intensity of each peak is then measured from the number of counts in the relevant energy channels (ROIs) of the spectrum.

6.3.2 Relative Amounts

In tissue, which does not vary greatly in composition, topology, thickness or relationship to the electron beam and detector, SEM-EDX can be used semi-quantitatively by comparing the numbers of counts in regions of interest. Both frozen-hydrated specimens (such as nectary trichomes; Robards 1985) and dried specimens (such as pollen tubes; Heslop-Harrison et al. 1985) can be analyzed semi-quantitatively.

Semi-quantitative analysis can also be used to examine differences in the ratio of two elements. In thin sections, the relative concentrations of two elements can be found because the ratio of intensities of two elements is directly proportional to the ratio of their concentrations (Russ 1974). The constant of proportionality varies between instruments and conditions, and can be determined experimentally using known solutions. Morgan (1985) and Morgan et al. (1975) give typical values (Table 2).

Table 2. Typical values for the constants of proportionality between the number of X-rays produced by an element and the concentration of each element. The relative concentrations of two elements can be calculated by taking the ratio of the number of X-rays counted from each element (i.e. the area under the energy graph, less background) and multiplying it by the ratio of the constants given. The values vary depending on the setup of the EDX analyzer. Values are taken from Morgan (1985)

Co	Na	Mg	P	S	K	Ca
1.0	6.7	2.5	1.3	1.2	1.0	0.9

6.3.3 Absolute Amounts

If the relative amounts of elements are known, using the methods of Morgan et al. (1975; Sect. 6.3.2), absolute amounts of elements can be determined if the concentration of one of the elements is known absolutely. Another quantitative technique, such as flame ionization, can be used to give the absolute value.

Quantitative analyses of biological specimens can also be carried out by comparing peaks from the specimen with those from a standard. Elements incorporated into resins are available commercially, either as sections of known thickness or blocks, or can be made in the laboratory. If cryofixation is used, standard solutions, usually including a cryoprotectant and organic matrix, can be frozen, sectioned and placed on support films or stubs before analysis. Often, the frozen tissue can be surrounded by the standard solution, and the two sectioned and analyzed together. Detailed quantitative methods of analysis of thin sections are given by Hall and Gupta (1979), Goldstein et al. (1981) and in summary by Morgan (1985).

Bulk specimens can also be analyzed quantitatively, but the analysis is more involved than that for thin sections, since many factors including excitation conditions, geometry, intensity ratios of elements and physical information about pure elements must be considered. The references mentioned above, and Newbury et al. (1986) give further details of the techniques.

7 Alternatives to EDX

7.1 Using X-Ray and EM Techniques

EDX analysis relies on the measurement of different energies of X-rays. Two further methods of analyzing characteristic X-ray energies are often used.

Energy and wavelength are related, so a wavelength dispersive X-ray detector (WDX) may be used. A crystal is used to diffract X-rays from the specimen, and the amount of diffraction depends on the X-ray wavelength. A detector which counts the number of X-rays entering is placed at an angle from the crystal, so the number of X-rays occurring in a narrow band of wavelengths (or energies) is counted. The approach reduces the peak spread, reducing the overlapping of peaks, and it allows peaks from all elements producing X-rays (not hydrogen or helium) to be counted. The energy resolution, peak-to-background ratio, and count rate can be much higher than an EDX analyzer will allow. Only one characteristic X-ray is counted at a time, while EDX analysis measures X-rays from all elements simultaneously.

EDX analysis measures the energy of the characteristic X-ray produced when an electron changes to a lower energy state. The excitation of the electron requires the same characteristic energy. The energy comes from the incident electron (Fig. 4), and the technique of electron energy loss (EEL) measures the characteristic loss of energy by the scattered, incident electron. EEL spectroscopy can be used to give the same quantitative information about a specimen as EDX analysis, but has advantages for the measurement of low atomic number elements.

X-ray emission can also be induced by other particles than electrons. For example, Reiss et al. (1983) have used proton induced X-ray emission (PIXE) to examine calcium gradients in pollen tubes.

There are many electron beam-specimen interactions other than X-ray production (Fig. 3). Some give emissions which are characteristic of the elements in the specimen. The number of backscattered electrons, i.e. elastically scattered electrons from the beam, increases with the atomic number of the elements in the specimen. Auger electrons are emitted from an atom following its ionization, and like X-rays, have an energy characteristic of the atom from which they are emitted which can be measured.

7.2 Cytochemistry

Cytochemistry, i.e. the use of specific staining reactions to detect the presence of elements, has been used widely in plant science. Cytochemical or histochemical techniques (described in other volumes of the Modern Methods in Plant Analysis series; and Pearse 1968; Gurr 1965) and extensions such as immunohistochemistry, may provide similar data to EDX analysis about the locations (although at potentially much lower resolution) and amounts of elements in plant tissues. It may be used in parallel with EDX analysis to confirm findings (e.g. Heslop-Harrison 1986), because of speed or cost considerations, or because EDX would give misleading results.

In some cases, cytochemical techniques may give more useful results than EDX analysis. Cytochemistry may be more sensitive and detect trace amounts of elements, and some reactions distinguish between conjugated elements and free ions. Organic molecules can be stained and identified. 'Wet' methods are usually used and may reduce the resolution of localization for reaction products, quantification may be difficult and laborious reactions may be required.

7.3 Bulk Analysis

Elemental analysis of plant tissues can also be determined using many other techniques, described in other volumes of this series and elsewhere. Such analyses are usually applied to bulk tissue samples. Examples include flame ionization and flame spectroscopy, mass spectrometry, nuclear magnetic resonance and physical separation of elements followed by characterization.

8 Conclusions

EDX analysis provides a fast, high resolution and simple method to examine the amounts and location of many different elements within a tissue simultaneously. It can be used at many different levels of sophistication although care is always required in specimen preparation and data analysis. Useful qualitative and semi-quantitative results can be easily obtained.

A range of different analytical techniques have been described above which are applicable to all plant tissues in studies of structure, function and development. The particular strategy for analysis must be chosen to suit the nature of the investigation and the type of specimen; EDX analysis is the optimal technique for many investigations of elements in plants.

The fixation techniques and images available for cryopreserved and sectioned specimens are becoming increasingly better and more accurate. As prices decline, EDX analysis systems will probably become more widely available. Up to now, most systems have relied on specialized computers, but instruments increasingly use IBM PC compatible microcomputers. Techniques of quantitative and semi-quantitative analysis are improving, and the use of software will make even the most complex analytical techniques increasingly available to the plant scientist.

Acknowledgements. I thank BP Venture Research Unit and the US Department of Agriculture Competitive Grant Number 5901–0410–9–0363–0 for support. I am grateful to Dr. B. Reger for help with the EDX analyses described, and Dr. T. Schwarzacher for her assistance with preparation of the figures.

References

Bald WB (1987) Quantitative cryofixation. Hilger, Bristol

Baldi BG, Franceschi VR, Loewus FA (1987) Localization of phosphorus and cation reserves in *Lilium longiflorum* pollen. Plant Physiol 83:1018–1021

Cameron IL, Hunter KE, Smith NKR (1984) The subcellular concentration of ions and elements in thin cryosections of onion root meristem cells: an electron-probe EDS study. J Cell Sci 72:295–306

Chong J, Ali-Kahn ST, Chubey BB, Gubbles GH (1983) Energy dispersive X-ray analysis of field peas, *Pisum sativum* with different cooking quality. Can J Plant Sci 63:1071–1074

Cosslett VE, Duncumb P (1956) Microanalysis by a flying spot X-ray method. Nature 177:1172–1173

Danilatos GD (1981) Design and construction of an atmospheric or environmental SEM (part 1). Scanning 4:9–20

Echlin P (ed) (1978) Low temperature biological microscopy and microanalysis. Royal Microscopical Society, Oxford

Echlin P, Lai CE, Hayes TL (1982) Low-temperature X-ray microanalysis of the differentiating vascular tissue in root tips of *Lemna minor* L. J Microsc 126:285–306

Echlin P, Hayes TL, McKoon M (1983) Analytical procedures for bulk frozen-hydrated biological tissues. In: Gooley R (ed) Microbeam analysis. San Francisco Press, San Francisco, pp 243–246

Edelmann L (1986) Freeze-dried embedded specimens for biological microanalysis. Scanning Electron Microsc 1986/IV:1337–1356

Elbers PF (1983) Quantitative determination of elements by energy dispersive X-ray microanalysis and atomic absorption spectrometry in egg cells of *Lymnaea stagnalis*. J Microsc 131:107–114

Fitzgerald R, Keil K, Heinrich KFJ (1968) Solid state energy dispersion spectrometer for electron microprobe X-ray analysis. Science 159:528–530

Goldstein JI, Newbury DE, Echlin PE, Joy DC, Fiori C, Lifshin E (1981) Scanning electron microscopy and X-ray microanalysis. Plenum, New York

Gurr E (1965) The rational use of dyes in biology. Williams and Wilkins, Baltimore

Hall TA (1986) The history and the current status of biological electron probe X-ray microanalysis. Micron and Microscopia Acta 17:91–100

Hall TA, Gupta BL (1979) EDS quantitation and application to biology. In: Hren JJ, Goldstein JI, Joy DC (eds) Introduction of analytical electron microscopy. Plenum Press, New York, pp 169–197

Heinrich KFJ (1981) Electron beam X-ray microanalysis. Van Nostrand, New York

Heslop-Harrison JS (1986) Ion localisation in the stigma and pollen tube of cereals. In: Mulcahy DL, Bergamini Mulcahy G, Ottaviano E (eds) Biotechnology and ecology of pollen. Springer, Berlin Heidelberg New York, pp 351–356

Heslop-Harrison JS, Reger BJ (1985) Chloride and potassium ions and turgidity in the grass stigma. J Plant Physiol 124:55–60

Heslop-Harrison JS, Reger BJ (1986) X-ray microprobe mapping of certain elements in the ovary of *Zea mays* L. Ann Bot 57:819–822

Heslop-Harrison JS, Heslop-Harrison J, Heslop-Harrison Y, Reger BJ (1985) The distribution of calcium in the grass pollen tube. Proc R Soc Lond B 225:315–327

Hyatt AD, Marshall AT (1985) An alternative microdroplet method for quantitative X-ray microanalysis of biological fluids. Micron and Microscopia Acta 16:39–44

Joy DC (ed) (1987) Analytical electron microscopy 1987. San Francisco Press, San Francisco

Kahn MA, Weber DJ, Hess WM (1985) Elemental distribution in seeds of the halophytes *Salicornia pacifica* var *Utahensis* and *Atriplex canescens*. Am J Bot 72:1672–1675

Kausch AP, Wagner BL, Horner HT (1983) Use of the cerium chloride technique and energy dispersive X-ray microanalysis in plant peroxisome identification. Protoplasma 118:1–9

Laeuchli A, Schwander H (1966) X-ray microanalyser study on the localization of minerals in native plant tissue sections. Experientia 22:503–505

Lanning FC, Eleuterius LN (1983) Silica and ash in tissues of some coastal plants. Ann Bot 51:835–850

Moore R, Cameron IL, Hunter KE, Olmos D, Smith NKR (1987) The locations and amounts of endogenous ions and elements in the cap and elongating zone of horizontally oriented roots of *Zea mays* L.: an electron-probe EDS study. Ann Bot 59:667–677

Morgan AJ (1979) Non-freezing techniques of preparing biological specimens for electron probe X-ray microanalysis. Scanning Electron Microsc II 635–648

Morgan AJ (1980) Preparations of specimens. Changes in chemical integrity. In: Hayat MA (ed) X-ray microanalysis in biology. University Park Press, Baltimore, Maryland, pp 65–165

Morgan AJ (1985) X-ray microanalysis in electron microscopy for biologists. Roy Microsc Soc Handb No 6. Oxford University Press, Oxford

Morgan AJ, Davies DA, Erasmus DA (1975) Analysis of droplets from isoatomic solutions as a means of calibrating a transmission electron analytical microscope (TEAM). J Microsc 104:271–280

Moseley HGJ (1913) The high frequency spectra of the elements. Phil Mag 26:1024

Newbury DE (ed) (1988) Microbeam analysis. San Francisco Press, San Francisco

Newbury DE, Joy DC, Echlin P, Fiori CE, Goldstein JI (1986) Advanced scanning electron microscopy and X-ray microanalysis. Plenum, New York

Oatley CW, Nixon WC, Pease RFW (1965) Advances in electronics and electron physics, ed L Marton. Academic Press, New York, p 181

Okoli BE, McEuen AR (1986) Calcium containing crystals in *Telfairia* Hooker (Cucurbitaceae). New Phytol 102:199–207

Pearse AGE (1968) Histochemistry: theoretical and applied. 3rd edn, Williams and Wilkins, Baltimore

Revel JP, Barnard T, Haggis GH (eds) (1984) The science of specimen preparation for microscopy and microanalysis. Scanning Electron Microscopy Inc., AMF O'Hare, Illinois

Reiss HD, Herth W, Schnepf E, Nobiling R (1983) The tip-to-base calcium gradient in pollen tubes of *Lilium longiflorum* measured by proton induced X-ray emission (PIXE). Protoplasma 115:153–159

Robards AW (1985) The use of low temperature methods for structural and analytical studies of plant transport processes. In: Robards AW (ed) Botanical microscopy. Oxford University Press, Oxford, pp 39–64

Russ JC (1974) The direct element ratio model for quantitative analysis of thin sections. In: Hall TA, Echlin P, Kaufman R (eds) Microprobe analysis as applied to cells and tissues. Academic Press, London, pp 269–276

Sakai WS, Sandford WG (1984) A developmental study of silicification in the abaxial epidermal cells of sugarcane *Saccharum officinarum* leaf blades using scanning electron microscopy and energy dispersive X-ray analysis. Am J Bot 71:1315–1322

Saubermann AJ (1988a) Quantitative electron microprobe analysis of cryosections. Scann Microsc 2:2207–2218

Saubermann AJ (1988b) X-ray mapping of frozen hydrated and frozen dried cryosections using electron microprobe analysis. Scanning 10:239–244

Saubermann AJ, Heymann RV (1987) Quantitative digital X-ray imaging using frozen hydrated and frozen dried tissues. J Microsc 146:169–182

Sigee DC (1988) Preparation of biological samples for transmission X-ray microanalysis: a review of alternative procedures to the use of sectioned material. Scann Microsc 2:925–935

Steveninck RFM van, Steveninck ME van (1978) Ion localization. In: Hall JL (ed) Electron microscopy and cytochemistry of plant cells. Elsevier/North Holland, pp 187–234

Stewart A, Nield H, Lott JNA (1988) An investigation of the mineral content of barley grains and seedlings. Plant Physiol 86:93–97

Sumner AT (1981) The distribution of quinacrine on chromosomes as determined by X-ray microanalysis. Chromosoma 82:717–734

Westermark U, Erikson I, Lidbrandt O (1988a) Lignin distribution in spruce (*Picea abies*) determined by mercurization with SEM-EDXA technique. Wood Sci Technol 22:251–257

Westermark U, Lidbrandt O, Eriksson I (1988b) Lignin distribution in spruce (*Picea abies*) determined by mercurization with SEM-EDXA technique. Wood Sci Technol 22:243–250

Wolf B, Schwinde A (1988) Investigations on ultrathin cryosections from cryoprotected and non-cryoprotected tissues by electron energy loss spectroscopy: advantages and limits of applications. Micron and Microscopia Acta 19:147–153

Zierold K, Hagler HK (1989) Electron probe microanalysis applications in biology and medicine. Springer, Berlin Heidelberg New York

Subject Index